Soft Order in Physical Systems

NATO ASI Series

Advanced Science Institutes Series

A series presenting the results of activities sponsored by the NATO Science Committee, which aims at the dissemination of advanced scientific and technological knowledge, with a view to strengthening links between scientific communities.

The series is published by an international board of publishers in conjunction with the NATO Scientific Affairs Division

A	**Life Sciences**	Plenum Publishing Corporation
B	**Physics**	New York and London
C	**Mathematical and Physical Sciences**	Kluwer Academic Publishers
D	**Behavioral and Social Sciences**	Dordrecht, Boston, and London
E	**Applied Sciences**	
F	**Computer and Systems Sciences**	Springer-Verlag
G	**Ecological Sciences**	Berlin, Heidelberg, New York, London,
H	**Cell Biology**	Paris, Tokyo, Hong Kong, and Barcelona
I	**Global Environmental Change**	

Recent Volumes in this Series

Series B: Physics

Soft Order in Physical Systems

Edited by

Y. Rabin

Bar-Ilan University
Ramat Gan, Israel

and

R. Bruinsma

University of California at Los Angeles
Los Angeles, California

Springer Science+Business Media, LLC

Proceedings of a NATO Advanced Research Workshop on
Soft Order in Physical Systems,
held in honor of Shlomo Alexander,
February 16–25, 1993,
in Les Houches, France

NATO-PCO-DATA BASE

The electronic index to the NATO ASI Series provides full bibliographical references (with keywords and/or abstracts) to more than 30,000 contributions from international scientists published in all sections of the NATO ASI Series. Access to the NATO-PCO-DATA BASE is possible in two ways:

—via online FILE 128 (NATO-PCO-DATA BASE) hosted by ESRIN, Via Galileo Galilei, I-00044 Frascati, Italy

—via CD-ROM "NATO Science and Technology Disk" with user-friendly retrieval software in English, French, and German (©WTV GmbH and DATAWARE Technologies, Inc. 1989). The CD-ROM also contains the AGARD Aerospace Database.

The CD-ROM can be ordered through any member of the Board of Publishers or through NATO-PCO, Overijse, Belgium.

Library of Congress Cataloging-in-Publication Data

Soft order in physical systems / edited by Y. Rabin and R. Bruinsma.
      p.   cm. -- (NATO ASI series. Series B, Physics ; vol. 323)
    "Published in cooperation with NATO Affairs Division."
    "Proceedings of a NATO Advanced Research Workshop on Soft Order in
  Physical Systems, held in honor of Shlomo Alexander, February 16-25,
  1993, in Les Houches, France"--Verso t.p.
    Includes bibliographical references and index.
    ISBN 978-1-4613-6046-9   ISBN 978-1-4615-2458-8 (eBook)
    DOI 10.1007/978-1-4615-2458-8
    1. Polymers--Congresses. 2. Chemistry, Physical and theoretical-
  -Congresses.   I. Rabin, Yitzhak. II. Bruinsma, R. III. North
  Atlantic Treaty Organization. Scientific Affairs Division.
  IV. NATO Advanced Research Workshop on Soft Order in Physical
  Systems (1993 : Les Houches, Haute-Savoie, France) V. Title.
  VI. Series: NATO ASI series. Series B, Physics ; v. 323.
  QC173.4.P65S64 1994
  530.4'13--dc20   94-5030
                                                             CIP

ISBN 978-1-4613-6046-9

© 1994 Springer Science+Business Media New York
Originally published by Plenum Press, New York in 1994
Softcover reprint of the hardcover 1st edition 1994

SPECIAL PROGRAM ON CHAOS, ORDER, AND PATTERNS

This book contains the proceedings of a NATO Advanced Research Workshop held within the program of activities of the NATO Special Program on Chaos, Order, and Patterns.

SPECIAL PROGRAM ON CHAOS, ORDER, AND PATTERNS

PREFACE

A humoristic view of the physics of soft matter, which nevertheless has a ring of truth to it, is that it is an ill-defined subject which deals with ill-condensed matter by ill-defined methods. Although, since the Nobel prize was awarded to Pierre-Gilles de Gennes, this subject can be no longer shrugged-away as "sludge physics" by the physics community, it is still not viewed universally as "mainstream" physics. While, at first glance, this may be considered as another example of inertia, a case of the "establishment" against the "newcomer", the roots of this prejudice are much deeper and can be traced back to Roger Bacon's conception about the objectivity of science. All of us would agree with the weaker form of this idea which simply says that the *final* results of our work should be phrased in an observer–independent way and be communicable to anybody who made the effort to learn this language. There exists, however, a stronger form of this idea according to which the above criteria of "objectivity" and "communicability" apply also to the process of scientific inquiry. The fact that major progress in the physics of soft matter was made in apparent violation of this approach, by applying intuition to problems which appeared to defy rigorous analysis, may explain why many physicists feel somewhat ill-at-ease with this subject.

The inadequacy of the notion that one can do physics in a way which can be taught to a computer is particularly clear to those of us who had the privilege to work with Shlomo Alexander. After discussing a scientific problem with Shlomo one quite often feels that Shlomo said something very deep and important but, at the same time, one fails to fully appreciate the connection between the question asked and Shlomo's answer. Only much later, sometimes after one spends weeks and months thinking about the problem, one suddenly grasps the delicate relationship between the problem he raised and Shlomo's comments. The reason is that that Shlomo's deep insights are the product of his unique physical intuition.

We can only speculate on the origins of this intuition: his unusual background as an experimentalist who turned into a high-powered theorist, his superb geometrical imagination, his deep feeling for symmetry arguments and last, but not least, the breadth of his interests which go far beyond physics (his collaboration with his wife Esther, on economics, is but one example). While the results of Shlomo Alexander's insights can be turned into simple and comprehensible formulas, his intuition and unique way of reasoning belong to the realm of art. This art has enriched us all in a way which makes Shlomo's impact on the physics of soft matter go even beyond his many beautiful works.

Yitzhak Rabin and Robijn Bruinsma

CONTENTS

2. Crystallography

3. Dynamics of Disordered Systems / Glasses

4. Percolation, Diffusion and Fractons

5. <u>Surfactants and Liquid Crystals</u>

SCIENCE AND SOCIETY

LE CENTRE DE PHYSIQUE ET l'ECOLE d'ETE DE PHYSIQUE

THEORIQUE DES HOUCHES

Une Perspective Historique

par le Professeur Yuval Ne'eman

Directeur de l'Institut Sackler de Hautes Etudes
Université de Tel-Aviv, Israel
President de l"Agence Spatiale Israelienne
Ancien President de l"Université de Tel-Aviv
Ancien Ministre de la Recherche, de la Science et de la Technologie
Ancien Ministre de l"Energie et de l'Infrastructure
Ancien President du Commissariat Israelien a l"Energie Atomique
Ancien Directeur de la Recherche au Ministère Israelien de la Défense
Membre de l'Académie des Sciences d'Israel
Membre Etranger, Académie Nationale des Sciences des États Unis
Academie Américaine des Arts et des Sciences
Academie des Sciences de New York

Mr le Maire, Méssieurs les Présidents

Nous fêtons ici aujourd'hui l'anniversaire de mon ami et collegue en Israel, le Professeur Shlomo Alexander. Cette manifestation d'amitie, de sympathie et d'appreciation nous touche tous, chercheurs Israeliens, membres de l'Academie des Sciences ou des cadres universitaires. Nous sommes fiers des travaux du Professeur Alexander - deuxieme generation dans la recherche en Physique en Israel, chose encore assez rare dans un état aussi jeune. Shlomo a aussi instruit et guidé une generation de chercheurs en physique de la matière condensée. Mr de Gènnes a décrit les resultats importants obtenus par Alexander, et je ne reviendrai pas sur cette liste impressionante. Par contre, avec Shlomo je profite de l'occasion pour citer les travaux de son épouse, Esther Alexander, en économie politique. J'ai eu le plaisir de travailler avec Esther en 1978-82, en tant que President du Commite d'étude du Projet Mediterranee-Mer Morte, quand Esther suivait nos travaux en tant que representante du Ministere de l'Energie. Plus tard, en 1990-92, c'est moi qui, en tant que Ministre de l'Enèrgie et de l'Infrastucture, lui ai demandé de rejoindre le Ministere avec les fonctions de Conseillère Economique du Ministre. Pendant cette periode de deux ans, j'ai admire ses analyses de la situation et surtout sa souplesse d'esprit. En une periode ou la mode est tout a fait celle du Capitalisme anti-etatiste de l'école de Chicago (y compris dans les anciens milieux socialistes Israeliens..) nous nous sommes trouvés tous deux nous éfforçant de faire comprendre a nos collègues qu'il est impossible d'absorber une augmentation de 10 pourcents de la population en deux ans (l'immigration des Juifs d'URSS) sans faire intervenir l'état dans l'économie nationale, par la création d'emplois etc..

Je voudrai aussi vous dire combien nous apprecions le fait que cette ceremonie se déroule aux Houches. Un proverbe Hebraique dit qu'avec trop d'arbres, on ne voit plus la forêt. Vous avez vécu avec cette Ecole de Physique depuis 1950, et la familiarité vous fait peut-être oublier ce qu'elle represente dans une perspective historique. On n'a pas encore écrit l'histoire de la communication scientifique et des institutions de hautes études et de recherche.   France, on a Montpellier et la Sorbonne au XII Siecle, les Universités. Sous François Ier il y a la création du Collège de France, un institut de recherche. Napoleon et Carnot creent les Grandes Ecoles, un nouveau modele pour les sciences mathématiques et la technologie. J'ose vous presenter Les Houches comme l'innovation suivante, la quatrième phase. Il n'est pas du tout surprennant que quelques années plus tard, l'Europe et les pays industrialisés copient ce système; aujourd'hui, il y a de part le monde une trentaine d'écoles en Physique, des centaines dans tous les domaines de la science.

En un age ou les domaines de recherche sont en plaine explosion et ou il est presque impossible de suivre tous les travaux - meme en un domaine tres restraint, il est essentiel d'avoir la possibilité de fournir aux jeunes chercheurs une introduction au probleme qu'ils auront a traiter. Le seul moyen de le faire - vu le rythme de l'avance du front - c'est de faire ce qu'a fait Cecile Morette en 1950: rassembler pour une à plusieurs semaines (preferablement) les chercheurs principaux de part le monde. Les étudiants y recoivent une demi-synthèse intérimaire, on leur fait le point sur les resultats les plus récents et on leur explique les difficultés. Ils peuvent poser toutes les questions et aux personnes les plus aptes a y répondre. L'autre fonction de ces écoles d'été ou d'hiver est la publication de ces demi-synthèses, tres interimaires, mais qui permettent aux chercheurs partout de se donner une perspective sur ce qu'il font - d'appercevoir la forêt, comme dans mon proverbe..

On a bien cité l'importance qu'avait eu l'Ecole des Houches, lors de sa création, pour la science en France, apres la dispersion causée par la Seconde Guerre Mondiale. Je voulais y ajouter l'aspect international. La création de l'Ecole de Physique des Houches a été une etape tres importante pour la recherche mondiale, un nouveau modèle, extremement utile. Mes felicitations, donc, a l'Université Scientifique et Medicale de Grenoble pour l'avoir prise sous sa tutelle.

Mr le Maire, en Sicile, a Erice, il y a une école du meme genre, fondée en 1962, qui a en meme temps transformé le pays, devenant un centre d'attraction du tourisme national et international: on vient tout simplement voir l'école.. Les Houches n'en a pas besoin, la nature l'a bien dotée. La neige et le ski font un levier plus que suffisant pour votre economie. Sachez quand meme, qu'en une region moins fortunée, comme la Sicile, c'est l'analogue de cette Ecole qui remplace le ski..

Merci encore - surtout a Pierre Gilles de Gennes - d'avoir initié cette rencontre. Bon succes dans la suite de ces cours!

The Physics Center and the Theoretical Physics Summer School at Les Houches: An Historical Perspective

Mr. Mayor and Honoured Guests,

Today's celebration commemorates the Sixtieth Anniversary of my friend and colleague in Israel, Professor Shlomo Alexander. This manifestation of friendship and appreciation on the part of Prof. de Gennes and our French colleagues touches us all, we Israeli researchers and Shlomo's colleagues, both at the Israel National Academy of Sciences and at the various Universitites and research establishments. We are indeed proud of Prof. Alexander's scientific achievements - a second generation leader in Israeli physics - in a young country in which this is still a rare case. Shlomo has also trained and guided a generation of researchers in the Physics of Condensed Matter. Prof. de Gennes has already enumerated the more important results obtained by Shlomo, and I shall not repeat this impressive list. Let me however use the opportunity, as Shlomo did, to praise his wife's achievements in Economics. Here I am expressing my professional views, based on my experiences in the Energy and Engineering fields: I have been fortunate to work with Esther, first in 1978-82, in my capacity of Chairman of the Steering Committee of the Med-Dead Project [1], with Dr Esther Alexander cooperating with us as the representative of the Ministry of Energy. Eight years later, in 1990-92, as Minister of Energy and Infrastructure, I asked her to work with me as the Minister's Economic Counsellor. For two years, I could base my policy on her Economic Studies and enjoyed her advice, learning to appreciate her non-conformal thinking. In a period in which it is fashionable to follow the Chicago School non-interventionist "free market" doctrines (even amongst the Israeli formerly socialist economic thinkers) we both found ourselves trying hard to convince our colleagues that it is impossible to absorb an addition of some ten percents to the population, within a two-years time-interval (the immigration of Jews to Israel, from the former USSR, in 1990- 1991) without Governmental intervention in the national economy, especially in generating employment.

I would also like to tell you how much we appreciate the fact that you chose this site for the ceremony. There is a Hebrew adage saying that when there are too many trees you do not perceive the forest. You have lived with this school since 1950, and familiarity makes you perhaps forget what it represents in a historical perspective. The history of Scientific Communication has not yet been written, together with the history of the Institutions of Higher Education. In France, it starts with Montpellier and the Sorbonne in the XIIth Century, the universities. Then, under Francis I, there is the creation of the College de France, a research institute. Napoleon and Carnot establish the Grandes Ecoles (Ecole Normale and Ecole Polytechnique), yet another model for the development of the mathematical and technological sciences. I dare describe to you Les Houches as the next innovation, the fourth stage. It is not at all surprising that within a few years, the countries of Europe and the industrialized world all immitated the system. There are nowadays some thirty similar schools just in Physics, probably a few hundreds in other scientific disciplines.

In an age in which the various areas of scientific research are in full explosive expansion,it has become almost impossible to follow all published works, even in a restricted field. And yet it is essential that young research students entering a field be supplied with an introduction to their area, including the latest results. The only way of achieving this aim is to do what Cecile Morette did in 1950: assemble for a few days or weeks the principal senior researchers in the given area, from all over the world. The students can thus receive a preliminary semi-synthesis; this should cover the recent results, explain the difficulties and point to the open problems. They can ask all questions - and get their answers from those who are most qualified to give them. At the same time, the school proceedings represent a published semi-synthesis, very informal and unpolished, but allowing colleagues everywhere to get a general view of the field - they can "perceive the forest", beyond the myriad isolated publications.

One does hear about the importance of the creation of Les Houches for the rebirth of French science, after its wartime collapse. My aim here has been to stress the international impact, beyond the specific problems of the France of 1950. The creation of the School of Physics at Les Houches represents an important advance in the organisation of world scientifc research, an innovative new model which has now become an essential element everywhere. I would like to congratulate the Scientific and Medical University of Grenoble and her President who is here with us for its having taken the School under its auspices.

Mr Mayor and honoured representatives of the region, I can tell you that in Sicily, at Erice, there is now a similar school, founded in 1962; over there it has transformed the region, having become a tourist attraction at the national and even the international levels. Les Houches needs no such touristic boost from science, nature has amply endowed it for this purpose. Snow and Skiing are quite successful as economic levers. And yet, you should know that elsewhere, without this natural endowment, a school like yours has even become an economic pillar of strength for the local community and the region. You too might decide some day to add the exploitation of this feature to your other assets.

Thanks again - especially to Pierre Gilles de Gennes - for initiating this meeting. I wish you all a successful completion of this session at the School.

Yuval Ne eman

FOCAL CONICS DOMAINS IN SMECTICS

P. Boltenhagen[1], M. Kleman[1] and O.D. Lavrentovich[2]

[1]*Laboratoire de Physique des Solides, Bât.510, Université de Paris-Sud 91405 ORSAY Cedex*
&
Laboratoire de Minéralogie-Cristallographie
Université Pierre et Marie Curie (Parie VI) et Paris VII
4 place Jussieu, case 115, 75252 PARIS Cedex 5

[2]*Kent State University, Liquid Cristal Institute*
Kent, Ohio,44242

Abstract: A lamellar phase constituted by equidistant flat layers presents typical structural defects called Focal Conic Domains (FCD), in which the layers (Dupin cyclides) are folded around two conjugated lines. The aim of this paper is to give a complete review of the geometrical and energetical aspects of FCDs. Also, some typical textures will be described. After a brief historical recall of the discovery of FCD's by G. Friedel, we describe the conformation of the layers in the FCD's. We consider the different fundamental types of structural defects of the lamellar phase, pointing out the topological differences between them. An analytical description of the Dupin cyclide is made, in particular we introduce their curvature, which is the starting point of the energy 's calculation of the FCD. We complete the description with a number of physical situations.

1 - INTRODUCTION

In the beginning of this century, Georges Friedel (G.F. and F.Grandjean, 1910 ; G.F., 1922) inferred from the observation at optical scales of line singularities in the since-called Smectic-A (SmA) phases that theses phases were lamellar. Diffraction methods were not yet invented, and the discovery of the lamellarity, which was some years later confirmed by his son Edmond (X ray diffraction), a student of Maurice de Broglie, was at that time based only on the geometrical properties of these singularities. These singularities were quite remarkable, made of pairs of conics, an ellipse E and one of the branches of an hyperbola H, situated in two orthogonal planes, and such that the apices of any of them is at the foci of the other. Georges Friedel and François Grandjean (1910) gave the name of 'confocal domain' to the region in space which is involved by the folding of the lamellae directly related to the pair of conics. In that time, physicists knew geometry better that it is known today, and G. Friedel immediately concluded from these observations that the system was indeed lamellar. For the sake of understanding what is important in this discovery, let us differentiate two steps. He was aware of Dupin's theorem (see Darboux, 1954) which states that, when the two focal surfaces of a family of parallel surfaces are degenerate to lines, these lines are cofocal conics, and that the family of parallel surfaces are cyclides (by definition the lines of curvature of a cyclide are circles). But this knowledge was probably shared by a large number of physicists, at that time, even among those whose names have not survived. What was fundamentally original in Friedel's discovery is that he gathered that these observed conics were <u>focal</u> lines, i.e. they were the geometrical locus of the centers of curvature of a family of parallel physical surfaces. This point being made, it was no great effort to conjecture that the molecules, which were known to have a preferential axis, were along these normals, and that they were organized in physical sheets.

The purpose of this paper is to give a complete review of the geometrical basis of the subject, supported by the description of a number of physical situations which have been recently studied; the subject of confocal domains, while still alive thanks to the renoval of the study of lyotropic phases, is probably the oldest subject in the physics of defects. Let us remark, that one of the main character of this theory, viz. that the full knowledge of the defects belonging to an ordered medium is enough to understand the nature (scalar, vectorial, etc...) of its 'order parameter', was already visible to its founders; but the importance of this discovery was not really appreciated before the topological theory of defects (Toulouse and Kléman, 1976 ; Volovik and Mineyev, 1977 ; Kléman, 1983). This is certainly not the only domain where the knowledge of focal domains is of great use : in other types of smectic phases, like SmC, SmC* (Bourdon et al., 1982), and even ordered smectics, and also in vesicles (see for example, Fourcade et al., 1992), whose topology often affects the shape of Dupin cyclides.

Our object has been treated a number of times in quite excellent papers ; apart the historical and still fundamental paper of G. Friedel (1922), it is worth mentionning the Bragg's paper (1933), which was the first to familiarize the english speaking community to the subject, and the paper of Bouligand 1972), who revisited the subject and contributed to its renewal. Since then, there have been a number of experimental and theoretical works on the subject; this outburst of a subject of such an old seating, 80 years after their discovery, is not without some special gusto.

This paper is written as an introductory review for the unlearnt. It is constructed as follows. The first part deals with some generalities of the geometry of Focal Conic Domains (FCD's). The second one is devoted to an analytical description of the sheets which constitute the domains. This is the starting point of the calculation of the energy of different types of FCD's, which is made in the third part. We complete the discussion with some experimental observations in lyotropic systems (fourth part). The fifth part is devoted to the problem of space filling: we are considering it in the case of FCD of the first and second species. The last part is devoted to the question of the nucleation and growth of FCD's.

2 - GEOMETRICAL ASPECTS OF FOCAL CONIC DOMAINS, GENERALITIES

A smectic A phase (SmA) or its lyotropic analog the L_α phase can be represented as a set of surfaces which are planar, parallel and equidistant in the ground state of the phase. Each deformed layer carries an energy per unit area f_s which reads, to quadratic order:

$$f_s = 1/2\,\kappa(\sigma' + \sigma'')^2 + \bar{\kappa}\,\sigma'\,\sigma'' \qquad (1)$$

In this expression σ' and σ'' are the principal curvatures of the surfaces, and κ and $\bar{\kappa}$ are the splay and the saddle-splay elastic constants. The mean curvature $\sigma' + \sigma''$ and the Gaussian curvature $\sigma'\,\sigma''$ are geometrical invariants of the surface (Darboux, 1954 ; Salmon, 1952). The second term related to the Gaussian curvature is generally omitted in many calculations on the ground that it does not enter in the minimization of the energy. This is a direct consequence of the Gauss-Bonnet theorem which tells us that the integral of the Gaussian curvature over a closed surface depends only on its topology:

$$\int \sigma' \, \sigma'' d\Sigma = 4\pi(1-N) \tag{2}$$

(N is the number of handles and "pores" on the surface (fig.1)). For example the integral (2) is equal to 0 for a torus and to 4π for a sphere. Thus this term is a constant and can be neglected in any smooth continuous deformation of the system (it cannot be measured in any deformation which implies a small 'virtual work'; there is therefore no condition of convexity of the free energy attached to it, and consequently no condition imposed on the sign of $\bar{\kappa}$); but it plays a important role in any change of the topology of the layers. We shall see its importance in the stability of confocal domains.

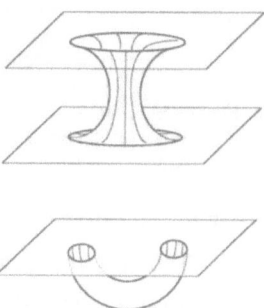

Fig. 1 - Pore and handle.

A total description of the energy of the system requires to add the independent contributions of all the layers and to take into account the energy of interaction betweeen the layers. This leads, to quadratic order, to the following expression for the volume energy density:

$$f = 1/2 \, K \, (\sigma ' + \sigma'')^2 + \bar{K} \, \sigma ' \sigma'' + 1/2 \, B \, \varepsilon^2 \tag{3}$$

where now $\kappa = K d_0$ and $\bar{\kappa} = \bar{K} d_0$, d_0 is the repeat distance of the smecic phase, and B is a modulus of compressibility. K and $|\bar{K}|$ have the same order of magnitude; the ratio $\lambda = \sqrt{K/B}$ has the dimension of a length, whose magnitude is comparable to d_0 by necessity.

The energy volume densities carried by the system for a pure curvature deformation $(f_c \approx K\sigma^2)$ or for a pure compression deformation $(f_\varepsilon \approx B)$ at large scales ($\sigma << d_0^{-1}$) have very different magnitudes $(f_c/f_\varepsilon \approx (\lambda\sigma)^2 << 1)$, telling us that the lamellar phase should deform in a set of quasi-equidistant and parallel surfaces ($f_\varepsilon \approx 0$). The confocal domains we shall dwell on do not constitute the most general solution to the problem of the geometry of parallel surfaces (it is in fact not the generic solution), but the most generally (although not unique) solution chosen by the smectic phases. The reason for the prevalence of this geometry will appear later, when we shall go into some detail into the geometry of focal conic domains (FCD's).

At this stage we want to state immediately that in each FCD the layers are folded around two conjugated lines, viz., an ellipse E and one branch H of a hyperbola, in such a way that they are everywhere perpendicular to the straight lines joining any point M' on the ellipse to any point M" on the hyperbola (fig.2). Any point M on the line M'M" is the orthogonal intersection with this line of a uniquely defined surface Σ_M, perpendicular everywhere to the 2-parameter family (the congruence) of lines M'M". All the parallel surfaces Σ_M

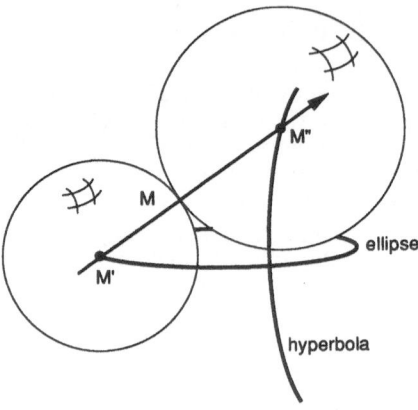

Fig. 2 - General geometrical aspect of Focal Conic Domain. The line M'M" is the perpendicular direction to the layer on the point M.

orthogonal to M'M" have the same centers of curvature, M' and M". $|\sigma'| = 1/M'M$ or $|\sigma''| = 1/M''M$ become infinitely large when M approaches either M' or M"; correspondingly, the Σ_M's are singular on M' and M" , where the energy density grows without limit.

There is a obvious analogy between the generic geometry of parallel surfaces and the geometry of wave fronts when light propagates in a medium of constant index of refraction; the normals to the layers (the congruence M'M") are the analogs of the light rays, and the two loci of the centers of curvature (the evolutes of the surface, to speak in mathematical terms) are the analogs of the two focal sheets (caustics) on which the intensity of light becomes infinite. In the special case we are considering, the caustics are degenerate into lines, so that the energy carried by the singularities scales like a length, rather than, generically, like a surface. As indicated in the introduction, Dupin's Theorem states that in the degenerate case the focal lines are a pair of conjugate ellipse and hyperbola.

One might wonder why the most usually observed focal domains are not those where the singularities are still further reduced, i.e. one focal sheet reduced to one point, the other one being sent at infinity (in such a case the layers are concentric spheres). These 'spherulites' are indeed observed, but only when the saddle-splay coefficient is favourable, i.e. $\overline{K}$ small enough compared to K, or even negative. This is stressing again the importance of the topology and its relationship to the elasticity coefficients. A spherolite is a special type of FCD. But, more generally, conventional focal conic domains of the G. Friedel type are not, by far, the only types of deformation sets in which the layers keep at constant distances one from the other: the giant screw dislocations observed in thermotropic SmA's by C. Williams (1975) are a remarkable example of a completely different topology: the visible singularities are made of a pair of helices which are the cuspidal edges of the evolutes of a helicoid generated by a straight line (The concept of 'virtual surface' was introduced for that purpose by Frank and Kleman see Kléman, 1983, chap. V, for illustration). All the layers are parallel to this helicoid, whose pitch is equal to the Burgers vector of the dislocation. Note that the physical part of the focal sheets is reduced to a line, the rest keeping 'virtual'; this phenomenon, which originates in the fact that some regions of space are multi-covered by the Σ_M's parallel to the helicoid, and that a choice has to be made which eliminates some parts of the Σ_M's which are close to their evolutes, has also (as above for FCD as opposed to generic FD) the advantage to construct domains of minimal singular energy. Other focal domains with virtual focal sheets have been more recently imagined by Fournier (1993a), in relation with boundary conditions on the sample; the same problem of the boundary conditions could also lead to focal domains with 'canal' surfaces (when only one of the focal surfaces is degenerate to a line, and the other one partly or not virtual). The first mention of virtual focal lines dates back to Steers *et al.* (1974), and refers to the problem of nucleation of focal domains near a free surface.

The Dupin cyclides offer a quite remarkable example in which the physical focal lines depends on criteria of energy minimization, beyond the fact that the singularities are lines. There are three simple cases, which we distinguish first by the sign of the corresponding

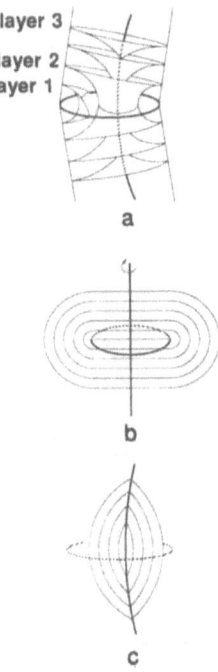

layer 3
layer 2
layer 1

a

b

c

Fig. 3 - a) Focal Conic Domain of the first specices (FCD-I)
b) Focal Conic of the third species (FCD-III)
c) Focal Conic of the second species (FCD-II)

Gaussian curvature $\sigma' \sigma''$ of the layers :

a)- if the physical part of the layer is located between M' and M", one gets a so-called *focal conic domain of the first species* (FCD-I, fig.3-a) this geometry yields $\sigma' \sigma'' < 0$. Dupin cyclides in a FCD-I show up features varying with the position of M on the segment M'M"; either Σ_M ends on the ellipse on two point singularities (layer 1, fig.3-a), or Σ_M is free of singularities and looks like a half-torus (layer 2, fig.3-a), or it ends on the hyperbola and has the form of a spheroid limited to its $\sigma' \sigma'' < 0$ part, with two conical indentations along the hyperbola (layer 3, fig.3-a). One see that in a FCD-I, the ellipse and the branch of hyperbola are both lines defects and thus visible. G. Friedel has described the different projections of a FCD-I observed in polarizing microscopy (fig.4).

b)- a FCD-I is bound by two half-cylinders; the continuation of the Σ_M's *outside* these cylinders has positive Gaussian curvature $\sigma' \sigma'' > 0$. This region consists of layers which are perpendicular to the half-infinite lines with origin at M". We call *focal conic domain of the third species* (FCD-III, fig.3-b) these domains, which are analytic continuations of a FCD-I and could therefore coexist with them, but which have not been observed, except in a few rare cases, in fact in the form of "complete" cyclides showing up both types of Gaussian curvatures. In a situation where positive Gaussian curvature is favoured, the system prefers to adopt the geometry of FCD-II, now described

c)- a so-called *focal conic domain of the second species* (FCD-II, fig.3-c) is obtained when the physical part of the normal to Σ_M is located along the half-line with origin in M'. In that case the cyclides look like a rugby-ball, with two conical indentations along the hyperbola. Their stacking can fill the entire space, i.e. the *inside* as well as the outside of the

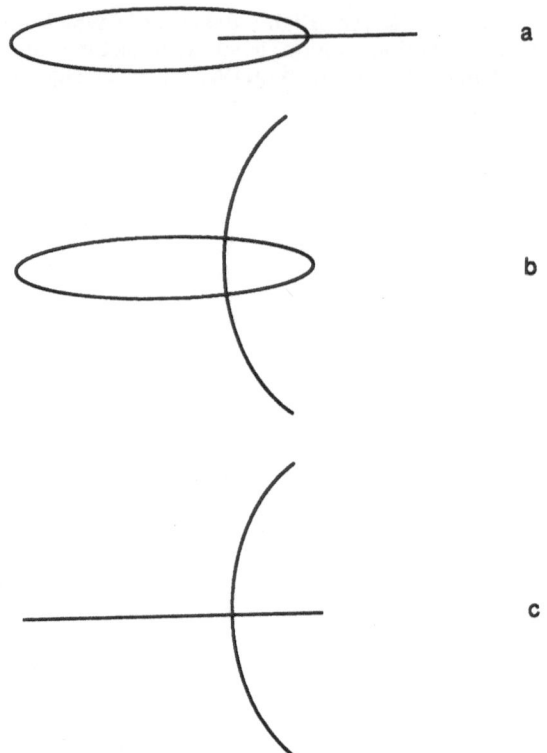

Fig. 4 -Projections of FCD-I observed in polarizing microscope
a) in the plane of the ellipse
b) in any plane
c) in the plane of the hyperbola

region of space occupied by a FCD-I having the same E and H , with which they cannot coexist physically. The ellipse is now virtual. In such a case, only the hyperbola will be visible under microscope.

3 - GEOMETRICAL ASPECTS, ANALYTIC DESCRIPTION

This section is devoted to an analytical approach of the geometry of Dupin cyclides and the energy of FCD-I and FCD-II. A more detailed account of the elements of surface theory which are used here, as well as more specific results, can be found in Darboux (1954). See also Weatherburn, 1955 ; Salmon, 1952 ; Willmore, 1982).

Let E and H be cofocal ellipse and hyperbola located in two perpendicular planes, and let the equation of the ellipse be, in standard notations:

$$z=0, \quad \frac{x^2}{a^2}+\frac{y^2}{b^2}=1 \qquad (4)$$

The equation of the cofocal hyperbola H then reads:

$$y=0, \quad \frac{x^2}{a^2-b^2}-\frac{z^2}{b^2}=1 \qquad (5)$$

Let M'(x',y',0) a point on E and M"(x",0,z") a point on H, and let us parameterize the conics in the usual way:

$$M' , \begin{cases} x' = a \cos u \\ y' = b \sin u \end{cases} \qquad 0 \le u \le 2\pi$$

$$M'' , \begin{cases} x'' = c \ ch \ v \\ z'' = b \ sh \ v \end{cases} \qquad -\infty \leq v \leq +\infty$$

where $c^2 = a^2 - b^2$.

The length M'M'' is :

$$M'M'' = \Delta = |a \ chv - c \cos u| = |ex' - e^{-1}x''| \qquad (6)$$

where $e = \dfrac{c}{a}$ is the eccentricity of the ellipse.

Let us now consider a point M on the line M'M'' and introduce the oriented lengths:
$$r' = MM' \qquad\qquad r'' = MM''$$
For the sake of clarity, we shall always orient the line from M' to M'', so that:
$$\Delta = r'' - r'$$
We introduce now the parameter r, defined as follows:

$$r' = \pm \{ex' - r\} \qquad r'' = \pm \{e^{-1}x'' - r\} \qquad (7)$$

where the same sign outside the brackets has to be chosen. (More precisely, sgn (r', r'') = sgn $\{e^{-1}x'' - ex'\}$). To each value of r is attached a surface $\Sigma(r)$. It is easy to show

that $\overrightarrow{M'M''}$ is perpendicular to $\Sigma(r)$ in M. Let indeed

$$S_r (M') \equiv (X-x')^2 + (Y-y')^2 + Z^2 - r'^2 = 0 \qquad (8a)$$

be the equation of a sphere centered in M' and passing through M, and similarly

$$S_r (M'') \equiv (X-x'')^2 + Y^2 + (Z-z'')^2 - r''^2 = 0 \qquad (8b)$$

be the equation of a sphere centered in M'' and passing through M. The set of spheres S_r (M') envelop a surface which is precisely $\Sigma(r)$. Indeed, the point of contact of S_r (M') with its envelop belongs at the same time to S_r (M') and to the derived surface:

$$\frac{dS_r (M')}{dM'} \equiv \frac{dS_r}{dx'} = \frac{\partial S_r}{\partial x'} + \frac{dy'}{dx'}\frac{\partial S_r}{\partial y'} = y'X + (e^2-1) x'Y - ey'r = 0 \qquad (9)$$

and it is a matter or simple algebra to show that M belongs to the intersection of S_r(M') and $\frac{dS_r(M')}{dM'}$; it obviously also belongs to the intersection of S_r(M'') and $\frac{dS_r}{dM''}$ (M''). Therefore $\Sigma(r)$ is the locus of the common envelop to S_r(M') and S_r(M''), and M' and M'' are its centers of curvature ; $\Sigma(r)$ is a surface which is everywhere perpendicular to a line joining some point M' on E to some point M'' on H : it is therefore the surface we are looking for. This surface is a cyclide (i.e. its lines of curvature are circles), since the complete contact of each S_r(M') with $\Sigma(r)$ is the circle C_r(M') along which S_r(M') and $\frac{d S_r}{d x'}$ (M') intersect. $\Sigma(r)$ is a Dupin cyclide.

Note also that M' is the apex of a cone of revolution whose basis is C_r(M') and which lies on H ; reciprocally, M'' is the apex of a cone of revolution whose basis is C_r(M'') and which lies on E. All these geometrical properties are well known.

We also introduce the curvatures $\quad \sigma' = \pm\dfrac{1}{r'} \ ; \quad \sigma'' = \pm\dfrac{1}{r''}$

In the FCD-I case, where M sits between M' and M'', we have r' > 0, r'' < 0 ; hence we have along a given normal (u,v) a set or r-values in the range:

$$c \cos u < r < a \operatorname{ch} v \qquad (10)$$

the whole set of r-values for the complete FCD-I domain being in the range [-c, + ∞[.

r = -c ; the Dupin cyclide is reduced to a point which is the apex J of the ellipse opposite to the physical branch of the hyperbola

-c< r <+c ; the Dupin cyclide has two singular points on the ellipse (obtained for r' = 0), none on the hyperbola

c< r <a ; the Dupin cyclide has no singular points and is homotopic to a torus. The mean Dupin cyclide $r_0^2 = ac$ is very special : according to a conjecture due to Willmore, it is a toroidal surface which is an absolute minimum of the curvature energy \F(1;2) κ \i(;;(σ'+ σ")² dΣ).

r>a ; singular points on the hyperbola only.

In the FCD-II case, where M sits beyond M", we have r',r">0 ; hence along a given normal {u,v}, we have

$$r > a \operatorname{ch} v \qquad (11)$$

and the whole set of r-values for the complete FCD-II domain is in the range [a,∞[. For r=a the Dupin cyclide is reduced to the apex F of the branch of H which is physical.
In all cases, it is possible to take

$$\sigma' = \frac{1}{c \cos u - r} = \pm \frac{1}{r'} \quad ; \quad \sigma'' = \frac{1}{a \operatorname{ch} v - r} = \pm \frac{1}{r} \qquad (12)$$

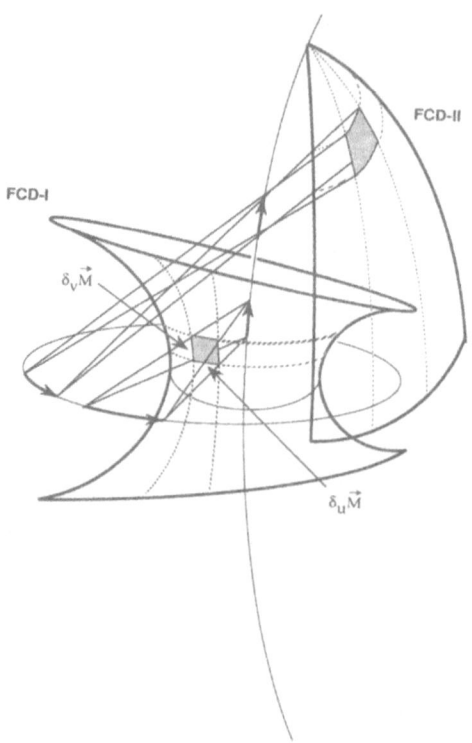

Fig. 5 - Infinitesimal elements of surface dΣ of a Dupin cyclide in the case of FCD-I and FCD-II.
$$d\Sigma = |\delta_v \vec{M}| |\delta_u \vec{M}|$$

To go further, we have to express the element of surface $d\Sigma(r) = AB\, du\, dv$ of the $\Sigma(r)$ cyclide (fig.5). Consider an infinitesimal variation δu of u at constant v. M is displaced from M to $M+\delta_u\, \vec{M}$, and M' from M' to $M'+\delta_u\, \vec{M'}$; r' is modified but M" and r" stay fixed. We have :

$$A\, du = |\delta_u\, \vec{M}| = \delta_u\left(\frac{r'}{\Delta}\, \overrightarrow{M'M"}\right) = \pm\, \frac{b\, \sigma"}{\sigma'-\sigma"}\, du$$

$$B\, dv = \pm\frac{b\, \sigma'}{\sigma'-\sigma"}\, dv$$

Hence :

$$d\Sigma(r) = \frac{b^2|\sigma'\, \sigma"|}{(\sigma'-\sigma")^2}\, du\, dv \qquad (13)$$

4 - ENERGY OF FCD's

Let us now calculate the elastic energy of a FCD. We shall split $F = \int f\, du\, dv\, dr$ into two parts:

$$F = \int f\, du\, dv\, dr = F_1 + F_2 \qquad (14)$$

with :

$$F_1 = {}^{-}_{+}\frac{1}{2}\, K\, b^2 \int \frac{du\, dv\, dr}{(c\cos u - r)(achv - r)} = {}^{-}_{+}\frac{1}{2}K\, (1-e^2)a \int \frac{du\, dv\, d\rho}{(e\cos u - \rho)(ch\, v - \rho)} \qquad (15)$$

$$F_2 = {}^{-}_{+}\, (\bar{K} + 2K)\, b^2 \int \frac{du\, dv\, dr}{(achv - c\cos u)^2} = {}^{-}_{+}\, \Lambda(1-e^2)a \int \frac{du\, dv\, d\rho}{(ch\, v - e\cos u)^2} \qquad (16)$$

The upper signs correspond to FCD-I, and the lower signs to FCD-II. Note that the K-term contributes to the 'topology', since it appears in F_2, which is an integral of the Gauss-Bonnet type. We shall note $\Lambda = \bar{K} + 2K$, $\rho = \frac{r}{a}$; e = eccentricity.

a) FCD-I : negative Gaussian curvature.
The F_2 term can be easily integrated and given an exact form:

$$F_2 = -\, 4\, \pi\Lambda a\, (1-e^2)\, \kappa\, (e^2) \qquad (17)$$

where a is the semi-major axis of the ellipse, e its eccentricity, and $\kappa(x)$ the complete elliptic function of the first kind. Note that F_2 is negative when Λ is positive, a fact which is always insured if $\bar{K} > 0$. This is the most commonly met physical situation, and it explains the prevalence of defects with negative Gaussian curvature. At this point it is worth mentioning that current models of membrane and layers with mechanical forces between molecules always foresee $\bar{K} > 0$ for monolayers, but that for bilayers the sign of $\bar{K}$ depends on the magnitude of the elastic coefficients (Petrov et al., 1984).

The F_1 term shows up singular behaviour near the singular lines, i.e. when $r \rightarrow c$ cos u or $r \rightarrow a$ ch u. We shall <u>assume</u> that the core radius does not depend of the layer (i.e. does not depend on r) ; this assumption is obviously greatly simplifying the situation (for example the layers with intersect the hyperbola far from the plane of the ellipse do show practically no singularity ; see Fournier (1993b) for a critical discussion ; we therefore introduce the following cut-offs on r :

$$r_{cutoff} = a \text{ ch } v - r_c \quad \text{near the hyperbola}$$
$$r_{cutoff} = c \cos u + r_c \quad \text{near the ellipse}$$

and get the following core term in the energy :

$$F_{1-sing} = 4 \pi K a (1-e^2) \kappa (e^2) \ln \frac{a}{r_c} \tag{18}$$

where r_c is typically of the order of the repeat distance of the layers.

The non-singular term writes :

$$F_{1non-sing} = 2\pi K a (1-e^2) \int_{-\infty}^{+\infty} \frac{\ln (ch^2v - e^2)}{(ch^2v - e^2)^{1/2}} dv \tag{19}$$

The total energy F scales like $a(1-e^2)$ and becomes in principle vanishingly small when $e \rightarrow 1$, i.e. for the so-called parabolic focal domains (PFD), where the two conjugate conics are two parabolae. But note that $\kappa(e^2)$ increases without limit for $e \rightarrow 1$, and that the transition from $e \neq 1$ to $e = 1$ is not analytic in the above equations. We treat therefore the PFD's apart.

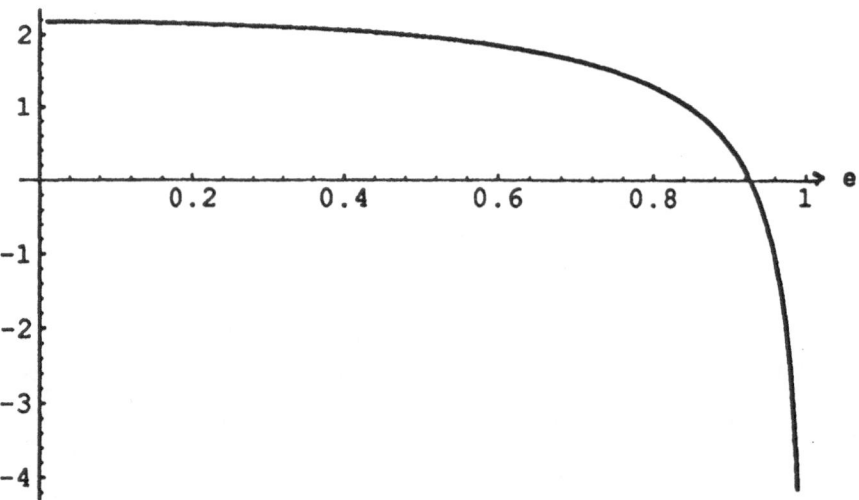

Fig. 6 - Plot of $F_2+F_{1non-sing}$ (eq.17 + eq.19) as a function of the eccentricity e. The sum is vanished for some value of the parameter Λ/K for each value of e.

A further remark concerns the sum of the two non-singular terms $F_2 + F_{1\text{non-sing}}$. Assuming Λ positive, we see that this sum vanishes for some value of the parameter $\frac{\Lambda}{K} = 2$ $+ \frac{\overline{K}}{K}$ for each value of the eccentricity e (see fig.6). Therefore, this sum becomes negative for any FCD of smaller eccentricity, but is in competition with the variation of $F_{1\text{-sing}}$ with e. In any case the general trend is towards FCD's of small eccentricity, and this might explain the frequent occurrence of toric focal domains (TFCD's) in thermotropic SmA's, where Λ is probably rather large.

Another further remark is that, when the total non-singular contribution to the curvature energy is very small, say $< Ba^3$, where B is a modulus of compressibility, it is probably necessary to take into account in the full energy to the compression term, since parallel layers are not in equilibrium if the layers interdistance is a constant (see Kleman and Parodi, 1974). Such a complete model of the energy of FCD's is all wanting.

b) TFCD's with negative Gaussian curvature
The equations above take very simple forms for $e=0$; we get :

$$F = 2\pi^2 \left[Ka \, \ln \frac{a}{r_c} - \Lambda \; a \right] + C \, 4\pi \, Ka \qquad (20)$$

where
$$C = \int\limits_{-\infty}^{+\infty} \frac{\ln \text{ chv}}{\text{chv}} \, dv = 2.17759...$$

Notice that for $\dfrac{\Lambda}{K} = \dfrac{2C}{\pi} \approx 1.386...$, the energy is reduced to its singular term. Notice that in the toroidal case the core radius is a well defined quantity in the sense that it is invariant along the ellipse. Then , taking $a \sim r_c$, i.e. reducing the TFCD to a pore, the total energy vanishes, in the approximation we are considering, which means that the compression term, which have been neglected up to now, becomes predominant. For $\dfrac{\Lambda}{K} > \dfrac{2C}{\pi}$, it is easy to see that F takes some minimum value for a particular value of a solution of $\dfrac{\partial F}{\partial a} = 0$. In the absence of B-term, we find:

$$a \sim \frac{r_c}{e} \, \exp\left(\frac{\Lambda}{K} - \frac{2\,C}{\pi} \right) \qquad (21)$$

Finally, for $\Lambda/K < \dfrac{2C}{\pi}$, the energy F(a) does not have a physical minimum -because $\ln \dfrac{ae}{r_c}$ cannot be negative-, and the TFCD's are unstable defects, little mobile because of the lattice friction. This is probably the most general case, while the case above $\left(\dfrac{\Lambda}{K} \geq \dfrac{2\,C}{\pi} \right)$ should probably be indicative of a precritical or critical regime, near a phase transition.

Similar considerations should hold for general FCD's ($e \neq 0$), but it is more difficult to put them quantitatively, since the core radius is not constant along the focal lines.

c) PFCD's with negative Gaussian curvature
This case has to be treated apart, since it does not follow analytically from the general case. We summarize the results.

Let P and Q be two running points on two confocal parabolae (fig.7) :

$$P = \begin{cases} x = 2f\,\alpha \\ y = 0 \\ z = -\dfrac{f}{2} + f\,\alpha^2 \end{cases} \qquad\qquad Q = \begin{cases} x = 0 \\ y = 2\,f\,\beta \\ z = \dfrac{f}{2} - f\,\beta^2 \end{cases} \tag{22}$$

The distance PQ reads :

$$PQ = f\,(1 + \alpha^2 + \beta^2) \tag{23}$$

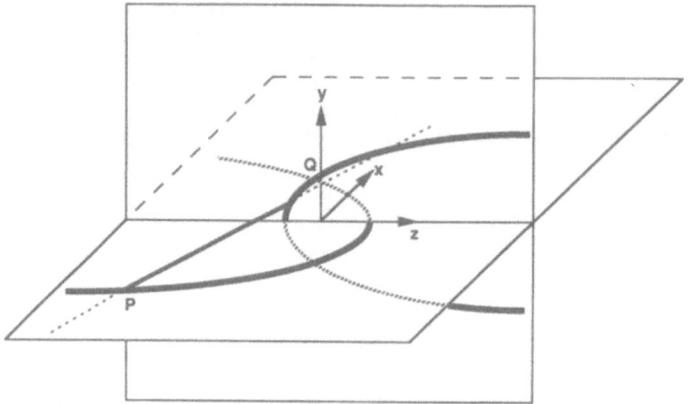

Fig. 7 - Parabolic Focal Conic Domain (PFCD).

and the radii of curvature (measured along the line oriented from P to Q) are :

$$PM = R_P = \frac{1}{\sigma_\alpha} = -\,f\,\alpha^2 - \frac{f}{2} + r \tag{24}$$

$$QM = R_Q = \frac{1}{\sigma_\beta} = -\,f\,\beta^2 - \frac{f}{2} + r$$

where:

$$-\left(\beta^2 + \frac{1}{2}\right) < \frac{r}{f} < \alpha^2 + \frac{1}{2}$$

Equality is forbidden since it yields an infinite curvature ; the excluded region near by is the core region.

Rosenblatt *et al.* (1977), who first discovered PFC's, have shown in a series of pictograms (see their article), that a PFCD-I extends over all space : the layers are practically planar and perpendicular to the common axis of the parabolae, at long distance from the foci ($|r/f| \gg 1$). When $|r/f| > 1/2$, $\Sigma(r)$ cuts one of the parabolae in two points, where the curvature increases without limit. The layers with $|r/f| < 1/2$ have no point singularity : they do not cut the parabolae, but rather envelop them, forming handles ; the Dupin cyclide $r=0$ is symmetric with respect to the plane $z=0$, with a twist of $\dfrac{\pi}{2}$ when passing from the region $z>0$ to the region $z<0$. Using the definition of the cyclides as envelops of spheres (see above eqs. 8 and 9 for the general case), we find that the cyclides can be expressed analytically by the following parametric equations :

$$\frac{x}{y} = \alpha \left\{ 1 + \frac{r}{f} - \frac{z}{f} \right\}$$

$$\frac{y}{f} = \beta \left\{ 1 - \frac{r}{f} + \frac{z}{f} \right\} \qquad\qquad\qquad (25)$$

$$z + r + \beta y - \alpha x = 0$$

The element of area $d\Sigma(r,\alpha,\beta)$ reads :

$$d\Sigma = \frac{4}{|\sigma_\alpha \, \sigma_\beta|} \, \frac{d\,\alpha \, d\,\beta}{(1+\alpha^2+\beta^2)^2} \qquad\qquad (26)$$

and the energy of a PFD-I is of the form :

$$F = - 4\Lambda \int \frac{d\alpha \, d\beta \, dr}{(1+\alpha^2+\beta^2)^2} + 2K \int d\alpha \, d\beta \, f^2 \, |\sigma_\alpha \, \sigma_\beta| \, dr \qquad (27)$$

We shall not develop here the full expression of F. It will be sufficient at this point to remark that :

– the topological term sums up to $-4\pi\Lambda f$ in the continuous limit (f>>d) for the layers $|r/f| < \frac{1}{2}$: each layer has a topological contribution which amounts to -4π. For $|r/f| \geq \frac{1}{2}$, the topological contribution diverges logarithmically with the size $2L$ of the PFCD measured along the z-axis, which scales (for L larger than f) like the maximum value of r ; we have finally :

$$F_{topol} \approx - 4\pi \, \Lambda \, f \left[1 + 2 \ln \frac{2\,L}{f} \right] \qquad\qquad (28)$$

– the K term yields a non-singular, finite, contribution for $|r/f| < \frac{1}{2}$, viz :

$$F_{Knon\text{-}sing} = 2 \, \pi^3 \, K \, f \qquad\qquad (29)$$

and a singular contribution for $|r/f| > \frac{1}{2}$. Since the effects of curvature decrease when r increases, the curvature of those surfaces is certainly not of a large order of magnitude in the region outside the core r_c. Taking r_c constant on segments of parabolae of the order of f, assuming that the singularities are negligible outside these segments, we expect :

$$F_{K\text{-}sing} \approx K \, f \ln^2 \frac{fL}{r_c^2} \qquad\qquad (30)$$

Again, the sum of the contributions in eqs. 28, 29 and 30 does not take into account the core energy. A more complete discussion of all those energies would be useful, and could probably be done without too much difficulty with the present days desk computer facilities.

d) FCD-II : positive Gaussian curvature

We choose the range of variation $r > achv$, introducing also a constant cut-off r_c, related to the maximal value of v on each $\Sigma(r)$ by the relation $r = achv_{max}+r_c$, and a size $r_{max} = +R$. This latter condition means that we limit the system by the cyclide parameterized by R. Since the layers have positive Gaussian curvature, this is closed surface of genus zero, with two singular points at opposite points on the hyperbola.

Let us first calculate the energy of the layer parameterized by r. The range of variation for v and u reads:

$$1 < chv < r/a \quad ; \qquad 0 < u < 2\pi \tag{31}$$

The elastic energy of the layer is split as above:

$$F_r = \int f\, dv\, du = f_{1,r} + f_{2,r} \tag{32}$$

with:

$$f_{1,r} = \frac{1}{2}\,\kappa\, b^2\, \frac{4\pi}{\sqrt{r^2 - e^2}\,\sqrt{r^2 - a^2}}\, \ln\, \frac{2(r-a)}{r_c}$$

$$\tag{33}$$

$$f_{2,r} = (2\kappa + \bar{\kappa})\, 4\pi \sqrt{\frac{r^2 - a^2}{r^2 - e^2 a^2}}$$

We will now restrict our calculation to the only experimentally observed case of rotationally symmetrical domains (e=0, a=b).
In this case the principal curvatures read (Fig.8):

$$\sigma' = \frac{1}{r} \quad ; \qquad \sigma'' = \frac{\cos\theta}{r\cos\theta - a} \tag{34}$$

and the infinitesimal element of surface dΣ is:

$$d\Sigma = 2\pi r\, (r\cos\theta - a)\, d\theta \tag{35}$$

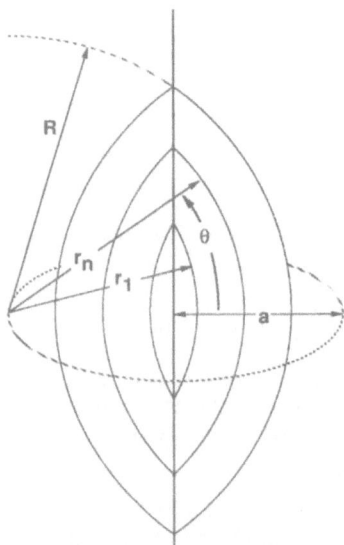

Fig. 8 - Focal Conic Domain of the second Species (FCD-II) in the case e=0. The number n (n=1,2,...) labels the bilayers. a is the radius of the circle, virtual in the case of FCD-II. r_n= a+ d(n - 1/2). The meridian cut of the domain is the intersection of two circles with radius R. θ is the angle between the normal the bilayer and the the plane containing the circle.

Equations (33) become:

$$f_{1,r} = \frac{1}{2} \kappa \, a^2 \frac{4 \pi}{r \sqrt{r^2 - a^2}} \ln \frac{2(r-a)}{r_c}$$

$$f_{2,r} = 4 \pi \, (2\kappa + \bar{\kappa}) \sqrt{\frac{r^2 - a^2}{r^2}}$$

(36)

Thus the elastic energy of the layer at r reads:

$$F_{\kappa,\bar{\kappa},r} = 4 \pi \, (2\kappa + \bar{\kappa}) \sqrt{\frac{r^2 - a^2}{r^2}} + \frac{2\pi\kappa a^2}{r\sqrt{r^2-a^2}} \ln \frac{2(r-a)}{r_c}$$

(37)

where $r = a/|\cos\theta_{max}|$
The total energy of the FCD-II reads:

$$F = \int F_{\kappa,\bar{\kappa},r} \frac{dr}{d}$$

(38)

Integrating the topological term over the whole volume of the FCD-II ($a < r < R$), we find for the total energy:

$$F = 4 \pi \Lambda \, R \left\{ \sqrt{1 - \frac{a^2}{R^2}} - \frac{a}{R} \arccos\frac{a}{R} \right\} + 2 \pi \, K \, a^2 \int \frac{\ln(2r-a)/r_c}{r\sqrt{(r^2-a^2)}} dr + W_c$$

(39)

W_c is a core energy. In this case it is possible to estimate it. We will assume that the energy density of the core is of the order of $\alpha K \sin^2 \omega$ (by analogy with a pure divergence deformation of a nematic medium), where ω is the angle between the normal to the bilayer (line PQ) and the defect line (line S (fig.9).) The energy of the line defect reads:

$$W_c = \int_{a+d/2}^{R} \alpha K \left(\frac{a^2}{r_n}\right) dr_n = \alpha K a^2 \frac{R-r_1}{R r_1}$$

(40)

r_1 denoting the first layer (see fig. 9).

This expression tells us that when $a = 0$ the core energy vanishes. In fact there is no line defect because when $a=0$, the FCDII becomes a system of concentric spheres, the line is reduced to a point.
We shall now estimate the energy of deformation of the matrix, corresponding to the introduction of the FCD-II in the homeotropic region (fig.10). The energy density of this deformation is of the form $w_B = \frac{1}{2} B \left(\frac{\partial u}{\partial z}\right)^2$, where u is the perpendicular displacement (in the z direction) of the layers from their initial position corresponding to the flat layers.

In a very simplified model, u of the layers in the homeotropic region can be taken as:

$$u = R_t \cos qx \cos hy \exp\left(-\frac{z}{L}\right)$$

(41)

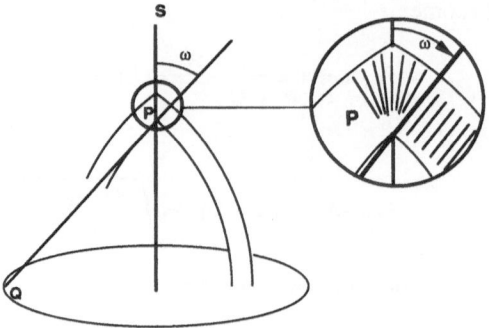

Fig. 9 - Description of the FCD-II core model. In the angle ω the deformation is of a pure divergence type. The density energy of the core line can be estimate by α K$\sin^2\omega$ per unit length.

with $q = \dfrac{\pi}{2R_t}$ and $h = \dfrac{\pi}{2R_l}$

The deformation must minimize the energy; this condition yields:

$$L = \frac{1}{\sqrt{\dfrac{K}{B}}\,(q^2 + h^2)} \qquad (42)$$

The integration of the energy density related to the deformation of the matrix over the ranges $-R_l < x < R_l$; $-R_t < y < R_t$; $-\infty < z < \infty$ take the form:

$$W_B = \frac{\pi^2}{8}\,B\,\sqrt{\frac{K}{B}}\,R_t^2\,\left(\frac{R_t}{R_l} + \frac{R_l}{R_t}\right) \qquad (43)$$

Notice that for a given number of layers, W_B has a minimum when $R_l = R_t$, *i.e* for spherical domain.

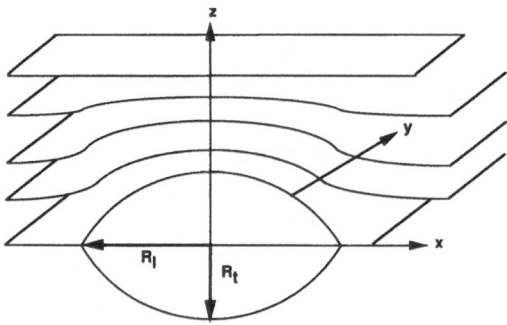

Fig. 10 - Deformation of the homeotropic region due to the introduction of FCD-II.

5 - OBSERVATIONS OF FCD's IN A LYOTROPIC LAMELLAR PHASE

Lyotropic lamellar phases are made by equidistant organic bilayers regularly separated by water or oil. Some of these lamellar phases show very interesting textures of FCD's. This is the case when the lamellar phase is between two phases of different topology. In the system studied by Boltenhagen *et al*.(1991,1992), both types of FCD I and II have been observed.

a) Oily streaks and FCD-I in lyotropic lamellar phase

Oily-streaks are the most usual structural defects in lyotropic lamellar phases. They appear as long bands with a complex inner structure, which subdivide the sample into so-called homeotropic regions where the layers are parallel to the boundaries of the preparation, and thus appears in black between cross nicols.

The 3-components system cetylpyridinium chloride/hexanol/brine, which we have studied in some details for its textures, shows up in the swollen region ($\phi_{brine} > 0.7$) a lamellar phase L_α located in the phase diagram between an isotropic sponge phase L_3 (made of a single connected bilayer randomly folded, which separates space into two connected regions of brine, according to the current model (Porte et al.,1989)) and an isotropic micellar phase L_1.

When the observations are carried in the lamellar phase for a set of compositions close to the L_3 phase, the predominant defects are oily streaks which are split into chains of FCD's of the first species (Boltenhagen et al, 1991). The ellipses are in the plane of the observations, with their long axes transverse to the oily streaks.

Fig.11 illustrates these characters. The plane which contains the ellipse is always parallel to the layers belonging to the homeotropic region. By changing the focusing of the microscope, it is possible to follow a line singularity, which lies in the vertical plane containing the major axis of the ellipse and passes through the focus of the ellipse. This line is one branch of hyperbola. Between crossed nicols, the focus of the ellipse is the center of four dark brushes (fig.11). These observations lead to the conclusion that these domains are FCD's of the first species.

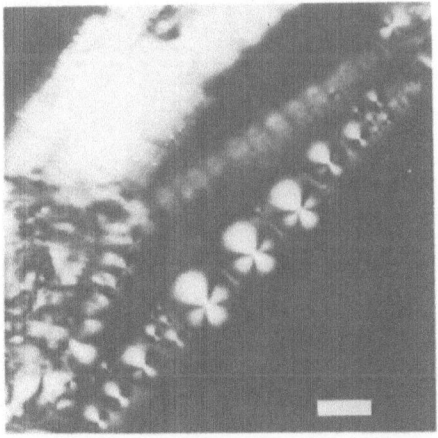

Fig. 11 - Oily streak which contains FCD-1. The oily streak is a chain of FCD-I. The gap between two domains is filled with edge dislocations CpCl/hexanol/Brine = 14.8/15.2/70. Sample thickness 130 μm. Bar 50 μm.

As first shown by Colliex et al (1974) (see also Kleman, 1983; Rault, 1975), an oily streak is a mode of instability of an edge dislocation of large Burgers vector. Dislocations of small Burgers vector appear in the sample as a result of the capillary process of introduction of the chemical between the glass plates, and gather together after a few hours, to form large Burgers vector dislocations which are split into FCD's as described above. The intervals between the FCD's are filled either with the segments of dislocations which have not transformed completely, or with FCD's of smaller size (see Fig.12). The largest FCD's have geometrical features which are directly related to the Burgers'vector β (β=nd, a multiple of the L_α phase repeat distance); most generally $\beta = b_1 - b_2$, meaning that the dislocation is the sum of two dislocations of opposite signs b_1, $-b_2$, whose Burgers vectors scale with the size of the sample. The largest FCD's have a major axis a=$(b_1+b_2)/2$, (i.e. scan the whole dislocation inner region), a minor axis $(b_1 b_2)^{1/2}$ and an eccentricity e = $(b_1+b_2)/(|b_1-b_2|)$. This geometry illustrates the topological relationships which exist between dislocations and disclinations (focal lines in FCD's are special disclination lines) - for a

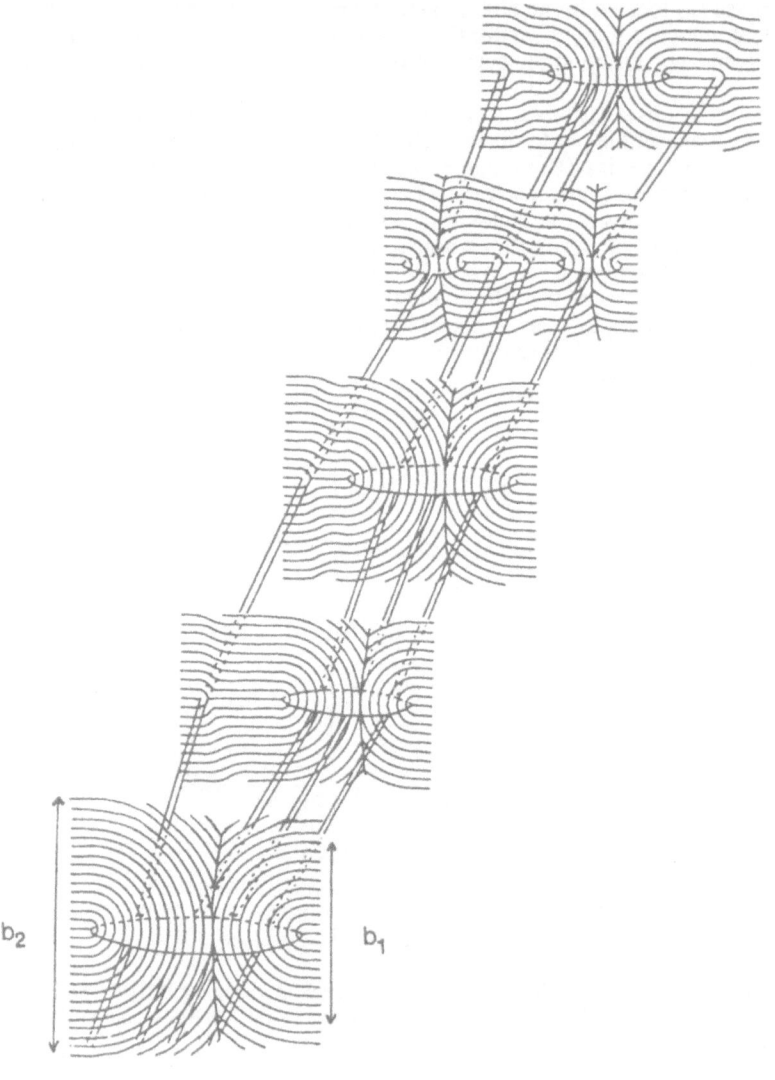

Fig. 12 - General structure of an oily streak. The oily streak is a chain of FCD-I joined by edge dislocations.

discussion of this latter point, which would bring us too far in this review, see Kleman (1983) and Bourdon et al (1982), where this question is developped for the specific case of FCD's. See also Williams et al (1975) for another example of an instability of (surface) edge dislocations versus FCD's.

The stability of the FCD splitting of the edge dislocation is theoretically studied in Boltenhagen et al (1991). An important element in this stability is the proximity of the sponge phase, the only region of the phase diagram where FCD-I's are visible.

To understand this point we must remember that the sponge phase contains a large quantity of pores or handles, which are of negative Gaussian curvature. Thus, a large

positive $\bar{K}$ stabilises the sponge phase (Porte et al, 1989). The measurement of the

geometrical parameter of the ellipse allows to estimate the ratio $\bar{K}/K$, which has indeed been found positive in the vicinity of the lamellar-sponge phase transition. So the existence of this particular splitting in FCD-I, which contains layers with negative Gaussian curvature, near

the sponge phase, is enhanced by a large positive value of $\bar{K}$.

b) Focal conic domains of the second species in lyotropic lamellar phase

In the same system, it has been observed for the first time FCD's of the second species (Boltenhagen et al, 1992). This is the case when the lamellar phase is near the micellar phase, which is made of globular or rod-shaped micelles. Observations with the polarizing microscope show textures which differ drastically from typical textures ever reported for smectic like phases. One observes birefringent elongated domains (fig.13), located at different levels of the surounding matrix, which is also birefringent. The meridian cut of the boundary of each of those domains has the shape of the intersection of two circles with the same radius R. In these domains the optical axis is distributed in a radial-like manner, so it is perpendicular to the domain boundary. These observations lead to the conclusion that these domains are FCD's of the second species, in which the layers are folded around two conjugated lines, viz., a circle and a straight line. The circle is virtual, and in this case, only the straight line is visible under microscope.

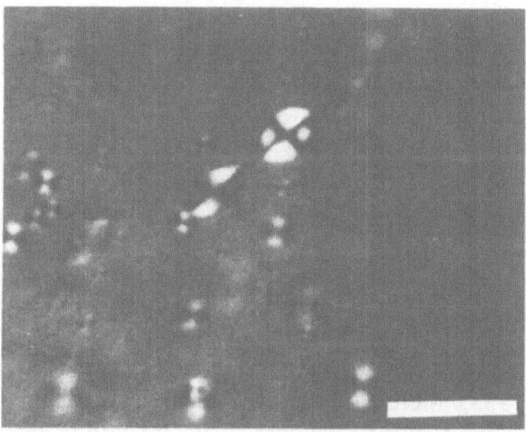

Fig. 13 - Focal Conic Domains of the second species - CpCl/hexanol/Brine = 12.5/7.5/80. Bar 50μm

These domains are not stable, and as time elapses, they relax into spherical ones. During this process, the virtual circle is reduced to a point and the number of layers constituting the domain is conserved.

Contrary to the case of FCD-I, layers in the FCD-II's have positive Gaussian

curvature; this is related to a negative value of $\bar{K}$, which has been estimated by several methods (Boltenhagen et al.,1992, and in preparation; Boltenhagen, 1993).

6- SPACE FILLING WITH FOCAL CONIC DOMAINS

Since the focal conic domains have a nontrivial shape which does not tile space, the appearance of conical FCD-I's or spherical FCD-II's raises the question of space filling with distorted smectic layers. Here we consider examples with usual FCD-I's and FCD-II's; the case of toric FCD-I's is trivial since a 2D periodic network of equal TFCD's can fill the space with gaps where the layers are striclty parallel and planar, which match smoothly with the negative Gaussian curvature parts of the tori. The case of PFCD's has been considered experimentally and theoretically by Rosenblatt et al. (see also Oswald et al. 1982); each PFCD fills all space, and the layers are pratically flat (see above) at some distance of the axis of the parabolae; there is therefore a pratically smooth matching from one PFCD to another, as soon as the axes are parallel.

a) Focal conic domains of the first species.

Usually in polarizing microscope one observes not isolated focal conic domains, but entire groups of contacting domains with a common apex. This geometry is known as the first of Friedel's law of FCD-I association: if two FCD-I's are in contact, they are tangent to each other along the common generator of two conical surfaces. Since the layers are everywhere perpendicular to the generators, they cross the line of contact without singularities (see Friedel 1922, Kleman 1982). In particular, this geometry arises in the defect texture of grain boundaries: instead of dislocations, the boundary is filled with ellipses where long axes are parallel and which have equal eccentricities. The hyperbolae have consequently parallel asymptotes, and the angle between the asymptotes measures the disorientation between the two grains. The "common apex" is at infinity on both sides. But the question arises, how it is possible for layers to cross smoothly from domain to domain not only along the lines of contact, but in the whole volume of a gap between them?

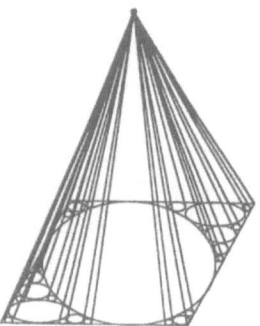

Fig. 14 - Filling of the pyramid by the iterative system of FCD-I's with common apex.

For a long time the only model was that one of iteration filling proposed by Bidaux et al. (1973). Let us imagine a part of space which has the shape of a pyramid with a characteristic size L close to the sample thickness, Fig.14. In general, the whole space can be filled by such sets of pyramids and complementary sets of tetrahedra representing parts of FCD-I's, see Bragg (1933). More specifically: one starts by placing inside the pyramid a domain of largest possible size $\sim$ L; the apex of the domain is located at the apex of the pyramid. Then the remaining gaps are filled with smaller conical domains, etc., down to molecular scales; the smallest domain has height L and base radius $a^* \sim \lambda \sim \sqrt{K/B}$. The result $a^* \sim \lambda$ is an intrinsic feature of the model where the filling is defined by the competition between dilation and pure curvature terms. However, in experimentally observed textures the iteration process stops often at scales much larger than molecular, $a^* \gg \lambda$. As one can see from Fig.15, there are regions with characteristic size $a^* \sim 10\mu$m between the large domains that are free from smaller domains, and hence are filled with layers of a special form. Sethna and Kleman (1982) have proposed that the gaps between the FCD-I's with common apex can be smoothly filled with spherical layers, i.e., with parts of...spherical FCD-II! The model is as follows:

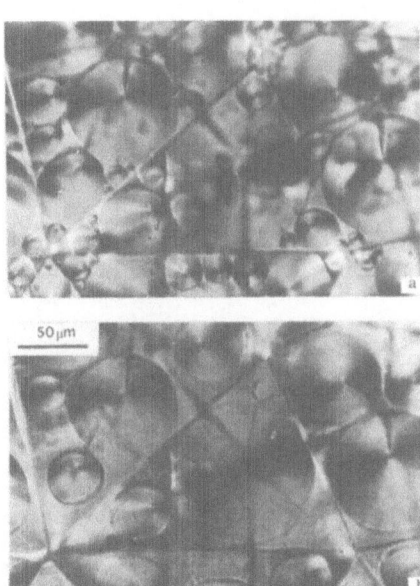

Fig. 15 - Typical polygonal texture of smectic A phase containig ellipses of FCD-I located at the cell boundary. In between the large FCD-I one observes clear gaps that are filled with spherically curved layers. The macroscopic size of the smallest domains is defined by the balance of the elastic constant K and the anisotropy of the surface energy; when one proceeds with temperature from the deep smectic A phase (a) towards the smectic B phase (b) the critical size increases and the smallest domains disappear.

Let us imagine a FCD-II with spherical concentric layers and drill out from it several circular cones with vertices at the center of the sphere (Fig.16 a). Now we fill these holes with FCD-I cones. It is clear that the remaining regions of the FCD-II match smoothly with these FCD-I's. The experimental realization of such a filling is illustrated by the texture of a spherical droplet (Fig.16 b): one can easily distinguish the cones of FCD-I embedded into the sphere with common apices in the center of the drop; the only defects in the textures are circles at drop surface bounding the basis of the FCD-I's and the straight radial lines ending at the center of the droplet. Similar geometry holds not only for droplets but also for the general case of pyramids.

Now the problem is to find a physical reason for a macroscopic cut-off of the iteration process. The new factor that explains why a^* does not necessarily coincide with λ is the anisotropy of the surface energy of the smectic phase. For a qualitative understanding it is sufficient to point out that there are two types of molecular anchoring at the sample surface. In the regions with spherical packing the molecules are normal to the surface; within the islands occupied by the bases of FCD-I's the molecules are tangentially anchored, see Fig.16 c. The surface energies $\sigma_{//}$ and $\sigma_{\perp}$ for these two types of molecular alignment are not equal in general. If $\Delta\sigma = \sigma_{//} - \sigma_{\perp} < 0$, then the appearance of a small FCD-I with radius a in between the larger domains saves surface energy $\sim (-\Delta\sigma)a^2$; the elastic enegy cost is $\sim Ka$. The balance of the surface and pure curvature terms results in a characteristic length (Lavrentovich 1986 and1987):

$$a^* \sim K/ (-\Delta\sigma) \qquad (44)$$

which can be very different from λ. On scales exceeding this critical lenght a^*, the space filling is realized by an iterative system of FCD-I's, obeying G.Friedel association rules. On scales smaller than a^* the remaining gaps cannot be filled with FCD-I's because the losses in elastic energy are larger than the surface energy gain. These gaps are filled with layers of spherical curvature. Fig.15 shows how the small domains disappear when one increases the elastic constant K (and thus a^*) approaching the smectic A- smectic B transition temperature

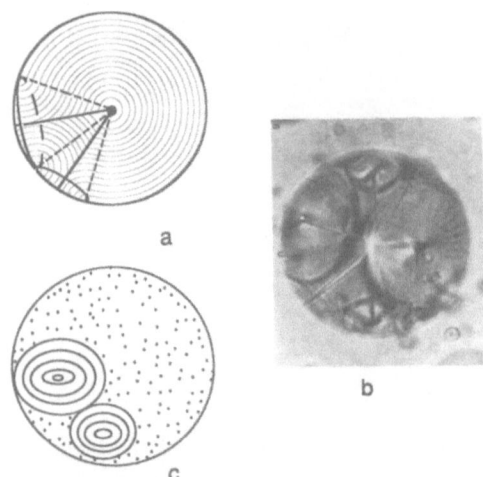

Fig. 16 - a- The scheme explaining the geometry of the Sethna-Kléman model: the interstities between the FCD-I's are filled with parts of the spherical FCD-II; **b-** Experimental realisation of the model within the smectic A droplet freely suspended in an isotropic matrix; **c-** The orientation of the molecules at the surface of droplets is different for region filled with FCD-I's and FCD-II layers.

(in this experimental case). Experimental determination of a^* allowed to estimate $\Delta\sigma$ for smectic A/glycerin interface to be of the order of $(\sim 10^{-2})\text{erg/cm}^2$, which is significantly smaller than the isotropic part of the interface energy σ ($\sigma \sim 10 \text{ erg/cm}^2$).

The confined geometries FCD-I just described above does not change the equilibrium shape of the smectic volume: the shape is fixed either by rigid glass plates (flat cells) or by isotropic surface tension $\sigma >> |\Delta\sigma|$ (smectic droplets suspended in an isotropic matrix such as glycerin). A more difficult question arises when one considers the nucleation of the smectic A from its isotropic melt, where σ is small and can be comparable with the energy of bulk distortions. It is well known that the smectic A phase appears as elongated "bâtonnets" first studied by Friedel and Grandjean in 1910. The "bâtonnets" are decorated by complicated sets of focal conic domains. Here again they appear starting with characteristic size $a^* \sim K/ (-\Delta\sigma)$ (see Fournier and Durand, 1991). Since the anisotropic and isotropic parts of the smectic/isotropic interfacial energy are comparable the appearance of FCD's can be accompanied by changes in the nuclei shape. As it was shown by Fournier et al.(1991), one should distinguish three regimes according to the magnitude of L with respect to a^*, which might be estimated for the SmA-isotropic melt interface as 0.1-1μm (Fournier et al. (1990)).

(a) For $L<<a^*$ the nuclei show an undistorted bulk structure and have an elongated shape given by the Wulf's construction known for three-dimensional crystals. The layers are strictly undistorted since both curvatures and dilation lead to enegy losses higher than the surface energy gain.

(b) For usual three-dimensional crystal, the distortions of the lattice are negligible whatever the system size may be; thus in the Wulf's construction the bulk structure is always assumed to be strictly ideal. In a SmA, which is a one dimensional crystal, the situation is different when the system is large enough. For $L>>a^*$ focal conic networks relax the surface energy anisotropy ($\Delta\sigma$) and the SmA bulk is curved to provide the equilibrium of the system as a whole. Moreover, large SmA nuclei behave like a three-dimensional liquid in the sense that their equilibrium shape is almost spherical: the surface isotropic energy $\sim\sigma L^2$ is higher than the elastic energy of any focal conic configuration inside the nuclei.

(c) The case $L \sim a^*$ is the most difficult to describe: here bulk elasticity is in real competition with both anisotropic and isotropic parts of the surface energy. In that respect it is interesting to note that in some substances the smectic A phase grows from the isotropic melt via long cylindrical structures (with radius 1-3 μm and length up to few centimeters) that originate from more or less rounded nuclei (see e.g., Meyer et al, 1990, Adamczyk, 1989). Very recently Naito et al (1993) have proposed a model using so-called Weingarten surfaces on which there exist a functional relation between the mean and Gaussian curvatures. The model allows the existence of both spherical and cylindrical nuclei. However, it does not consider the anisotropy of the surface tension. This point might be important, since the main difference between "bâtonnets" filled with focal conic domains and spherical filaments is apparently the difference in the surface orientation. In "bâtonnets" the tangential orientation of molecules is preferable while in filaments the surface orientation seems to be normal. The complete description of the smectic nucleus shape remains a puzzling theoretical problem.

b) Focal conic domains of the second species.

Although it is not clear whether iterative space filling of a smectic phase with spherulites has been observed —the electron microscopy studies of the lyotropic phases performed by Boltenhagen et al (1993) in the vicinity of the L_α-L_1 phase transition revealed a filling with spherulites that have different sizes ranging from macroscopic (a few micrometers) to microscopic (tens and hundreds Angström) but this filling might well be explained in terms of pretransitional effects— it appears interesting to consider such a possibility.

There are some important differences in packing of space with FCD-I's and FCD-II's. First of all, a FCD-I with conical shape has volume $L \times a^2$ and extends through the whole sample (length L). For example, in Figs. 14-16, each FCD-I has its base located at the sample surface. This is why the anisotropy of the surface energy is so important for the scenario with FCD-I's. In contrast, the FCD-II has a closed shape with characteristic volume a^3, where a might be much smaller than L. Thus the physical limit for the iteration process is defined solely by the bulk properties of the lamellar phase.

The largest spherulites have a macroscopic size $R \gg \lambda$ (defined e.g. by the sample size or by the shear rate) and distort the lamellar matrix. The energy W_B of the layer compressibility outside the spherulites of radius R scales like $B\lambda R^2$ if a mean separation is larger than R^2/λ and like BR^3 if the separation is smaller than R (in the latter case the dilatation ε of the layers is of the order of unity). These distortions can be relaxed by smaller FCD-II's lying in between the large FCD-II's since the geometry of the FCD-II implies only curvature deformations and energy $\sim \Lambda a$. The iteration process will interrupt at scales a^* that do no provide a sufficient energy gain when substituting curvature by dilation. To define a^*, let us introduce the number g of spheres of radius $R \geq a$ packed in a volume L^3, $g = (L/a)^\gamma$. The residual volume (that is not occupied by FCD-II's) is $V(a) \sim L^\gamma a^{3-\gamma}$. Then the total free energy scales as :

$$F(\rho) \cong \Lambda L^\gamma a^{1-\gamma} + BL^\gamma a^{3-\gamma}. \tag{45}$$

Here γ, contrarily to the Appolonius exponent which is met in the problem of 2D-circles filling (Bidaux et al. (1973)), is not an universal constant. But, in some experimental cases (concrete) a relevant value of γ is approximately 2.8 (Omnès 1985). This value yields indeed a minimum of $F(\rho)$ when $\Lambda > 0$ and a physical limit of iterations :

$$a^* \cong \sqrt{(\gamma-1)\Lambda/(3-\gamma)B} \tag{46}$$

Thus the iterations go down to a very small scale $\sim \lambda$ as in the Bidaux et al model for FCD-I's filling. It is difficult to imagine some other cut-offs of the iteration because in balance of the curvature and dilatation elastic terms the only characteristic length is that defined by λ.

The space filling spherulitic state is a (meta)stable one if $\overline{K}$ is negative and large enough. The fusion of two neighboring spherulites is hindered by the necessity of creating an energetically unfavorable "passage" with negative Gaussian curvature between them (see Fig.17). However, with positive $\overline{K}$ the situation will be different.

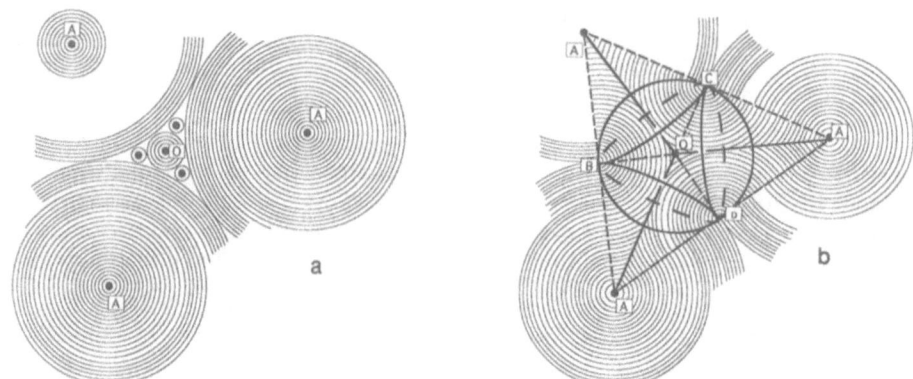

Fig. 17 - Iterative filling of the space with FCD-II's centered at points A and O (a) and fusion of the FCD-II's via the FCD)I's ABOC, ACOD, ABOD (b).

7. NUCLEATION AND GROWTH OF FOCAL CONIC DOMAINS

The considerations given above do not provide answers to the questions of nucleation and growth. Experiments (see, e.g. Fig. 15) show the possibility of transitions between a domain-free state (a = 0) and focal conic textures composed of FCD's with macroscopic size (a > a*).

Let us restrict ourselves to the simplest case when a focal conic domain in the form of a TFCD is supposed to appear in a flat cell of fixed thickness h from initially uniform stack of layers under the action of the magnetic field H applied across the cell. If the diamagnetic anisotropy $\chi_{//} - \chi_{\perp} = \chi_a$ of the smectic liquid crystal is negative, it is natural to expect a nucleation of TFCD's since the reorientation of layers decreases the diamagnetic coupling with the field. Such a phenomenon has been observed by J.C. Dabadie et al. (1990) in ferrosmectics. On the other hand, the TFCD can be smoothly embedded into a system of parallel layers without dilation ; hence the elastic energy cost of the domain appearance could be relatively small.

The physical mechanisms that govern the behavior of the domain is different for domains smaller or larger than the cell thickness h (Fig. 18). Let us introduce a dimensionless order parameter ρ, which is the ratio of the domain radius to the cell thickness, $\rho = a/h$. The case of arbitrary ρ was considered by Lavrentovich and Kleman (1993) ; here we consider only two limiting cases : nucleation ($\rho \ll 1$) and expansion ($\rho \gg 1$) of the TFCD.

a) Nucleation of TFCD's.

For $\rho \ll 1$ the deformations of layers are restricted to a region of volume a^3 and do not modify the orientation at the boundaries. The nucleation is defined by the balance of the elastic and field terms. The elastic energy scales linearly with TFCD radius, $\sim h\rho$, while the field term should scale as $\sim h^3\rho^3$, which is the volume of significant layers curvature. More

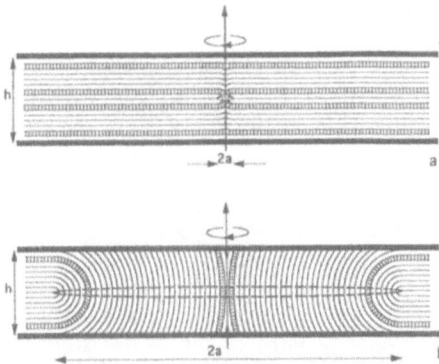

Fig. 18 - The structure of the smectic cell with TFCD for two extreme situations: the radius a of the domain is much smaller (a) or much larger (b) than the cell thickness h. The small domain does not change significantly the boundary anchoring: maximal angle of inclination is arctg($2a/h$) << 1; the volume of distorsion is ~ a^3. In contrast, the large domain occupies the volume ~ ha^2 and the surface orientation of molecules is pratically tangential.

precisely, the difference between the free energies of the domain and uniform state is represented as :

$$\Delta F(\rho<<1)=A_1\rho + A_3\rho^3 +...,$$

$$A_1 = 2\,\pi^2\,K\,h\,(\beta-2-\bar{K}/K), \qquad A_3 = \frac{\pi^2}{6}\,|\chi_a|\,H^2\,h^3 \qquad (47)$$

where $\beta = \ln(2a/rc) \approx$ const and $(-\bar{K}/K)$ appears because of positive Gaussian curvature of the TFCD.

The behavior of the system is determined by the signs and values of the coefficients of expansion. The coefficient A_3 is always negative since $\chi_a < 0$. If $\bar{K}$ is large enough, the coefficient A_1 might be also negative and the appearance of the TFCD will be inevitable even without external field. However, for the stable smectic A phase A_1 should be always positive, i.e., $\bar{K}$ is not too high. The balance of the linear positive term and cubic negative term leads to the first-order character of the instability : $\Delta F(\rho)$ goes through a maximum $\Delta F_c = \Delta F(\rho_c)$ at some critical radius ρ_c, which defines the critical TFCD-nuclei. Only the TFCDs with $\rho > \rho_c$ transform the metastable uniform state into a stable defect state. The TFCDs with $\rho < \rho_c$ (embryos) are unstable and decay.

The problem is that the energy barrier $\Delta F_c \cong A_1^{3/2}/(3\sqrt{3A_3}) \sim 10^{-9}$ erg separating the metastable uniform ($\rho = 0$) and stable domain state is too high to be surmounted by thermal fluctuations ($k_B T \sim 4.10^{-14}$ erg) for plausible H (10^2 kGs in experiments with typical thermotropic smectics). Thus the TFCD nucleation from initially uniform stack of SmA layers is a very rare event (except for large $\bar{K}$). A possible solution is that the nucleation starts from a nonuniform state. The main idea is simple : a bulk dust particle or surface irregularity dilates (or compresses) the smectic layers and thus the initial state is characterized by some nonzero dilation energy. The nucleation of the TFCD means the substitution of the dilation by curvature deformations which generally have less energy. The last circumstance should decrease the barrier ΔF_c. In fact, there are direct experimental confirmations that the field-induced nucleation occurs at surface or bulk irregularities (see, e.g., Hinov (1988) ; Jakli and Saupe (1992). Fournier, Warenghem and Durand (1993)).

Let us consider, as an example, a heterogeneous nucleation caused by a dust particle whose shape is close to a circular bicone of height 2l and base radius R (Fig. 19). It creates deformations which are relaxed by a set of dislocations loops. The energy of a dislocation loop with radius r_0 is (Kleman, 1974) $f_0 = \pi \lambda Bd^2 r_0/\xi$, where ξ is a core radius and d is Burgers vector taken equal to one layer thickness. Thus the energy of n = 2l/d dislocation loops of radius R is approximately:

$$F_{disl} = \pi BdlR \frac{\lambda}{\xi} = \pi K \frac{ld}{\lambda\xi} R \qquad (48)$$

(The interaction energies scale also like R).

The nonzero elastic energy F_{disl} of the initial state drastically changes the ΔF behavior in the initially distorted region, because it enters as a new negative contribution (-A'_1) into the ΔF expansion :

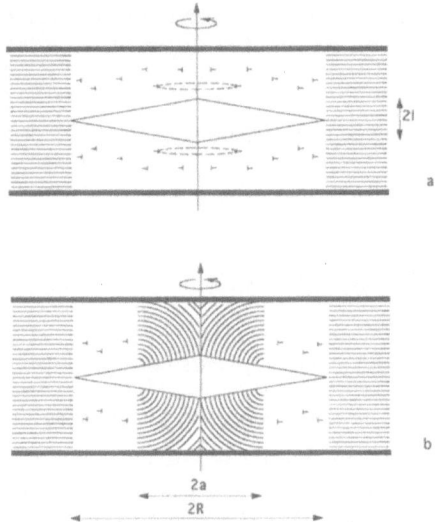

Fig. 19 - Dust particle of biconical shape in the SmA cell: (a) initial state with distorsions relaxed by dislocation loops; (b) nucleation of TFCD.

$$\Delta F((\rho<<1) = (A_1 - A'_1)\rho + A_3\rho^3+...$$

$$= 2 \pi^2 K h (\beta-2-\bar{K}/K-\frac{ld}{2\pi\lambda\xi})\rho+A_3\rho^3+... \qquad (49)$$

Now the barrier $\Delta F^* \cong 2(A_1 - A'_1)^{3/2}/3\sqrt{3}|A_3|^{1/2}$ is significantly reduced by the contribution A'_1 from the dislocation energy. The barrier even completely disappears when $\beta - 2 - \bar{K}/K < \frac{ld}{2\pi\lambda\xi}$. With $d \sim \xi \sim \lambda \sim 30$ Å, $\bar{K} = 0$, $\beta \sim 4 - 10$ the last situation is satisfied for particles with height as small as $2l \sim (20 - 60)d$, i.e. $2l \sim 10^3$ Å. It simply means that the irregularity can nucleate the TFCD even when the external field is absent.

b) Expansion of TFCD.
The behavior of the system differs principally for $\rho >> 1$ and $\rho << 1$, because of the confined nature of the cell. First of all, the volume of the reoriented layers for $\rho >> 1$ scales

as a^2h (Fig. 19b) rather than as a^3 (Fig. 18a) ; thus the driving term $\pi a^2 h \times \chi_a H^2/2$ is now proportional to ρ^2 rather than to ρ^3. Furthermore, the layers in the TFCD region are practically reoriented along the field (Fig. 18b) and the anchoring term $2\pi a^2 \times W_a \propto \rho^2$ should be considered. Here W_a is so-called anchoring coefficient wich is the work needed to reorient the layers from equilibrium direction. W_a has the same dimensionality as the surface energy anisotropy $\Delta\sigma$, but is not necessary equal to $\Delta\sigma$ in smectics (see below).

The domain expansion is defined mainly by the competition between the surface ($\sim \rho^2$) and the field ($\sim \rho^2$) energies, because the elastic term retains its behavior $\propto Ka \propto \rho$ and can be neglected since $\rho \gg 1$. Balance of these energies defines a "saturation" field H_{sat} above which the TFCD expands infinitely :

$$H_{sat} = 2 \sqrt{\frac{W_a}{|\chi_a|h}} . \tag{50}$$

It is worth taking note that H_{sat} for a modest W_a is significantly smaller than the thresholds of the well-known Helfrich (1970) undulation instability and Parodi's (1972) dislocation instability. These two instabilities are defined by the balance of the field energy

and the dilation energy, $H_{sat} = \sqrt{\frac{2\pi\lambda B}{|\chi_a|h}} = H_p$. Thus the ratios :

$$\frac{H_{sat}}{H_{HH}} \sim \frac{H_{sat}}{H_p} \sim \sqrt{\frac{W_a}{\lambda B}} \tag{51}$$

can be small, ~ 0.06, with $W_a \sim 0.1$ erg/cm^2, $\lambda = 30$ Å and $B \sim 10^8$ dyn/cm^2. For strongly anchored SmA, however, one can expect surface layer melting $W_a \sim Bd$, and thus the ratios (51) will be close to unity.

As it follows from recent experiments (Li and Lavrentovich (1993)), when the anchoring is strong, the field-reoriented state occurs as TFCDs with $\rho \ll 1$, but, starting with $\rho = 1$, these domains transform into stripe domains with constant width. The reason is that the anchoring cost of stripe domain is smaller than that for TFCD domains with $\rho \gg 1$. We shall not enter into the details of these experiments and their interpretation.

REFERENCES

Adamczyk A., 1989, Mol. Cryst. Liq. Cryst., 170:53

Boltenhagen P., Lavrentovich O.D. and Kleman M., 1991, J. Phys. II France 1:1233

Boltenhagen P., Kleman M et Lavrentovich O.D., 1992, C. R. Acad. Sci. Paris, 315, Série II:931-935

Boltenhagen P., O.D. Lavrentovich and Kleman M., 1992, Phys. Rev.A, Vol. 46, Num. 4, R 1743-R1746

Boltenhagen P., Thèse ,Orsay, (1993), " Quelques aspects des défauts dans les phases lamellaires gonflées".

Bouligand Y., J. Physique, 1972, 33:525

Bourdon L., Kleman M. and Sommeria J., 1982, J. de Phys. 43:77

Bragg W., 1933, Transaction Faraday Soc. 29:1056

Colliex C., Kleman M. and Veyssié M., 1974, Eight. Inter. Congress on Electron Microscopy, Camberra, 1:718

Dabadie J.C., Fabre P., Veyssié M., Cabuil V. and Massart R., 1990, J. Phys.: Condens. Matter., 2 SA291

Darboux G., Théorie Générale des Surfaces (Gauthier Villars, Paris, 1988), republished by Chelsea Pub. Cy, New York, 1954

Fourcade B., 1992, J. Phys. II France 2 :1705-1724

Fourcade B., Mutz M. and Bensimon D., 1992, Phys. Rev. Lett. 68:2551

Fournier J.B., a) 1993, Phys. Rev. Lett. 70:1445 ; b) to be published

Fournier J.B., Dozov I. and Durand G., 1990, Phys. Rev. A 41:2252

Fournier J.B. and Durand G., 1991, J. de Phys. II, 1:845

Fournier J.B., Warenghem M. and Durand G., 1993, Phys. Rev. E 47:1142

Friedel G., 1922, Annales de Physique 18: 273

Friedel G. and Grandjean F., 1910, Bull. Soc. Franç. Minér. 33:192, 409

Friedel G. and Grandjean F., 1910, C.R. Hebd Séan. Acad. Sci. 151:762

Hinov H.P., 1988, Liq. Cryst.,3:1481

Jakli A. and Saupe A., 1992, Mol. Cryst. Liq. Cryst. 222:101

Kleman M. , 1977, J. de Physique 38:1511

Kleman M., 1983, Points, Lines and Walls, John Wiley and Sons, Chichester

Kleman M., 1974, J. Phys. (Paris) 36:595

Kleman M. and Parodi O., 1975, J. de Physique 36:671

Lavrentovich O.D., 1986, Sov. Phys. JETP 64:984

Lavrentovich O.D., 1987, Mol. Cryst. Liq. Cryst., 151:417

Lavrentovich O.D. and Kleman M., 1993, Phys. Rev. E, 48

Li Z. and O.D.Lavrentovich, submitted

Meunier et Billard, 1969, J. Mol. Cryst. Liq. Cryst. 7:421

Meyer R.B., Jones F.and Palffy-Muhoray P., July 1990, in Proceedings of the Thirteenth International Liquid Crystal Conference, Canada, (unpublished).

Naito H., M. Okuda and Ou.Yang Zhong, 1993, Phys. Rev. Lett., 70:2912

Omnés R., 1985, J. Physique France 46:139

Oswald P., Behar J. and Kleman M., 1982, Phil. Mag. A 46:899

Petrov A.G. and Bivas I., 1984, Progress in Surface Sci., Ed. S. Davidson 16:389

Porte G., Appell, J., Bassereau P. and Marignan J., 1989, J. Phys. France 50:1335

Pratibha R. and N.V. Madhusudana, 1992, J. Phys.II France 2:383

Rault J., 1975, Comptes Rendus Acad. Sci. (Paris) B280:417

Rosenblatt C.S., Pindak R., Clark N.A. and Meyer R.B., 1977, J. Physique 38:1105

Salmon G., 1952, "Analytic Geometry of Three Dimensions" Chelsea, New York

Sethna J.P. and Kleman M., 1982, Phys. Rev. A26:3037

Steers M., Kleman M. and Williams C.E., 1974, J. Phys. Lett. 35:L-21

Toulouse G. and Kleman M., 1976, J. Physique Lett. 37:L-149

Volovik G.E. and Mineyev V.P., 1977, Zh. Eksp. Teor. Fiz 72:2256

Williams C., 1975, Phil Mag. 32:313

Williams C. and Kleman M., 1975, J. de Phys., 36 C1 315

Willmore T.J., 1982, "Total Curvature in Riemannian Geometry" (Ellis Horwood Ltd, Chichester,)

Weatherburn C.E., 1955, "Differential Geometry in Three Dimensions", Cambridge Univ. Press, Cambridge

ON POLYMER BRUSHES AND BLOBOLOGY: AN INTRODUCTION

A. Halperin

Laboratoire Leon Brillouin
CEA-Saclay
91191 Gif sur Yvette Cedex
France

I-INTRODUCTION

Polymer adsorption has long been a focal point of polymer science[1,2,3,4]. For many years, such phenomena were discussed in terms of uniform adsorption models. In these models all monomers are taken to be uniformly attracted to the surface[1,5]. A different scenario was proposed by Alexander in 1977 under the name of Polar Head Adsorption[6]. In this case the polymer head group is strongly adsorbed while the other monomers are non adsorbing or weakly attracted to the surface. The proposed term did not survive. Chains attached by their headgroups to a surface, a line or a point are currently refered to as *grafted* or *tethered* chains[7,8]. On the other hand, the concept of terminally anchored chains proved to be of enduring value. It enables a unified description of diverse systems such as branched polymers, mesophases of block copolymers and adsorbed layers of terminally functionalized chains. Furthermore, assemblies of tethered chains are of inherent interest because of their distinctive properties. In turn, these are traceable to the stretched configurations adopted by the tethered polymers when the grafting density is high and the chains crowd each other. In the following I will not attempt to present an exhaustive review of this field since such reviews already exist[7,8]. Rather, this review attempts to present a didactic introduction to blob arguments as applied to densely grafted layers known as polymer brushes.

The blob concept was invented by Daoud and deGennes[9,10] in 1975 following the discovery of the relationship between polymers and critical phenomena[11]. Rigorous theoretical implementation of this relationship, using renormalization group and field theory methods[12], is typically difficult and often impractical. "Blobological" arguments provide a complementary approach[10]. Both the advantages and the short comings of blobology are significant. Blob arguments are technically undemanding. Consequently one may obtain results at a fraction of the effort required when more rigorous techniques are employed. Furthermore, because of its simplicity, blobology enables theoretical description of complicated and otherwise intractable problems. On the negative side, the information provided by blob arguments is incomplete. The method is effective in uncovering the asymptotic scaling behavior. However, it cannot be used to determine numerical prefactors and to study crossover between different regimes. Finally, blobology, as a method, tends to rely on intuition and is less than systematic. The blobological part of this review is motivated by this last observation. In particular, the need for a more systematic exposition of blobology. Since a complete review of blobology is impossible within the available space the reader is assumed to be familiar with deGennes's monograph[10]. The first two sections, II and III, are concerned with blobs in general. The relationship between blobs and the view of polymers as critical objects is discussed in section II. A systematic catalogue of blobs in current use is presented in section III. Different types of blobs arguments, as illustrated by examples concerning brushes, are presented in the next six sections. Section IV presents the blobological description of flat brushes immersed in both good and poor solvents. The generalization of this picture to non-planar geometries is discussed in section V. These two sections are primarily concerned with semi-dilute polymer solutions in good solvents, when the blob size is set by the monomer volume fraction. Many-chain systems in which this prescription is inapplicable are considered in section VI. Most of section VII concerns the mean field description of chains at the periphery of a finite brush. From a blobological point of view this is an example of a problem for which a full blobological argument is difficult to formulate. However, this difficulty may be circumvented by blobological conjectures based on mean field results as illustrated by the Raphael-deGennes analysis of the "overspill effect"[13]. Examples illustrating the different roles played by blobs in the discussion of dynamical problems are given in section VIII. Finally, a number of current developments, concerning brush research, are discussed in section IX. Because of its didactic aspirations, this review makes no claim to completeness. Neither brush research nor blobology are exhaustively covered and the choice of subjects reflects my personal taste. Yet, it is hoped that the text does present a coherent introduction to the methodology of blob arguments. While partial overlap with earlier reviews[7,8] and with deGennes monograph[10] was unavoidable, an attempt was made to go beyond the limits set in these three references.

II - BLOBS AND CRITICAL PHENOMENA

Isolated, neutral and flexible polymers are critical objects[10]. By this we mean that the dimensions of the volume occupied by a group of g monomers, R_g, obey $R_g \sim g^\nu$ irrespective of the choice of g i.e., this behavior is independent of scale. In particular, the overall chain size, R, diverges as $R \sim N^{\nu}$ when the polymerization degree, N, is increased. A flexible chain in an athermal good solvent is characterized by $\nu \approx 3/5$, an ideal coil in a melt by $\nu = 1/2$ while a collapsed chain in a precipitant by $\nu = 1/3$. On the other hand, for rigid linear chains we have $\nu = 1$. In the language of critical phenomena the behavior of each of these systems is determined by a different fixed point[14,15,16]. As we shall see, in such situations the blob incorporates the whole chain. In other circumstances we encounter cross-over phenomena. The ν values characterizing the small length scale behavior differ from those found on large length scales. This behavior occurs when more than a single critical point is embedded in the critical surface[15,16]. The physical trajectory of the system, specifying its approach to criticality, initially approaches a mixed fixed point before it is deflected towards the fixed point which determines the true critical behavior of the system. Each of these fixed points is associated with a different universality class i.e., a different ν. A blob is defined by the *cross-over* between the two universality classes which occurs as g is increased. In particular, a blob is a *maximal group of monomers exhibiting the small scale behavior*. The small scale behavior corresponds to that of an isolated, unperturbed chain under appropriate conditions (for example, solvent quality). In other words, a blob is a chain segment behaving as an isolated chain in a suitable state. Each blob is allocated an energy of kT.

These concepts are best illustrated by the behavior of a self avoiding chain confined to a slit[17,18] of width H. On length scales smaller than H the chain does not experience confinement. Consequently, its properties are essentially those of a free three dimensional chain with $\nu \approx 3/5$. This corresponds to the mixed fixed point in our preceding discussion. On large length scales the chain behaves as a two dimensional self avoiding random walk with $\nu \approx 3/4$. This ν is associated with the fixed point which determines the critical behavior of the chain. One can view such a chain as a string of blobs of size H. Each blob consists of g monomers such that $g^{3/5}a \approx H$. The chain as a whole behaves as a string of N/g blobs exhibiting the statistics of a two dimensional self avoiding random walk. Its in-plane radius is thus $R \approx (N/g)^{3/4}H$. The excess free energy due to the confinement, as given by the kT per blob prescription, is $(N/g)kT \approx (a/H)^{5/3}kT$.

The usefulness of the blob concept stems precisely from our ability to view the polymer as a string of blobs. This specifies the configurational statistics of the chain on short and long length scales, its dimensions, free energy and, ultimately, its dynamics. The blob method works because the configurations of the polymer reflect, simultaneously, its whole physical trajectory as the chain approaches criticality. While the critical point is approached by increasing N, the behavior of finite chain segments of length $g < N$ is essentially that of a finite chain of this length. The finite chain segment corresponds thus to an appropriate subcritical point on the physical trajectory. Since all length scales, $1 < g < N$, exist within the polymer, the whole physical trajectory is reincarnated and may be observed simultaneously. It is also important to stress that while blobs are related to cross-over phenomena, they are not useful in their analysis. The cross-over effects occur for fractions of blobs or for few ($\approx 1 - 3$) blobs. Only situations involving single blobs or many blobs are amenable to this type of analysis.

When the fixed points of the polymeric system are known, it is straightforward to formulate the corresponding blob picture. This is, by itself, of interest as a way of interpreting such results and extending them. The opposite route, identifying the fixed points via blobological arguments, is a more delicate undertaking. While this route is not technically demanding, the absence of systematic procedure makes it difficult to implement. A strong intuitive input is always required. Yet, the pursuit of the blobological approach is worthwhile. When it does work, it enables the analysis of complicated systems by use of technically simple methods. While a fully systematic approach to blobology is impossible to formulate, it is possible to suggest useful guidelines. As we have discussed, the blob picture involves three ingredients: (1) The chain statistics within the blob as characterized by ν_b. This small scale behavior is determined by the mixed fixed point. (2) The blob size, ξ, or equivalently, the number of constituting monomers, g. These are related via $\xi \approx g^{\nu_b} a$. (3) The configuration of the string of blobs as specified by ν_s. In turn, ν_s is determined by the fixed point of the system. It determines the overall chain dimension when it is viewed as a string of blobs, $R \approx (N/g)^{\nu_s} \xi$. It is often convenient to begin the construction of the model with the single blob.

III - BLOBOLOGY: A BRIEF REVIEW

Blobs are chain segments exhibiting the behavior of a single unperturbed chain[10]. The choice of the unperturbed reference state depends on the circumstances: solvent quality, spatial dimensionality, role of surface interactions etc. For flexible chains the primary choice is between two models: A random walk (RW), and a self avoiding random walk (SARW). The definition of a particular blob involves two ingredients: Size, ξ, and the number of constituting monomers, g. The two are related via $\xi \approx g^{\nu_b} a$ where a is the monomer size and ν_b is determined by the reference state. For a RW $\nu_b = 1/2$ while for three and two dimensional SARW $\nu_b \approx 3/5$ and $\nu_b \approx 3/4$ respectively.

It is helpful to distinguish between screening blobs and single chain blobs. The screening blobs typically occur only in semi-dilute solutions involving good solvents and are thus usually a multi-chain effect[10,19]. These familiar blobs reflect the competition between two types of repulsive monomer-monomer interactions[20]. Intrachain interactions promote buildup of correlations. In turn, these give rise to to a cross-over from Gaussian to SARW statistics. The cross-over occurs when the interaction energy due to repulsive binary contacts, $kT(v/a^3)g^2a^3/\xi^3$ is comparable to kT i.e., $(v/a^3)g_{co}^{1/2} \approx 1$. For $g < g_{co} \approx (a^3/v)^2$, the effect of the binary interactions is too weak to perturb the ideal chain behavior. SARW behavior is expected in longer segments, $g > g_{co}$, when the binary interaction energy is of order kT. Note that for athermal solvents, with $v \approx a^3$, $g_{co} \approx 1$. The asymptotic behavior of an *isolated chain* of length $N > g_{co}$ is that of a SARW. This brings us to the role of chain-chain interactions. Intrachain correlations are weakened by interactions with monomers of different chains. Such interchain interactions result in screening of the excluded volume effect thus leading to RW statistics. The screening blobs are defined by the onset of the screening i.e., the cross-over between the SARW behavior within the blob and the RW statistics on larger length scales. The onset of screening is identified with g for which the binary interaction energy is of order kT. Since SARW statistics tend to lower the number of such contacts, their number per site scales as $\sim \phi^{9/4}$ rather than as $\sim \phi^2$. Accordingly, the screening blob is defined by $kT(v/a^3)\phi^{9/4}\xi^3 \approx kT$ where $\phi \approx ga^3/\xi^3$. Since this form obtains past the overlap threshold, c^*, for $\phi > Na^3/R_F^3 \approx N^{-4/5}$, it reflects a major contribution of interchain monomer-monomer interactions. Clearly, this criterion is only meaningful for $\xi > g_{co}^{1/2}a$. It is equivalent to the familiar definition[10] involving the combination of $\phi \approx ga^3/\xi^3$ and $\xi \approx g^{3/5}a$. Both yield $\xi \approx \phi^{-3/4}$ and $g \approx \phi^{-5/4}$. Excluded volume interactions are screened out beyond ξ. Accordingly, a string of screening blobs is expected to behave as an ideal Gaussian chain. A free polymer chain behaves thus as a Gaussian string of N/g blobs of size ξ and its overall span is $R \approx (N/g)^{1/2}\xi \approx N^{1/2}a\phi^{-1/8}$. However, in other situations, to be discussed later, this string of blobs can be stretched. In such cases the associated elastic penalty is Gaussian. It assumes a particularly simple form, $kT(R/R_o)^2 \approx kT(R^2/Na^2)\phi^{1/4}$, when all blobs are of equal size. Such is the case for a semi-dilute solution of free, linear chains. As we shall discuss later, it is necessary to allow for spatial variation of ξ when considering adsorbed layers, branched polymers etc. In all cases described in terms of screening blobs the free energy density and the osmotic pressure scale as kT/ξ^3.

Screening blobs are normaly a multichain effect. A single, isolated coil will not usually exhibit such screening induced structure[19]. However, single chains can support blobs induced by other types of interactions: confinement[17,18], tension[10,21], adsorption to a surface[22], etc. In discussing these single chain blobs it is sometimes helpful to consider ξ and ν_b separately. The blob size, ξ, can be set by geometrical confinement or by a variety of interaction energies. On the other hand, ν_b for a given ξ may depend on the solvent quality and other factors. Chains constrained to narrow slits[17,18] and capillaries[10,18] with impenetrable walls exemplify geometrical confinement. The finite (or smallest) spatial dimension sets the blob size. In the two cases noted above ξ is set, respectively, by the slit width and the diameter of the capillary.

On length scales smaller than ξ, the chain behavior is similar to that of the reference state. On larger scales the behavior is of a system of reduced dimensionality. Thus, a flexible chain immersed in a good solvent and confined to a slit of width H may be envisioned as a two dimensional SARW of confinement blobs of size $\xi = H$. In a capillary of diameter D the chain behaves as a one dimensional string of blobs of size $\xi = D$. Note that geometrical confinement is of interest beyond this limited scope since other situations may be discussed in these terms. For example, the adsorption of a single chain onto a flat attractive surface involves confinement to a virtual slit[23]. As we shall illustrate later, such geometrical interpretations can be a useful tool for the construction of blob descriptions.

Confinement blobs arise because of boundary conditions imposed by rigid, repulsive surfaces. In turn, these may be attributed to infinite repulsive energies between the monomers and the walls. The more prevalent blobs involve a variety of finite interaction energies. A blob is defined when the *interaction energy associated with the chain segment is of order* kT. On larger scales the interaction energy is large enough to perturb the reference state. The interaction energy may be due to a variety of sources such as binary monomer-monomer interactions, electrostatic interactions, tension or attraction to a surface. In the electrostatic[10] case we consider a weak polyelectrolyte in a salt free solvent. A small fraction of the monomers, $\rho \ll 1$, carry monovalent charges. The electrostatic energy of a group of g monomers is thus $(\rho g e)^2/\epsilon\xi$ where ϵ is the dielectric constant of the solvent. Setting this energy equal to kT defines the electrostatic blob. To obtain ξ and g it is necessary to know ν_b. In principal, both RW and SARW behaviors are possible. However, for water soluble polyelectrolytes the backbone is normally hydrophobic thus experiencing a poor solvent environment. This suggests that $\nu_b = 1/2$ is more appropriate for these cases. The combination of $(\rho g e)^2/\epsilon\xi \approx kT$ and $\xi \approx g^{1/2}a$ yields $g \approx \rho^{-4/3}(\epsilon a kT/e^2)^{2/3}$ and $\xi \approx \rho^{-2/3}(\epsilon a kT/e^2)^{1/3}a$. The electrostatic repulsions are too weak to perturb the Gaussian statistics within the blob yet strong enough to induce strong stretching of the string of blobs, $\nu_s = 1$. The length of the chain is thus $L \approx (N/g)\xi \approx N\rho^{2/3}(e^2/\epsilon a kT)^{1/3}a$. Note that this approach is only meaningful for $\rho \ll 1$ when $g \gg 1$.

Pincus blobs[10,21] are induced by subjecting the chain to a tension f. The blob size is set by the requirement $\xi f \approx kT$. The chain statistics below this scale are unmodified while on larger scales the string of blobs is fully stretched. Both $\nu_b \approx 3/5$ and $\nu_b = 1/2$ are possible. SARW behavior is expected for large ξ and in good solvents when $\xi \approx g^{3/5}$ and the blob contains $g \approx (kT/fa)^{5/3}$ monomers. The length of the stretched chain is thus $L \approx (N/g)\xi \approx Na(fa/kT)^{2/3}$. When $\nu = 1/2$ occurs L obeys the Gaussian force law, $L \sim f$. As for electrostatic blobs, the description is only valid in the limit of $g \gg 1$ or, in this case, of $fa \ll kT$.

The adsorption of an isolated chain at an attractive surface immersed in a good solvent can be described in terms of adsorption blobs[22]. The interaction with the surface is due to adsorption energy of $-\delta kT$ per monomer in contact with the interface. The chain as a whole is expected to adopt a flattened, pancake like configuration. However, sufficiently short chain segments retain their isotropic configurations i.e., their dimensions parallel to the adsorbing surface are comparable to their dimensions in the perpendicular direction, both scaling as $\sim g^{3/5}a$. Such undeformed chain segments nevertheless experience g^{Φ} contacts with the surface. $\Phi \approx 3/5$ is the cross-over exponent associated with the special surface adsorption transition[24,25]. Each adsorption blob is defined by the condition that the adsorption energy per blob, $-\delta g^{\Phi}kT$, is of order kT or $\delta g^{\Phi} \approx 1$. Accordingly, $g \approx \delta^{-5/3}$ and, since $\xi \approx g^{3/5}a$ in good solvent conditions, we obtain $\xi \approx \delta^{-1}a$. The chain as a whole behaves as a two dimensional SARW of adsorption blobs and its in plane radius is $R \approx (N/g)^{3/4}\xi \approx N^{3/4}\delta^{1/4}a$. Note that while the two previous examples correspond to chains confined by cylindrical capillaries, this example is reminiscent of a chain confined to a slit. The adsorption behavior of star polymers may be analyzed in terms of such adsorption blobs[26].

Free, isolated, collapsed chains may be viewed as a spherical arrays of close packed collapse blobs[27]. Collapse blobs, like screening blobs, occur because of binary monomer-monomer interactions. However, screening blobs occur in good solvents when the binary interactions are repulsive while collapse blobs result from attractive monomer-monomer interactions in a poor solvent. When the temperature is not too far below the θ temperature, the associated interaction energy is weak enough so as not to perturb the Gaussian behavior of short chain segments. Quantitatively the collapse blob is defined by the combination of $vg_c^2/\xi_c^3 \approx 1$ and $\xi_c \approx g_c^{1/2}a$ where vkT is the second virial coefficient. Since at the vicinity of the θ temperature $v \approx v_o(\Delta T/\theta)$ where $\Delta T = \theta - T$, the collapse blobs consist of g_c monomers such that $g_c^{1/2}|\Delta T|/\theta \approx 1$. This picture is, again, valid only in the limit of $N \gg g_c \gg 1$ or, in this case, $\Delta T \ll \theta$. The radius of the collapsed globule, within this picture, is $R \approx (N/g_c)^{1/3}\xi_c$. The density within these collapse blobs, $g_c a^3/\xi_c^3$, is comparable to that of the polymer precipitate. The boundary of the collapsed globule is rather sharp and endowed with a surface tension of $\gamma \approx kT/\xi_c^2$

By way of summary, let us review the various scenarios encountered with regard to ν_b and ν_s. In a good solvent the flexible polymer chain behaves as a SARW and ν_b, in three dimensions, is $\approx 3/5$. This is only true for large enough blobs, with $g > g_{co}$. If $g < g_{co}$ the repulsive interactions within the blob are too weak and RW statistics, $\nu_b = 1/2$, are expected. This is also the case in θ and poor solvents. The possible configurations for the string of blobs are richer. A free collapsed chain may be envisioned as a close packed, spherical array of collapse blobs. The close packing in this case is due to the attractive binary interactions which are of negligible importance within the blob. In this case $\nu_s = 1/3$ as is apparent from $R \approx (N/g)^{1/3}\xi_c$. A free chain in a semi-dilute solution may be viewed as an ideal string of screening blobs with $R \approx (N/g)^{1/2}\xi$ and $\nu_s = 1/2$. For an isolated chain in a slit as well as for the single adsorbed chain, the string of blobs exhibits the behavior of a two dimensional SARW with $\nu_s \approx 3/4$. As opposed to the previous example, the excluded volume interactions are not screened in this case. A one dimensional behavior, $\nu_s = 1$, is expected for an isolated polyelectrolyte chain as well as for a chain under tension. Both behave as chains confined to cylindrical capillaries. Altogether, we can summarize the various blob ansatzs in a table form (Table I) keeping in mind that other types of blobs may be defined.

Table I. Blob types.

Blob Type	Interaction	ξ/a	ν_b	ν_s
screening	binary repulsions	$\phi^{-3/4}$	$3/5, 1/2$	$1/2, 1$
collapse	binary attractions	$|\Delta T|/\theta$	$1/2$	$1/3, 1$
electrostatic	electrostatic repulsions	$\rho^{-2/3}(\epsilon a k T/e^2)^{1/3}$	$1/2, 3/5$	$1/2, 1$
Pincus	tension	kT/f	$1/2, 3/5$	1
adsorption	monomer-wall attraction	δ^{-1}	$3/5$	$3/4, 1/2$
confinement	monomer-wall repulsions	H	$1/2, 3/5$	$3/4, 1/2$

IV - TETHERD CHAINS: THE FLAT BRUSH

Following our brief review of blobs in general we now present illustrative examples of blobological arguments concerning polymer brushes. It is convenient to follow historical development and begin with a discussion of a flat grafted layer. In this case it is possible to obtain the blob picture by minimizing an appropriate free energy. As we shall see the results can be interpreted in terms of confinement blobs. In turn, this enables the generalization of the blob picture to more complicated geometries. Since our primary interest is in the methodology of blob arguments we focus the discussion on the simplest case: Neutral, flexible chains which are irreversibly grafted to a flat surface with a constant surface density σ^{-1}. Each terminally anchored chain is allocated an area σ which is further assumed to be small enough so as to ensure chain overlap. Under these conditions the crowded chains are stretched along the normal to the surface. Such layers of densely tethered chains are known as brushes. The surface is further assumed to be repulsive. This assumption enables us to disregard complications due to the adsorption of non terminal monomers[6,28].

Our discussion[8] is founded on the Alexander model[6,29,30]. This Flory type model is based on two assumptions: (i) The concentration profile is step like with a constant monomer volume fraction $\phi = Na^3/L\sigma$ where N is the polymerization degree, a is the monomer size and L is the thickness of the layer. (ii) the chains are uniformly stretched with their ends straddling the boundary of the layer, at a distance L from the surface. The equilibrium state of the layer corresponds to a minimum of the free energy per chain, F_{chain}. In turn, F_{chain} incorporates two contributions: An elastic term, F_{el}, allowing for the loss of configurational entropy upon stretching and an interaction term, F_{int}, accounting for monomer-monomer interactions. We initially consider the behavior of a brush immersed in a good solvent[31]. In this case F_{int} is due to repulsive, binary monomer-monomer interactions. The overlap threshold in this regime is $\sigma^* \approx R_F^2$ where $R_F \approx N^{3/5}a$ is the Flory radius of the swollen coil.

To proceed it is necessary to obtain an explicit expression for F_{chain}. The two simple approaches[10] available are the Flory approximation and the blob picture. Within the Flory approximation the polymers are viewed as ideal, uncorrelated chains. Accordingly, the number of monomer-monomer interactions per site scales as $\sim \phi^2$. Since the volume per chain is σL, $F_{chain}/kT \approx (v/a^3)\phi^2\sigma L$ where vkT is the second virial coefficient. The elastic penalty of a Gaussian chain is $F_{el}/kT \approx L^2/R_o^2$ where $R_o \approx N^{1/2}a$ is the unperturbed radius of the coil. Altogether

$$F_{chain}/kT \approx (v/a^3)\phi^2\sigma L + L^2/R_o^2$$

and the equilibrium condition, $\partial F_{chain}/\partial L = 0$, yields

$$L/a \approx N(a^2/\sigma)^{1/3} \quad ; \quad F_{chain}/kT \approx N(a^2/\sigma)^{2/3}.$$

Both F_{int} and F_{el} are overestimated within this approach because self avoidance correlations are neglected. The Flory approximation is nevertheless very useful: (i) It typically yields the correct equilibrium dimensions because the two overestimates cancel when the equilibrium condition is invoked. (ii) It provides a useful upper bound for F_{chain}. (iii) Most important, the Flory approximation is simple to apply irrespective of difficulties such as complicated geometries. While it is a good idea to use the Flory approximation as a first step in the analysis of a problem, it is important to understand its limitations. Since the free energy density is overestimated, so is the osmotic pressure. As a result erroneous results are obtained for related properties such as the force law for brush compression. The overestimate of F_{int} and F_{el} can result in wrong predictions for equilibrium dimensions when F_{chain} includes extra terms and cancellation of errors may not be relied upon. Such is the case for aggregates of block copolymers[8] where F_{chain} must allow for a surface energy contribution.

Within the blob picture the brush is viewed as a slab of semi-dilute solution. It is thus envisioned as an array of close packed blobs of size $\xi \approx \phi^{-3/4}a$ consisting of $g \approx \phi^{-5/4}$ monomers. The interaction free energy, as given by the kT per blob ansatz, is $F_{int}/kT \approx N/g \approx N\phi^{5/4}$. The elastic free energy of a Gaussian string of blobs is $F_{el}/kT \approx L^2/R_o^2(\phi)$ where $R_o(\phi) \approx (N/g)^{1/2}\xi \approx N^{1/2}a\phi^{-1/8}$ is the unperturbed radius of the string. Altogether F_{chain} is[32]

$$F_{chain}/kT \approx \phi^{9/4}\sigma L + (L/R_o)^2\phi^{1/4}.$$

Both F_{int} and F_{el} are smaller by a factor of $\phi^{1/4}$ in comparison to the expressions used in the Flory approximation. F_{int} is lower since the number of monomer-monomer contacts decreases because of the self avoidance. The extra $\phi^{1/4}$ factor in F_{el} arises because the Gaussian chain consists now of blobs rather than monomers. Within this picture, the equilibrium state of the brush is specified by

$$L/a \approx N(a^2/\sigma)^{1/3} \qquad F_{chain}/kT \approx N(a^2/\sigma)^{5/6}.$$

Thus, F_{chain} as calculated blobologically is indeed lower than the F_{chain} obtained via the Flory approximation while the brush thickness obtained by the two methods is identical.

Within the Alexander model, a chain in a brush behaves as if confined to a cylindrical capillary of cross section σ and height L (Figure 1). Blobologically, this is an especially important point since the diameter of this virtual capillary, $\sigma^{1/2}$, sets the blob size, ξ. In terms of our preceding discussion this follows because at equilibrium $F_{el} \approx F_{int}$ or $L^2/(N/g)\xi^2 \approx N/g$. In turn, this leads to $L \approx (N/g)\xi$ and to $\xi \approx \sigma^{1/2}$. Thus, the grafted chain may be envisioned as a *fully stretched string of blobs*. It is similar to the one dimensional chain of confinement blobs created upon constraining a linear polymer to a cylindrical capillary. However, since the string is Gaussian and the capillary is virtual, it is necessary to allow for traverse fluctuations giving rise to a random walk component parallel to the wall. Accordingly, the chain span in this direction is

$$R_\perp \approx (N/g)^{1/2}\xi \approx N^{1/2}(\sigma/a^2)^{1/12}a$$

This picture, of a chain confined to a capillary, is of great utility. As we shall discuss, it guides the generalization of the blob picture to non-planar geometries and the analysis of dynamical properties. Before we address this issues, it is useful to consider the structure of a brush immersed in a poor solvent.

First it is important to note that the overlap threshold for a brush in a collapsed state is smaller than that found in a good solvent. Since the dimensions of an isolated collapsed globule scale as $R \sim N^{1/3}$, $\sigma^* \sim N^{2/3}$ instead of $\sigma^* \sim N^{6/5}$ in a good solvent. Thus, upon decreasing the solvent quality, a dense brush with $N^{2/3}a^2 < \sigma < N^{6/5}a^2$ may be transformed into a collection of non-interacting collapsed globules. This and other scenarios obtained in this regime are discussed further in section IX. When $\sigma < N^{2/3}a^2$ the brush thickness may be obtained by use of the capillary picture described above[33,34]. However, in this case the blob size is determined by the solvent quality[27] as specified by $\theta - T$ rather than by $\sigma^{1/2}$. The layer thickness is set by the condition that the capillary of length L is close packed with collapse blobs, $\sigma L \approx (N/g)\xi_c^3$, or

$$L \approx (N/g_c)(\xi_c^2/\sigma)\xi_c \approx N(a^2/\sigma)\xi_c$$

This result may also be obtained by minimizing the appropriate free energy[35]. In this case F_{chain} must also allow for "three blob" interactions

$$F_{chain}/kT \approx L^2/R_o^2 - (\xi_c^3/L\sigma)(N/g_c)^2 - (\xi_c^6/L^2\sigma^2)(N/g_c)^3$$

where $R_o \approx (N/g_c)^{1/2}\xi_c$ is the radius of an ideal string of collapse blobs. The second term allows for the *attractive* two blob contacts and the last term reflects the contribution of repulsive ternary interactions between the blobs. While this F_{chain} reproduces the results obtained by other methods, it is somewhat questionable since the volume fraction of the blobs in the equilibrium state is of order unity. Finally, a

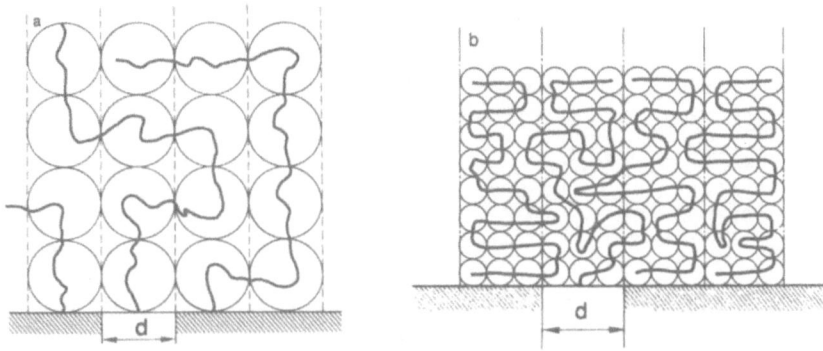

Figure 1. Schematic cross sections in brushes immersed in good (a) and poor (b) solvents. In the good solvent case the blob size is set by the grafting distance. The solvent quality determines the blob size in a precipitant.

few concluding remarks: The grafted chains in a brush are stretched even in the collapsed state and $L \sim N$ irrespective of the solvent quality. In marked distinction, isolated chains shrink from $R \sim N^{3/5}$ to $R \sim N^{1/3}$ as the solvent quality decreases. As in the good solvent case the lateral dimensions of the chain are determined by lateral Gaussian fluctuations leading to $R_\perp \approx (N/g_c)^{1/2}\xi_c$. On the other hand the density of the collapsed brush is much higher than that of a swollen brush. It is also expected to have a sharp interface associated with a line tension of $\gamma \approx kT/\xi_c^2$.

V - CURVED BRUSHES AND SPATIALY VARYING BLOBS

The Alexander model as outlined above is untenable for extended brushes grafted onto curved surfaces. In such cases the volume available to the chains grows with the distance from the surface. Consequently, the concentration profile expected is monotonously decreasing rather than step like. In the spherical case, first considered by Daoud and Cotton[36], one may generalize the Alexander model by envisioning chains confined to conical rather than cylindrical capillaries. However, it turns out that a slightly different geometrical construction is of wider applicability, allowing for a unified description[8] of spherical, cylindrical[37] and flat brushes. The grafted layer is envisioned as a stratified array of blobs. The blobs within a given sub-layer are of equal size. Any single grafted chain contributes one blob to each of the sub-layers. This last requirement allows for the stretched configurations of the densely grafted chains. For a flat grafted layer the area of all sub-layers, S, is equal. Thus, in a layer comprising of f chains the blob size is determined by $f\xi^2 \approx S$ or $\xi \approx \sigma^{1/2}$. In the case of cylindrical layers of length H the area of a given strata depends on its altitude. In particular, $S \approx rH$ where r is the radius of the strata. As a result $f\xi^2 \approx rH$ or $\xi \approx (rH/f)^{1/2}$. Finally, in spherical layers $S \approx r^2$ leading to $f\xi^2 \approx r^2$ or $\xi \approx r/f^{1/2}$. Knowing ξ it is easy to obtain the layer thickness, L, and the free energy per chain, F_{chain}. L is determined by mass conservation. Since grafted chains are stretched

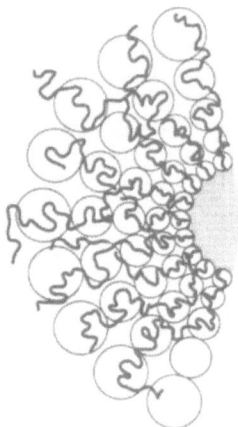

Figure 2. A spherical grafted layer. The hatched core is surrounded by a corona of swollen grafted chains. The resulting blob structure is self similar, $\xi \sim r$.

along the normal to the grafting site N is given by[38]

$$N \approx \int_{R_{in}}^{R_{in}+L} (\xi/a)^{5/3}\xi^{-1}dr$$

where R_{in} is the radius of the grafted surface and $(\xi/a)^{5/3}$ is the number of monomers in a blob of size ξ. This integral sums up the number of monomers in the blobs along the radial trajectory of the chain. For a flat layer, where $\xi \approx \sigma^{1/2}$, it leads to the recovery of $L \approx (N/g)\xi \approx N(a^2/\sigma)^{1/3}a$. In the limit of $L \gg R_{in}$, L of a cylindrical layer is $L \approx (fa/H)^{1/4}N^{3/4}a$ while in the spherical case $L \approx f^{1/5}N^{3/5}a$. The kT per blob prescription leads to the following expression[38] for F_{chain}

$$F_{chain}/kT \approx \int_{R_{in}}^{R_{in}+L} \xi^{-1}dr.$$

This integral simply counts the number of blobs along the trajectory of the chain. For flat layers this reduces to the familiar $F_{chain}/kT \approx N(a^2/\sigma)^{5/6}$. For spherical layer $F_{chain}/kT \approx f^{1/2}\ln(R_{in} + L)/R_{in}$. Extended cylindrical brushes, $L \gg R_{in}$, have $F_{chain}/kT \approx (fL/H)^{1/2}$. This together with $L/a \approx (fa/H)^{1/4}N^{3/4}$ leads to $F_{chain}/kT \approx (fa/H)^{5/8}N^{3/8}$.

VI - SINGLE CHAIN BLOBS AND CHAIN-CHAIN INTERACTIONS

Our preceding discussion focused on systems described in terms of close packed screening blobs. As was previously stressed, such a picture is inherently associated with multichain interactions. Single chain blobs may also occur in systems incorporating many chains. In such situations it is necessary to allow for the effect of chain-chain interactions. The precise method varies with the details of the system. In the following we present two illustrative examples concerned, respectively, with confinement and with Pincus blobs.

Consider first the role of chain-chain interactions in the case of confinement to a slit[17] of width $H \ll R_F$. As was discussed in sections II and III, a single confined chain in a good solvent may be viewed as a string of blobs of size H. The chain segments within the blobs obey three dimensional SARW statistics and the number of monomers in a blob is accordingly $g \approx (H/a)^{5/3}$. Excluded volume interactions between the confinement blobs give rise to a two dimensional SARW behavior of the blob string and the in-plane radius of the chain is thus

$$R_\parallel \approx (N/g)^{3/4} H \approx N^{3/4} (a/H)^{1/4} a.$$

This description is valid up to the overlap threshold, $\phi^* \approx Na^3/R_\parallel^2 H$. For $\phi > \phi^*$ it is necessary to modify the picture so as to allow for chain-chain interactions. Since the confinement blobs experience excluded volume interactions, we may view the confined layer as a two dimensional semi-dilute solution consisting of strings of blobs. As in the three dimensional case, the solution is envisioned as a close packed array of screening blobs. However, in this regime the screening blobs are two dimensional, consisting each of n confinement blobs. Their in-plane dimension, $\xi_\parallel \approx n^{3/4} H$, is such that n is set by $\phi \approx nga^3/\xi_\parallel^2 H$ i.e, $n \approx \phi^{-2}(a/H)^{8/3}$ and $\xi_\parallel/a \approx \phi^{-3/2}(H/a)^{1/2}$. On length scales larger than $\xi_\parallel$ excluded volume interactions are screened out and the string of $\xi_\parallel$ blobs obeys RW statistics. The in-plane radius of the chain is thus

$$R_\parallel \approx (N/ng)^{1/2} \xi_\parallel \approx N^{1/2}(a/H)^{3/2} \phi^{-1/2}.$$

Accordingly, the free energy density is $(n+1)kT/\xi_\parallel^2 H$ where one kT is assigned to the screening blob and to each of the confinement blobs. Clearly, this regime can only last while $\xi_\parallel > H$. Three dimensional semi-dilute solution behavior is recovered when $\xi_\parallel \approx H$ or $\phi \approx \phi^{**} \approx (H/a)^{1/2}$.

The second example concerns chain-chain interactions between strings of Pincus blobs. This situation occurs upon stretching of a swollen brush. In the following we focus on a brush consisting of polymers having each end grafted to a different plate and in contact with a reservoir of good solvent. The Alexander model is rigorously applicable to this system. In the absence of external forces, the brush will attain its equilibrium structure as described in section IV. Each grafted chain is in effect confined to a capillary of cross section σ. Equivalently, the chain may be viewed as a fully stretched string of blobs of size $\sigma^{1/2}$. Most important, the brush may be viewed as a slab of semi-dilute solution with the caveat that each string of blobs is characterized by $\nu_s = 1$ rather than by $\nu_s = 1/2$ as is the case in semi-dilute solutions of free homopolymers. This picture is no longer valid when the brush is stretched to

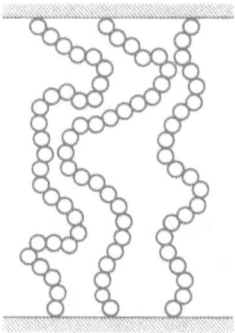

Figure 3. A stretched brush comprising of chains depicted as strings of Pincus blobs. In distinction to the semi-dilute solution behavior the blobs are not close packed.

$L > N(a^2/\sigma)^{1/3}a$. Since the unperturbed brush consists of fully stretched strings of blobs, further extension can be attained only if the blob structure is modified. The new regime, first discussed by Rabin and Alexander[39], is different in two respects: (i) The screening blobs are replaced by Pincus blobs i.e., the blob size is now determined by the tension, f, rather than by the monomer volume fraction, ϕ. (ii) The blobs are no longer close packed (Figure 3). This last point is of special importance since it results in the failure of the familiar semi-dilute solution prescription for the free energy density, kT/ξ^3. Instead, one may argue that the Pincus blobs interact as hard spheres. A Flory type approximation suggests that each blob experiences an interaction energy of $kT\Phi_B$ where $\Phi_B \approx (N/g)\xi^3/L\sigma \approx \xi^2/\sigma$ is the blob volume fraction. The interaction energy per chain is thus

$$F_{int}/kT \approx (N/g)\Phi_B \approx N(a^2/\sigma)(\xi/a)^{1/3}$$

which scales as $\sim \phi$ rather than as $\sim \phi^{5/4}$. The elastic free energy, as given by the kT per Pincus blob, is

$$F_{el}/kT \approx N(fa/kT)^{5/3} \approx (L/R_F)^{5/2}.$$

The analysis of Rabin and Alexander was elaborated by Barrat[40] in order to explain shear induced swelling of brushes immersed in good solvents. The stretching of brushes in poor solvents was considered by Halperin and Zhulina[27]. Ross and Pincus[41] considered the role of lateral inhomogeneities in this problem. Finally, in this regime shear is expected to induce brush shrinkage as was shown by Williams[42].

VII - BLOBOLOGICAL CONJECTURES BASED ON MEAN FIELD RESULTS: THE OVERSPIL EFFECT

Most of this section is concerned with the behavior of chains at the periphery of finite, flat brushes. Since such chains are less crowded, their stretching is correspondingly weaker. This gives rise to a novel *negative* line tension. The theoretical description involves a mean field, Flory type theory. While blobs are not invoked, this case is nevertheless of interest from the perspective of brush research as well as that of blobological methodology. The methodological interest stems from possible difficulties in the formulation of blob pictures allowing for spatial variations of ξ. Occasionally, such difficulties may be circumvented by using geometrical arguments as has been discussed earlier. In other situations one may use results obtained by mean field arguments as guidelines for formulating blobological conjectures. In the following example a mean field theory is used to obtain a line tension[13], τ. The analysis suggests that τ is proportional to the osmotic pressure, π, and to an area specified in terms of a characteristic length, λ, that is $\tau \approx \pi\lambda^2$. Mean field and blobological arguments are usually in agreement with regard to dimensions. Thus λ, as obtained via the mean field theory, is likely to remain unmodified by a blobological analysis. On the other hand, the free energy density as obtained by mean field theories is typically overestimated. To allow for the effect of correlations it is necessary to replace $\pi \sim \phi^2$ by $\pi \sim \phi^{9/4}$. Thus, given a mean field result of the form $\tau \approx \phi^2\lambda^2$ one is lead to the conjecture $\tau \approx \phi^{9/4}\lambda^2$. Clearly this argument can be used only when the mean field result is a product of factors of known blobological behavior.

This particular case concerns the behavior of brushes grafted onto finite albeit large plates. The behavior of the brush at the central regions of the plate is essentially indistinguishable from that of an infinite flat brush. However, the description of the peripheral region must allow for "edge" effects (Figure 4). In particular, a larger volume is available to chains grafted at the periphery of the plate because of the possibility of an "overspill". Consequently, such chains are less stretched. As a result, the boundary of the plate is associated with a negative "line tension", $-\tau$. This effect is of importance in the discussion of micelles formed by rod-coil and crystalline-coil block copolymers. It has recently been analyzed by Raphael and deGennes[13] . Their discussion of τ concerns a semi infinite brush grafted to the xy plane and extending from $x = 0$ to $x = \infty$. At $x = 0$ the chains are allowed to spill sideways into the ungrafted region. This spill produces a tilt in the grafted chains i.e., the average magnitude of the x component of end to end vector, $\langle R_x \rangle$, is non zero. This tilt decays and eventually vanishes as x increases. As we shall see, τ is determined by the decay length, λ, characterizing the width of the tilted region. The Raphael-deGennes analysis is based on the Alexander model for ideal grafted chains as extended to allow for chain tilt. Two ingredients are involved:(i) A relationship between the local tilt

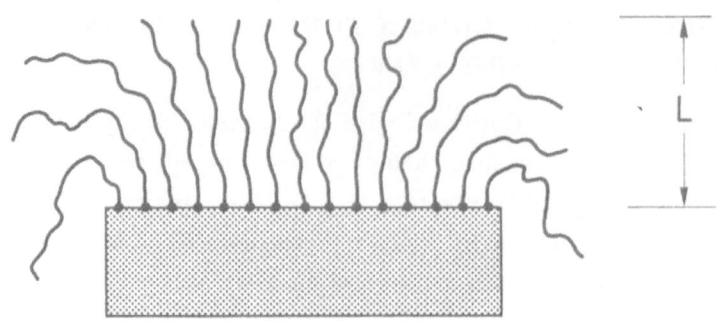

Figure 4. The "overspill" effect occurs at the periphery of flat, finite brushes. The peripheral chains are less stretched compared to the inner chains thus giving rise to a negative line tension.

and the surface number density of monomers, Γ and (ii) An extra term supplementing the Alexander free energy to allow for lateral chain deformation.

To relate the tilt and Γ, consider a rectangular strip of width $\sigma^{1/2}$ extending from x to $x + \Delta x$. The tilt is assumed to vary monotonically and only in the x direction. Only chains grafted within a distance $\langle R_x(x) \rangle$ from the x boundary contribute to Γ. These chains occupy a tilted prism such that half the prism invades the strip. The prism basal area is $\langle R_x(x) \rangle \sigma^{1/2}$ and since each chain consists of N monomers and occupies an area of σ, the number of monomers invading the strip is $N \langle R_x(x) \rangle / 2\sigma^{1/2}$. Similarly, $N \langle R_x(x + \Delta x) \rangle / 2\sigma^{1/2}$ monomers, belonging to grafted chains within a distance of $\langle R_x(x + \Delta x) \rangle$ inside the $x + \Delta x$ boundary, reside outside the strip. This suggests defining a "polarization"

$$P(x) = N \langle R_x(x) \rangle / 2\sigma$$

such that the total number of monomers within the strip, $\Gamma \Delta x \sigma^{1/2}$, is equal to $\sigma^{1/2}[\Gamma_o \Delta x + P(x) - P(x + \Delta x)]$ where $\Gamma_o = N/\sigma$ is the surface number density in the untilted layer. In the limit of small Δx this leads to

$$\Gamma = \Gamma_o - \partial P/\partial x.$$

To allow for uniform tilt, F_{chain} is supplemented by a Gaussian lateral stretching term, $\langle R_x \rangle^2 / Na^2 = \sigma^2 P^2 / N^3 a^2$. The free energy density is $f = (\Gamma_o/N) F_{chain}$. The resulting expression for f is generalized to allow for non uniform tilt by replacing Γ_o by Γ. f is then approximated by its expansion around $P = 0$ to second powers in P and in $P_x = \partial P/\partial x$. The final expression for $\Delta f = f - f_o$ is

$$\Delta f = (\kappa/2)P^2 - |f'|P_x + (1/2)f''P_x^2$$

where $\kappa/kT \approx N^{-3}(\sigma/a^2)$, $|f'|/kT \approx (a^2/\sigma)^{2/3}$ and $f''/kT \approx N^{-1}(\sigma/a^2)^{1/3}a^2$. To obtain the decay length, λ, we consider a semi infinite layer with an imposed $P = P_o$

at $x = 0$ and $P = 0$ at $x = \infty$. The physical $P(x)$ should minimize the corresponding τ as given by integrating Δf with respect to x. The solution of the appropriate Euler-Lagrange equation yields $P = P_o \exp(-x/\lambda)$ with $\lambda^2 = f''/\kappa$ or

$$\lambda \approx N(a^2/\sigma)^{1/3}a \approx L.$$

To obtain the line tension due to the overspill it is argued that in this case P near $x = 0$ is specified by $P(x) = -P_o \exp(-x/\lambda)$. Minimization of the corresponding expression for τ yields $P_o \approx L\Gamma_o$ and

$$-\tau a/kT \approx N^2(a^2/\sigma)^2 \approx \lambda^2\phi^2.$$

This result may be interpreted as the osmotic work done by a brush of unit width as it overspills, $\pi\lambda^2$, where $\pi/kT \approx \phi^2$ is the osmotic pressure in a brush consisting of ideal chains. Since the blob picture leads to $\pi \approx kT/\sigma^{3/2}$ with no modification in L, this suggests that the blob or scaling form of τ is specified by two length scales, λ and $\xi \approx \sigma^{1/2}$, and is given by $\tau a/kT \sim N^2\sigma^{-13/6}$.

VIII - A LITTLE ON BLOBOLOGICAL DYNAMICS

Blobs can assume two very different roles in the discussion of dynamical processes. A blob may be considered as a hydrodynamically impenetrable object. In other situations it specifies the elementary mesh unit of a network, permanent or transient, formed by interpenetrating polymers. Both roles may be illustrated by suitable examples concerning the dynamics of brushes. The first role is appropriate when discussing the expulsion of a chain out of a brush immersed in a good solvent[43]. In particular, we focus on the outward motion of a chain once its bond to the surface was severed. We aim to study the N dependence of the characteristic time of this process, τ_{exp}. As noted previously, within the Alexander model a grafted chain is, in effect, confined to a cylindrical capillary. Consequently, the length of the minimal path traversed by the chain scales linearly in N and the motion is reptative in nature. However, in marked contrast to the tubes familiar from the discussion of reptation in the bulk[10], the confining capillary is virtual and the reptative character is expected even in the absence of entanglements. Furthermore, the virtual capillary is finite and comparable in length to the span of the confined chain. As a result, end effects, having no counterpart in the reptation of free chains, become important. As the chain is expelled it is gradually released from confinement. In turn, the associated gain in configurational entropy gives rise to an outwards directed force. Thus, the chain expulsion is not a purely diffusive process. A complete description of the expulsion process may be obtained by solving the appropriate Langevin equation. However, since our interest is limited to the scaling behavior of τ_{exp}, we may resort to a simpler approach. In particular, we can estimate τ_{exp} for both purely diffusive and purely driven motions. As we shall see the τ_{exp} associated with the driven motion is much shorter thus providing a good approximation for the true τ_{exp}.

Let us first estimate the diffusive τ_{exp}. The outward motion of the chain is accompanied by a change in its configuration. The fully embedded chain is envisioned as a string of blobs of size $\xi \approx \sigma^{-1/2}$. As the chain moves outwards, the expelled monomers form a single, larger blob. When the chain approaches complete release, its configuration is essentially that of a free chain with $R_F \approx N^{3/5}a$. These configurational changes are associated with corresponding changes in the frictional properties of the chain. Because blobs behave as hydrodynamically impenetrable spheres, the friction coefficient of the fully embedded chain scales as $\zeta_{chain} \sim L$. When the chain is almost free $\zeta_{chain} \sim R_F$. The instantaneous ζ_{chain} for intermediate configurations is thus bounded by $6\pi\eta L \leq \zeta_{chain} \leq 6\pi\eta R_F$ where η is the viscosity of the solvent. Because the diffusion coefficient is given by $D \approx kT/\zeta_{chain}$, it scales as $L^{-1} \leq D \leq R_F^{-1}$. Finally, since the diffusive τ_{exp} is given by $D\tau_{exp} \approx L^2$, τ_{exp} is bound by $L^3 \leq \tau_{exp} \leq L^2 R_F$ or $N^3(a^2/\sigma) \leq \tau_{exp} \leq N^{13/5}(a^2/\sigma)^{2/3}$.

To obtain τ_{exp} for the driven process we consider a stationary expulsion process, when the driving force is balanced by the frictional force. The excess free energy of the confined chain, as given by the kT per blob ansatz, is $F_{chain}/kT \approx h/\sigma^{1/2}$ where h is the penetration depth of the chain, $L \geq h \geq 0$, and $\sigma^{1/2}$ is the blob size. The driving force is accordingly $-\partial F_{chain}/\partial h = -kT/\sigma^{1/2}$. It is assumed that both forces act only on the embedded segment i.e., the expelled and the embedded segments are decoupled. The frictional coefficient of an embedded segment consisting of $h/\sigma^{1/2}$ impenetrable blobs is $\zeta(h) \approx (h/\sigma^{1/2})\sigma^{1/2}\eta \approx h\eta$. The outward motion is thus described by $\zeta(h)dh/dt \approx -\partial F_{chain}/\partial h$ or $\eta h dh/dt \approx -kT/\sigma^{1/2}$ which leads to

$$\tau_{exp} \approx \sigma^{1/2}L^2 \approx N^2(a^2/\sigma)^{1/6}.$$

Since the driven τ_{exp} is significantly shorter we conclude that this is the dominant process for this geometry.

This analysis has been extended to allow for the "de-grafting" dynamics in the case of aggregated diblock copolymers[44]. The resulting Kramers type rate theory enables the analysis of the relaxation dynamics of copolymeric micelles[45]. Related studies were reported by Ligoure and Liebler[46] and by Johner and Joanny[47] who respectively considered copolymer adsorption out of non-selective and selective solvents.

The internal modes of the brush may be described by an approach due to deGennes[3]. This method enables the analysis of collective, elastic deformations of brushes of various geometries. Our presentation focuses on the breathing modes of a flat brush as described by the Alexander model. In this case the layer is transitionally invariant and the only relevant coordinate is z, the distance from the surface. The polymer layer is characterized by local monomer concentration, $c(z)$, a mesh size, $\xi(z)$, and an elastic modulus, $E(z)$, related by[10]

$$E(z) \approx kT/\xi^3(z).$$

The local dynamics are described by balancing the elastic restoring force and the viscous friction due to the solvent. The collective breathing modes are thus expected to obey

$$\frac{\partial}{\partial z}[E(z)\frac{\partial u}{\partial z}] \approx \frac{\eta}{\xi^2(z)}\frac{\partial u}{\partial t}$$

where $u(z)$ is the relevant component of the displacement field. For small longitudinal deformations the stress, $P(z)$, is proportional to the strain, $\partial u/\partial z$, and thus $P(z) = E(z)\partial u/\partial z$. The elastic restoring force, $\partial P(z)/\partial z$, is balanced by the viscous friction. The viscous friction arises because the breathing motion result in changes in the volume of the layer and, consequently, in solvent flow within it. The associated resistance to flow is comparable to that of array of closely packed capillaries of radius equal to the mesh size. Thus, the flow through a slab of unit area and of thickness dz at height z is equivalent to the flow through $1/\xi^2(z)$ close packed capillaries of radius $\xi(z)$ and length dz. Since the resistance of a single capillary is proportional to $\xi^4(z)$ and the capillaries are packed in parallel, the resistance of the array scales as $[\xi^4(z)\xi^{-2}(z)]^{-1} \approx \xi^{-2}(z)$. The local viscous friction is consequently proportional to $(\eta/\xi^2(z))\partial u/\partial t$.

The analysis is directed at finding eigenmodes of the form $u(z,t) = u_n(z)\exp(-t/\tau_n)$ subject to the boundary condition

$$E\frac{\partial u}{\partial z}|_{z=L} = 0$$

at the free boundary of the layer. The preceding discussion allows for a more general case when ξ is dependant on z . However, for a flat brush within the Alexander model $\xi(z) \approx \sigma^{1/2}$ and the description of the breathing modes is reduced to a diffusion type equation

$$(kT/\eta\sigma^{1/2})\partial^2 u/\partial z^2 = \partial u/\partial t.$$

Note that $kT/\eta\sigma^{1/2}$ is the diffusion coefficient of a single, impenetrable blob. Accordingly, the characteristic time associated with the longitudinal breathing modes of a flat brush is

$$\tau_{longitudinal} \sim L^{-2} \sim N^2(\sigma/a^2)^{1/3}.$$

A similar approach has been utilized by Fredrickson and Pincus in an analysis of the dynamics of a squeezed brush[48].

IX - SOME RECENT DEVELOPMENTS IN BRUSH RESEARCH: COLLAPSE AND MIXED SOLVENT BEHAVIOR

Brush research continued to flourish after the publication of the earlier reviews. An exhaustive summary of current developments is beyond the scope of the present review. Rather, I will focus on few notable results, experimental as well as theoretical, obtained in two areas. One is the collapse of brushes in poor solvents. The other is the behavior of brushes in mixed solvents. Brush collapse was studied by Auroy and Auvray[49] using small angle neutron scattering (SANS). Their experiments demonstrated the existence of a sharp interface between the dense brush and the bulk. This and the characteristic thickness of the layer are in good agreement with theory. The nature of the collapse for $\sigma < N^{2/3}a^2$ is not yet fully resolved. However, within mean field theory and in the limit of infinite chains, the collapse appears to take place as a second order phase transition. This picture is suggested by the work of Marko[50] on the basis of the analysis of Zhulina et al[51]. A second effect is the onset of lateral nonuniformity as the solvent quality decreases[49]. This last effect occurs in brushes of lower grafting density and is due to the dependence of σ^* on solvent quality. As noted earlier $\sigma \sim N^{6/5}$ in good solvents while in precipitants $\sigma \sim N^{2/3}$. Thus, for $N^{2/3}a^2 < \sigma < N^{6/5}a^2$ the brush when swollen by a good solvent, is expected to be laterally uniform. However, no chain-chain crowding is expected in a precipitant. The Alexander model and the Semenov type SCF theories are accordingly inapplicable to this regime. It was first studied by Lai and Binder[52] using computer simulations. Their results show in-plane aggregation resulting in lateral inhomogeneities. The segregation of the insoluble chains is arrested by the grafting. However, limited segregation, involving neighboring chains, is possible at the price of a deformation penalty. This regime was also studied by Grest[53] using simulation techniques and by Yeung et al[54] using SCF methods. A physical model for the in-plane aggregation was recently proposed by Williams[55]. It is based on the distinctive deformation behavior of collapsed chains[56]. In particular, a collapsed chain with its ends constrained to a separation larger than R_c exhibits a "ball and chain" configuration. This configuration involves a coexistence of a collapsed globule and a stretched chain segment. Sparsely grafted chains can lower the surface free energy per chain by aggregating laterally. This is attained at the price of a deformation penalty. As in the case of copolymeric micelles, the resulting structure consists of a dense central core and an external corona of stretched chains. However, in the present case the core and the corona are chemically identical and the stretched configurations of the coronal segments result from the grafting constraints.

The behavior of brushes immersed in mixed solvents involves two related issues: preferential solvation and the interplay between the brush and a prewetting transition of the bare surface. The first issue concerns the question whether a mixture of two solvents may be considered as a single effective solvent of intermediate quality. For dilute coils the effect of preferential solvation appears to be weak[57] and the single solvent picture holds. A different scenario is found in brushes. This effect was studied by Auroy and Auvray[49] using SANS. The initial aim of the study was the elucidation of the collapse behavior using mixtures of good and poor solvents. However, the brush thickness and the fraction of the better solvent within the brush varied non-linearly with the bulk composition. These results agree with theoretical discussions based on the Alexander model[58] and the SCF approach[50]. Subsequent study of brushes in a mixture of two *poor* solvents revealed a more dramatic effect[59]. While the brushes were collapsed in each of the neat solvents, significant swelling was observed in brushes immersed in a mixture of the two. This regime was not explicitly discussed theoretically. It is however easy to rationalize for solvent mixtures with positive mixing enthalpy. The effect may be attributed to screening of repulsive interactions between the different solvent molecules due to the presence of a third component, the polymer chains. A slightly more quantitative argument is possible upon noting that the collapse equilibrium is set by a balance of binary attractions and ternary repulsions. For simplicity we focus on the symmetric case, when the two solvents are equally poor. In such a case the Flory type free energy per chain is approximately

$$F_{chain}/kT \approx \sigma L[\frac{1}{2}v\Phi^2 + \frac{1}{6}w^2\Phi^3 + \Delta(1-\Phi)^2\phi_1\phi_2]$$

where Φ is the monomer volume fraction, $\Phi = Na^3/\sigma L$, v and w are the second and third virial coefficients, Δ is the interaction parameter of solvents 1 and 2 and the volume fraction of the two solvents is expressed as $\Phi_i = (1-\Phi)\phi_i$. The solvent-solvent interactions give rise to a quadratic term in Φ^2, $\Delta\Phi^2\phi_1\phi_2$. In turn, this gives rise to a modified second virial coefficient, $(\frac{1}{2}v + \Delta\phi_1\phi_2)\Phi^2$. When the enthalpy of mixing of 1 and 2 is positive, this correction can overcome the negative v associated with the poor solvent and lead to a positive effective v.

Brushes in mixed solvents are expected to exhibit an even richer behavior in presence of a prewetting transition. The prewetting transition involves a binary mixture and a wall exerting preferential attraction to one of the components. When the bulk is homogeneous but approaching phase separation this preferential attraction can result in the formation of a finite prewetting layer at the wall. This layer is richer in the prefered component. The formation of the prewetting layer takes place as a phase transition of order determined by the range of the interactions etc. In the presence of a brush, the system is characterized by two coupled order parameters specifying the monomer volume fraction and the excesses of the prefered solvent. Some of the rich behavior expected of this system was analyzed by Johner and Marques[60] and by Marko et al[61]. Two scenarios are of special interest. In both the prefered solvent is a precipitant. When the surface attraction is not too strong the growth of the prewetting layer may be arrested by the unfavorable interactions with the brush. In this limit one expects to find a finite prewetting layer "trapped" within the brush. A more interesting scenario is expected in the opposite limit, of strong surface attraction, when the prewetting layer is capable of attaining the thickness of the brush. As the prewetting layer thickens, the chain segments in the interior region of the brush stretch outwards so as to form a dilute exclusion zone. By so doing they avoid unfavorable interactions with the poor solvent at the expense of an elastic deformation. Eventually, when the prewetting layer extends further, the chains must collapse so as to minimize the overlap with the precipitant. From a different perspective, one may view a brush interacting with a wetting layer as an example of a brush experiencing a step like potential. The potential can be attractive or repulsive, depending on the solvent quality, resulting respectively in shrinking and swelling of the brush.

Finally, recent SANS experiments by Auroy and Auvray[62] probed the behavior of brushes immersed in a solution of homopolymers in a good solvent. Their results confirm the theoretical expectations. In particular, they observed the shrinking of the brush due to increased screening. Also observed was the associated decrease in the correlation length or blob size. Both results are of special interest as direct confirmation of the blob picture.

X - CONCLUDING REMARKS

As was stated in the introduction, this review makes no claims to completeness. Many subjects, concerning both blobology and brushes were not discussed. Among the blobological topics of importance, no attention was given to the structure of uniformly adsorbed layers[3,5] and the structure of randomly branched chains and gels[63]. Polyelectrolytes is another neglected subject[64]. Hopefully the main aims was attained, and this review can serve to complement deGennes monograph in introducing blobology. It also provides a limited update of the earlier reviews on brushes[7,8].

Acknowledgements: It is a pleasure to acknowledge the generous hospitality and insightful comments of M. Daoud

REFERENCES

1. (a) A.J. Silberberg, *J. Phys. Chem.* **66**, 1872 (1962). (b) E.A. diMarzio and F.L. McCrackin *J. Chem. Phys.* **43**, 539 (1965). (c) C.A.J. Hoeve *J. Chem. Phys.* **44**, 1505 (1966).
2. D.H. Napper, *Polymeric Stabilization of Colloidal Suspensions* , Academic Press, New York (1983).
3. P.G. deGennes, *Adv. Colloid Interface Sci.* **27**, 189 (1987).
4. M. Cohen-Stuart, T. Cosgrove and B. Vincent, *Adv. Colloid Interface Sci.* **24**, 143 (1986).
5. P.G. deGennes, *Macromolecules* **14**, 1637 (1981).
6. S. Alexander, *J. Phys.(France)* **38**, 977 (1977).
7. S.T. Milner, *Science* **251**, 905 (1991).

8. A. Halperin, M. Tirrell and T.P. Lodge, *Adv. Polym. Sci.* **100**, 31 (1992).
9. M. Daoud, J.P. Cotton, B. Farnoux, G. Jannink, H. Benoit, R. Duplesix, C. Picot and P.G. deGennes, *Macromolecules* **8**, 804 (1975).
10. P.G. deGennes, *Scaling Concepts in Polymer Physics*, Cornell University Press, Ithaca (1979).
11. P.G. deGennes, *Phys.Lett.* **38A**, 339 (1972).
12. (a) J. des Cloizeaux and G. Jannink, *Polymers in Solution*, Clarendon Press, Oxford (1990). (b) K.F. Freed, *Renormalization Group Theory of Macromolecules*, John Wiley, New York (1987).
13. E. Raphael and P.G. deGennes, *Physica* **A177**, 294, (1991).
14. The minimal necessary backgrounds may be found in Ch. 8 of ref. 15 or in Ch. 5 of ref. 16.
15. J.M. Yeomans, *Statistical Mechanics of Phase Transitions*, Clarendon Press, Oxford (1992).
16. J.J. Binney, N.J. Dowrick, A.J. Fisher and M.E.J. Newman, *The Theory of Critical Phenomena*, Clarendon Press, Oxford (1992).
17. M. Daoud and P.G. deGennes, *J. Phys.(France)* **38**, 85 (1976).
18. L. Turban, *J. Phys.(France)* **45**, 347 (1984).
19. Screening blobs can occur in a single chain upon confinement to a droplet. See F. Brochard and A. Halperin *C. R. Acad. Sci. Paris* **II302**,1043 (1986)
20. These are effective monomer-monomer interactions involving monomer-monomer, monomer-solvent and solvent-solvent interactions[10].
21. P. Pincus, *Macromolecules* **9**, 386 (1976).
22. E. Bouchaud and M. Daoud, *J. Phys.(France)* **48**, 1991 (1987).
23. P.G. deGennes, *J. Phys.(France)* **37**, 1445 (1976).
24. (a) J. Eisenriegler, K. Kremer and K. Binder, *J. Chem. Phys.* **77**, 6296 (1982) (b) J. Eisenriegler, *J. Chem. Phys.* **79**, 1052 (1983).
25. P.G. deGennes and P. Pincus, *J. Physique Lett.* **44**, 241 (1983).
26. A. Halperin and J.F. Joanny, *J. Phys.(France)* **II-1**, 623 (1991).
27. (a) C. Williams, F. Brochard and H.L. Frisch, *Ann. Rev. Phys. Chem.* **32**, 433 (1981) (b) I.M. Lifshits, A.Y. Grosberg and A.R. Khokhlov, *Rev Mod. Phys.* **50**, 6831 (1978) and references therein.
28. A. Halperin, *Macromolecular Reports* **A29**(suppl.2), 107 (1992).
29. P.G. deGennes in *Solid State Physics* Suppl. 14, L.Liebert ed. Academic Press, New York (1978).
30. P.G. deGennes, *Macromolecules* **13**, 1069 (1980).
31. Our discussion is limited to monomeric solvents. The behavior of brushes in polymeric solvents has been discussed in reference 29 and in: (a) L. Leibler, *Makromol. Chem., Macromol. Symp.* **16**, 1 (1988). (b) M. Aubouy and E. Raphael, *J. Phys.(France)* **II-3**, 443 (1993).
32. This free energy is actually valid only for brushes up to equilibrium thickness. The appropriate free energy for stretched brushes is discussed in section IX.
33. A. Halperin, *J. Phys.(France)* **49**, 547 (1988).
34. T.M. Birshtein and E.B. Zhulina, *Polym. Sci. USSR*, **25**, 2165 (1983).
35. A. Halperin and E.B. Zhulina, *Macromolecules* **24**, 5393 (1991).
36. M. Daoud and J.P. Cotton, *J. Phys.(France)* **43**, 531 (1982).
37. A.L.R. Bug, M.E. Cates, S. Safran and T.A. Witten, *J. Chem. Phys.* **87**, 1824 (1987).
38. C.M. Marques, L. Leibler and J.F. Joanny, *Macromolecules* **21**, 1051 (1988).
39. Y. Rabin and S. Alexander, *Europhys. Lett.* **13**, 49 (1990).
40. J.-L. Barrat, *Macromolecules* **25**, 832 (1992).
41. R.S. Ross and P. Pincus, *Europhys .Lett.* **19**, 79 (1992)
42. D.R.M. Williams, *Macromolecules* in press.
43. A. Halperin and S. Alexander, *Europhys.Lett.* **6**, 329 (1988).
44. A. Halperin, *Europhys.Lett.* **8**, 351 (1989).
45. A. Halperin and S. Alexander, *Macromoelcules* **22**, 2403 (1989).
46. C. Ligoure and L. Leibler, *J. Phys.(France)* **51**, 1313 (1990).
47. A. Johner and J.F. Joanny, *Macromolecules* **23**, 5299 (1990).
48. G.H. Fredrickson and P. Pincus, *Langmuir* **7**, 786 (1991)

49. P. Auroy and L. Auvray, *Macromolecules* **25**, 3134 (1992).
50. J. Marko, *Macromolecules*, **26**, 313 (1993).
51. E.B. Zhulina, O.V. Borisov, V.A. Pryamitsin and T.M. Birshtein, *Macromoelcules* **24**, 140 (1991).
52. P.-Y. Lai and K. Binder, *J. Chem. Phys.* **91**, 586 (1992).
53. G.S. Grest and M. Murat, *Macromolecules* **26**, 3106 (1993).
54. C. Yeung, A. Balazs and D. Jasnow, *Macromolecules* **26**, 1914 (1993).
55. D.R.M. Williams, *J. Phys.(France)* in press.
56. A. Halperin and E.B. Zhulina, *Europhys.Lett.* **15**, 417 (1991).
57. L. Gargallo and D. Radic, *Adv. Colloid Interface Sci.* **21**, 1 (1984).
58. P.-L. Lai and A. Halperin, *Macromolecules* **25**, 6693 (1992).
59. P. Auroy and L. Auvray, *Langmuir* in press.
60. C. Marques and A. Johner, *Phys. Rev. Lett.* **69**, 1827 (1992).
61. J. Marko, A. Johner and C. Marques, *Macromoelcules* in press.
62. P. Auroy and L. Auvray, (private communication).
63. M. Daoud and A. Lapp, *J. Phys: Condens Matter* **2**, 4021 (1990).
64. see for example (a) P. Pincus, *Macromoelcules* **24**, 2912 (1991). (b) E.B. Zhulina, O. V. Borisov and T.M. Birshtein, *J. Phys.(France)* **II-2**, 63 (1992).

THE ADHESION BETWEEN ELASTOMERS

Elie Raphaël and P.-G. de Gennes

Collège de France
Laboratoire de Physique de la Matière Condensée
URA n° 792 du C.N.R.S.
75231 Paris Cedex 05, France

I. INTRODUCTION

The phenomenon of *adhesion* concerns the interaction of two condensed phases brought into contact with each other. It involves a surprisingly large variety of materials - ranging from synthetic polymers to living cells and tissues. It is thought, for example, that the adhesive properties of external cellular membranes determine the main steps of an organism's development.

The complexity of the phenomenon of adhesion, as well as the diversity of materials capable of adhesive interaction, mean that a whole series of monographs would be required to constitute a comprehensive treatise. The purpose of this paper is more modest. We aim to present simple views on polymer adhesion to readers who are not familiar with this field. We choose to do so on a specific example, namely the adhesion between two cross-linked elastomers (a cross-linked elastomer consists of long, flexible chain-like molecules which are interconnected at various points by cross-links to form a molecular network; the polymer medium is locally fluid but the macroscopic flow of the material is prevented by the cross-links). This choice is motivated by a number of reasons: (a) the possibility of describing the fracture of the adhesive junction between the two elastomers in terms of a simple model, (b) the existence of controlled experiments that can be compared with the predictions of the model, (c) the possibility of introducing concepts that are of interest for other polymer adhesion problems, and, finally, (d) the fact that adhesion between elastomers is a technologically important field. Several texts on polymer adhesion are avaible (Wu, 1982; Kinloch, 1987; Lee, 1991; Vakula and Pritykin, 1991). A good reference for the results of the last few years is the review article by Brown (1991).

The problem we are interested in is represented on figure 1, where two chemically incompatible cross-linked elastomers A and B are in close contact. The interface between the two elastomers is strengthened by grafting some extra A chains (adhesion promoters) to the surface of the B elastomer. These chains - referred to as the *connectors* - cross the interface and penetrate into the bulk A elastomer. Note that since the two polymers (A) and (B) are incompatible, each connector crosses the interface only once. This situation is referred to as the one-stitch problem (Raphaël and de Gennes, 1992). The many-stitch problem has been investigated recently by Hong Ji and de Gennes (1993)). As a crack grows along the interface, the connectors are progressively pulled-out from the elastomer A. This suction process gives rise to a fracture energy that is larger than the work of adhesion W due to intermolecular interactions (typically of the van der Waals type). The aim of the present study is to analyze the effect of chain pull-out on the adhesion of the two elastomers.

The paper is organized as follows. Section II constitutes a brief introduction to linear elastic fracture mechanics. In section III we consider in detail the pull-out process.

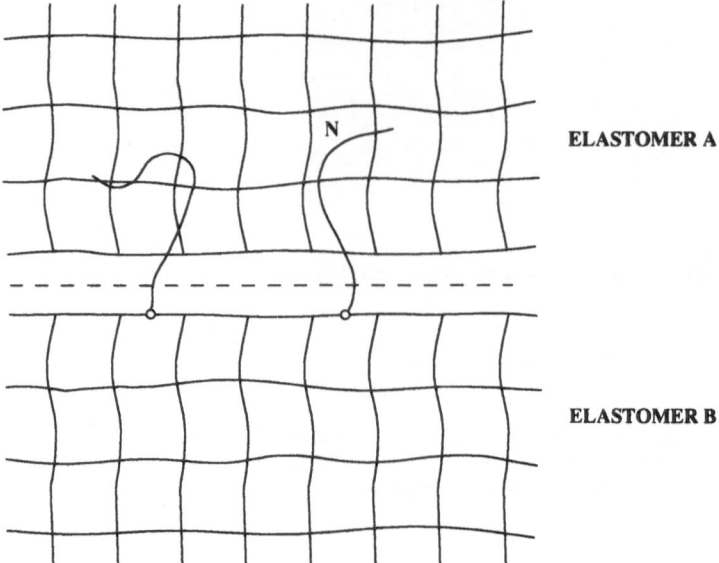

Figure 1. The interface between the two elastomers is strengthened by grafting some extra A chains (degree of polymerization, N) to the surface of the B elastomer.

The problem of a steadily growing crack along the interface between the two elastomers is analyzed in section IV. The paper ends with a discussion where comparison with the experimental results is made.

II. LINEAR ELASTIC FRACTURE MECHANICS

Let us consider a homogeneous, isotropic solid body. Under the action of applied forces, the solid body exhibit deformation. A point of initial position vector $\mathbf{r}$ (with components (x, y, z)) has, after the deformation, a new position $\mathbf{r'} = \mathbf{r} + \mathbf{u}$ where $\mathbf{u}(x, y, z)$ is the displacement field. The strain tensor u_{ik} is defined as (Landau and Lifshitz, 1986)

$$u_{ik} = \frac{1}{2} \left(\frac{\partial u_i}{\partial x_k} + \frac{\partial u_k}{\partial x_i} + \frac{\partial u_n}{\partial x_i} \frac{\partial u_n}{\partial x_k} \right) \tag{2.1}$$

(with $x_1 = x$, $x_2 = y$ and $x_3 = z$). In eqn (2.1) we have used the summation convention to suffixes occurring twice in an expression. Two neighboring points separated by a distance dl before the deformation are, after the deformation, separated by a distance dl' :

$$dl'^2 = dl^2 + 2\,u_{ik}\,dx_i\,dx_k \tag{2.2}$$

In the deformed body, internal stresses arise which tend to return the body to its original state. The force $\mathbf{F}$ per unit volume is given by

$$F_i = \partial\sigma_{ik}/\partial x_k \tag{2.3}$$

where σ_{ik} is the stress tensor.

For an linear elastic material the stress and strain tensors are related by (Hooke's law)

$$\sigma_{ik} = \frac{E}{1+\nu} \left(u_{ik} + \frac{\nu}{1-2\nu} u_{nn}\delta_{ik} \right) \tag{2.4}$$

where E is the Young's modulus and ν the Poisson's ratio.

Consider now the problem of a crack embedded in a linear elastic material (fracture mechanics within the confines of materials that obey Hooke's law is known as *linear elastic fracture mechanics*) . The crack extends in the negative x-direction with its tip at x = 0. The crack may be stressed in three different modes: (a) the cleavage or tensile-opening mode (mode I), (b) the in-plane shear mode (mode II), and (c) the antiplane shear mode (mode III), as depicted in figure 2. The superposition of the three modes describes the general case of loading. The mode I is technically the most important since it is the most commonly encountered and usually the one which most often results in failure (Kinloch, 1987). The following discussion will therefore be confined to this situation.

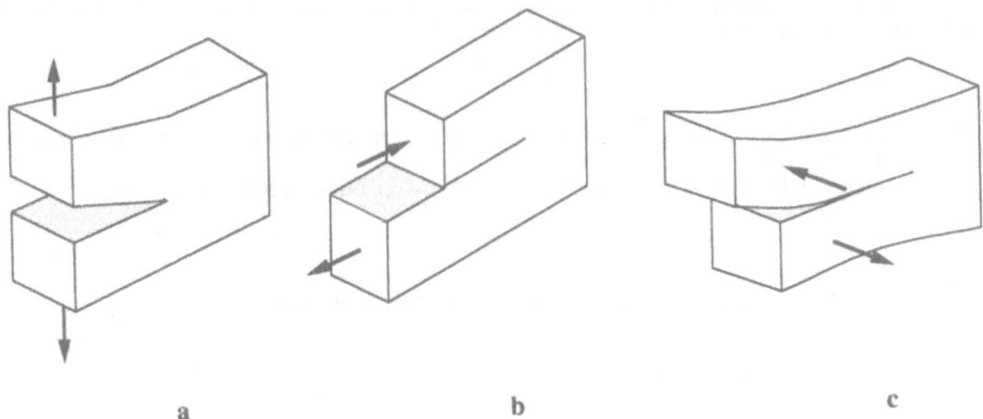

a   b   c

Figure 2. Modes of loading. (a) Cleavage mode: mode I. (b) In-plane shear mode: mode II. (c) Antiplane shear mode: mode III.

For a Mode I crack the tensile stress $\sigma(x) \equiv \sigma_{yy}(x,y = 0)$ and the crack displacement $u(x) \equiv u_y(x,y = 0)$ are respectively given by (Kanninen and Popelar, 1985)

$$\sigma(x) = \frac{K_I}{\sqrt{2\pi x}} \qquad x > 0 \qquad (2.5)$$

$$\sigma(x) = 0 \qquad x < 0 \qquad (2.6)$$

and

$$u(x) = \frac{4K_I}{E^*\sqrt{2\pi}}\sqrt{-x} \qquad x < 0 \qquad (2.7)$$

$$u(x) = 0 \qquad x > 0 \qquad (2.8)$$

The material parameter E^* is given by $E^* = E$ for *plane stress* conditions and $E^* = E/(1 - v^2)$ for *plane strain* conditions. These conditions are defined as follows:

$$\sigma_{zz} = 0 \qquad \text{plane stress} \qquad (2.9)$$

$$\sigma_{zz} = v\,(\sigma_{xx} + \sigma_{yy}) \qquad \text{plane strain} \qquad (2.10)$$

and $\sigma_{xz} = \sigma_{yz} = 0$ for both cases. In practice the state of stress near the crack tip varies from plane stress in a very thin specimen to plane strain near the center of a wide plate.

The quantity K_I is referred to as the *stress intensity factor*. It is a function of the applied loading and the geometry of the cracked body. Since the level of K_I uniquely defines the stress field around the crack, Irwin (1964) postulated that the condition

$$K_I \geq K_{Ic} \tag{2.11}$$

represented a fracture criterion (i.e., fracture occurs when the value of K_I exceeds the critical value K_{Ic}). K_{Ic} is a material property and is often termed the *fracture toughness*.

Another approach to obtain a criterion for fracture was obtained by Griffith (1920) and is based on an energy balance. Consider a Mode I crack of area A embedded in a linear elastic material. Let us imagine an infinitesimal propagation of the crack by an amount dA. In this process, part of the elastic energy U stored in the system is released. The work done by the external applied forces is dW. Griffith's hypothesis is that fracture occurs if and only if the *strain energy release rate* $G_I = $ (dW - dU)/dA is larger than a critical value G_{Ic}

$$G_I \geq G_{Ic} \tag{2.12}$$

The critical value G_{Ic} is called the *fracture energy*, or the critical strain energy release rate. It is a material parameter.

As shown by Irwin (1964), a simple relationship exists between K_{Ic} and G_{Ic} :

$$G_{Ic} = (K_{Ic})^2/E^* \tag{2.13}$$

Thus a critical K_{Ic} criterion is equivalent to a critical G_{Ic} criterion.

III. CHAIN PULL-OUT PROCESS

As a crack grows along the interface between the two elastomers, the connectors are progressively pulled-out from the material (the connectors are assumed not to break, but to slip out by a viscous process). The aim of the present section is to describe in detail this pull-out process. Let us first consider what happens when the two elastomers are separated by a uniform air gap of thickness h (figure 3). We will come back to fracture propagation in section IV.

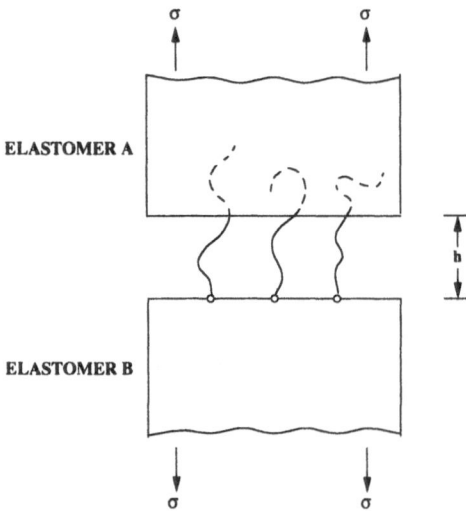

Figure 3. The two elastomers submitted to a uniform tensile stress σ.

The partially pulled-out chains are assumed to form single chain fibrils. The free energy e(h,n) of a fibril containing n monomers is given by

$$e(h,n) \cong \gamma_A \, a^2 \, n + kT \frac{h^2}{a^2 \, n} \tag{3.1}$$

The first term corresponds to the energy cost for exposing the n monomers to the air (a is a monomer size and γ_A is the interfacial energy between the bulk polymer A and the air). The second term in eqn (3.1) is a stretching term. Minimization of eqn (3.1) with respect to n gives

$$n(h) \cong \frac{h}{a} \left(\frac{\gamma_A \, a^2}{kT} \right)^{-1/2} \tag{3.2}$$

Since in practice $\gamma_A \, a^2 \cong kT$, the fibril is almost fully stretched. The corresponding value of the free energy is

$$e(h) \cong kT \frac{h}{a} \left(\frac{\gamma_A \, a^2}{kT} \right)^{1/2} \tag{3.3}$$

Equation (3.3) shows that there is a minimum force f* required for a fibril to exist (Raphaël and de Gennes, 1992):

$$f^* \cong \frac{kT}{a} \left(\frac{\gamma_A \, a^2}{kT} \right)^{1/2} \tag{3.4}$$

Suppose that a uniform external tensile stress σ is applied to the elastomers (figure 3). The energy g_{area} per unit area (as a function of the distance h between the two elastomers) has the form shown on figure 4, where W denotes the thermodynamic work of adhesion of the two elastomers in the absence of connectors:

$$W = \gamma_A + \gamma_B - \gamma_{AB} \tag{3.5}$$

(W is due to intermolecular interactions, typically of the van der Waals type). For $h > a$, the energy $g_{area}(h)$ is linear (see eqn (3.3))

$$g_{area}(h) \cong \left[\Sigma \, \frac{kT}{a} \left(\frac{\gamma_A \, a^2}{kT} \right)^{1/2} - \sigma \right] h + (const) \qquad (h > a) \tag{3.6}$$

where Σ is the number of connectors per unit interface area. As long as σ is smaller than the critical stress

$$\sigma^* = \Sigma \, f^* \cong \Sigma \, \frac{kT}{a} \left(\frac{\gamma_A \, a^2}{kT} \right)^{1/2} \tag{3.7}$$

the energy g_{area} is minimal for h = 0 and the system remains closed. But as soon as σ becomes greater than σ^*, the energy minimum is at h = +∞ and the system opens out. It is true that there remains an energy barrier but, in the fracture process to be discuss below, the fracture tip acts as a nucleation center and removes this barrier. Thus σ^* appears as a threshold stress for opening.

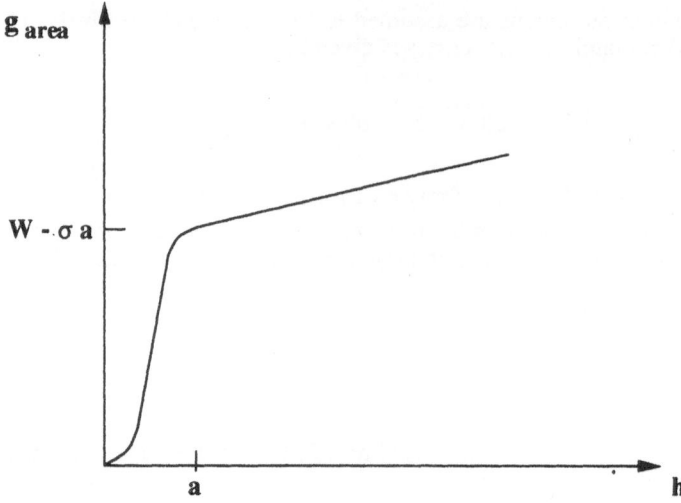

Figure 4. Energy per unit area as a function of the distance between the two elastomers.

For $\sigma > \sigma^*$, the connectors are progressively pulled-out of the elastomer. During this suction process, the energy is partly dissipated in viscous losses (caused by the slippage of the connectors in the elastomer) and partly stored in forming longer fibrils:

$$\sigma\, dh \cong [f_v\, ds + e'(h)\, dh]\, \Sigma \qquad (3.8)$$

($e'(h)$ represents the derivative of $e(h)$ with respect to h). Here $ds \cong (\gamma_A\, a/kT)^{-1/2} dh$ is the length of chain pulled-out when the distance between the two elastomers increased by dh and f_v is the friction force experienced by one connector

$$f_v \cong N\frac{h_f - h}{h_f}\, \zeta_0\, \frac{ds}{dt} \qquad (3.9)$$

In eqn (3.9), N is the degree of polymerization of the connector and ζ_0 is a monomer friction coefficient. The factor $(h_f - h)/h_f$ expresses the fact that the pull-out process becomes easier when only a small portion of the connector length remains to be pulled-out. For mathematical simplicity, we will hereafter ignore this correction. From eqns (3.7), (3.8) and (3.9) we arrive at the constitutive law

$$dh/dt = Q^{-1}\,(\sigma - \sigma^*) \qquad\qquad \sigma > \sigma^* \qquad (3.10)$$

$$= 0 \qquad\qquad \sigma < \sigma^*$$

where

$$Q \cong \Sigma\, N\, \zeta_0\, (\gamma_A\, a/kT)^{-1} \qquad (3.11)$$

The suction process ends when the connectors are completely pulled-out of the elastomer. This occurs for $n(h) \cong N$ which corresponds to a maximal value of the opening

$$h_f \cong a\, N \left(\frac{\gamma_A\, a^2}{kT}\right)^{1/2} \qquad (3.12)$$

IV. STEADY STATE CRACK GROWTH

IV.1. The Cohesive Zone

We now consider the problem of a steadily growing Mode I crack along the interface between the two elastomers. The crack extends in the negative x-direction with its tip at x = 0 (see figure 5). It propagates with a constant velocity V. We assume that the two elastomers have similar elastic properties and describe the materials outside the cohesive zone as a linear elastic material with Young's modulus E and Poisson's ratio ν.

The linear elastic analysis of section II predicts infinite stresses at the crack tip (eqn (2.5)). In reality, this divergence is relaxed by the dissipative pull-out process which takes place at the crack tip. The pull-out process is expected to occur in an approximately planar *cohesive zone* directly ahead of the crack tip. A thorough investigation of cohesive zone models can be found in Fager et al. (1991). See also Xu et al. (1991).

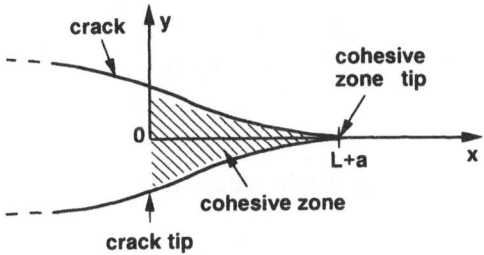

Figure 5. A schematic diagram of a cohesive zone ahead of a crack. The cohesive zone is defined by 0 < x < L+*a* , where *a* is a molecular size.

The adhesive junction between the two elastomers is an example of a *weak adhesive junction* (de Gennes, 1989a). When a fracture propagates along such a junction, the dissipation tends to be localized in a thin ribbon ahead of the crack tip.

If we assume the cohesive zone 0 < x < L + *a* to be small compared with the crack length, the applied loading can be simulated by the prescription of the elastic K [†] field far away from the crack tip

$$\sigma(x) = \frac{K}{\sqrt{2\pi x}} \qquad\qquad x \gg L + a \qquad\qquad (4.1)$$

$$u(x) = \frac{4K}{E^*\sqrt{2\pi}}.\sqrt{-x} \qquad\qquad x \ll 0 \qquad\qquad (4.2)$$

The elastic field associated with the cohesive zone can be described in terms of a source function $\Phi(x)$ (with 0 < x < L + *a*) defined by:

[†] For notational simplicity, the subscript Ic of K_{Ic} and G_{Ic} will be systematically dropped throughout the remainder of this article.

$$\sigma(x) \;=\; \frac{(1-v)}{2}\, E^* \int_0^x dy\; \Phi(y)\,(x-y)^{-1/2} \qquad\qquad x > 0 \qquad\qquad (4.3)$$

$$= 0 \qquad\qquad x < 0$$

$$u(x) \;=\; 2(1-v) \int_x^{L+a} dy\; \Phi(y)\,(y-x)^{1/2} \qquad\qquad x < L + a \qquad\qquad (4.4)$$

$$= 0 \qquad\qquad x > L + a$$

This formulation was first introduced by Cottrell (1969) and recently implemented by one of us (de Gennes, 1989a, 1989b). It can be shown (Hui and Raphaël, 1993) that eqns (4.3) and (4.4) are equivalent to the standard formulation

$$u(x) \;=\; \frac{4K}{E^*\sqrt{2\pi}}\sqrt{L+a-x}$$

$$-\;\frac{2}{\pi E^*} \int_0^{L+a} dt\; \sigma(t)\, \mathrm{Ln}\left|\frac{\sqrt{(L+a-x)}+\sqrt{(L+a-t)}}{\sqrt{(L+a-x)}-\sqrt{(L+a-t)}}\right| \qquad (0 < x < L+a) \qquad (4.5)$$

(see e.g. Fager et al., 1991)

For $x \gg L + a$, eqn (4.3) should reduce to eqn (4.1). The source function $\Phi(x)$ therefore satisfies

$$K - \sqrt{\pi/2}\,(1-v)\, E^* \int_0^{L+a} dy\; \Phi(y) \;=\; 0 \qquad\qquad (4.6)$$

Knowing K, the fracture energy G of the interface may be derive from the Irwin equation (2.13)

$$G \;=\; \frac{K^2}{2\mu(1+v)} \qquad\qquad \text{plane stress} \qquad\qquad (4.7a)$$

$$G \;=\; \frac{(1-v)\,K^2}{2\mu} \qquad\qquad \text{plane strain} \qquad\qquad (4.7b)$$

where μ is the shear modulus, $\mu = E/\,2(1+v)$.

In order to determine the source function $\Phi(x)$, we distinguish two regions within the cohesive zone. For $0 < x < L - a$, we adopt the constitutive law eqn (3.10) relating the opening rate $dh/dt = 2\,du/dt$ and the normal stresses acting on the cohesive zone. Inserting eqns (4.3) and (4.4) into eqn (3.10) and assuming plain strain conditions we obtain the fundamental equation (de Gennes, 1990)

$$\int_0^x dy\; \Phi(y)\,(x-y)^{-1/2} \;-\; \sigma^*/\mu \;=\; \lambda \int_x^{L+a} dy\; \Phi(y)\,(y-x)^{-1/2} \quad (0 < x < L-a) \qquad (4.8)$$

where the eigenvalue λ is defined by

$$\lambda = V/V^*$$ (4.9)

$$V^* = \frac{\mu}{2(1-\nu)Q} \cong \frac{\mu}{2(1-\nu)\Sigma N \zeta_0} \left(\frac{\gamma_A a}{kT}\right)$$ (4.10)

Equation (4.8) is an integral equation for the source function $\Phi(x)$. It must be supplemented by the boundary conditions

$$\sigma(x = L - a) = \sigma^*$$ (4.11)

$$2u(x=0) = h_f$$ (4.12)

For $L - a < x < L + a$, we assume that $\Phi(x)$ is unaffected by the connectors (Raphaël and de Gennes, 1992). Since in the absence of connectors the fracture energy is simply given by W, we have

$$\Phi(x) = \left[\frac{W}{4\pi \mu a^2 (1-\nu)}\right]^{1/2} \qquad (L - a < x < L + a)$$ (4.13)

VI.2. The Quasi-Static Limit

Let us first consider the quasi-static limit $V \to 0$. For $\lambda = 0$, equation (4.8) reduces to

$$\int_0^x dy \, \Phi(y) \, (x - y)^{-1/2} = \sigma^*/\mu$$ (4.14)

which leads to

$$\Phi(x) = \frac{\sigma^*}{\pi \mu x^{1/2}} \qquad 0 < x < L_0 - a$$ (4.15)

(the subscript 0 in L_0 refers to the limit $V \to 0$). Using eqns (4.6), (4.13) and (4.15) the applied stress intensity factor in the quasi static limit is found to be

$$K_0 = 2\sqrt{2/\pi} \, \sigma^* (L_0^{1/2} + \frac{1}{2} B^{1/2})$$ (4.16)

where

$$B = \frac{\pi \mu W}{(1-\nu)(\sigma^*)^2}$$ (4.17)

(we have neglected terms of order a/L_0).

The length $L_0 + a$ of the cohesive zone can be determined by using the boundary condition eqn (4.12)

$$h_f = 4(1 - \nu) \int_0^{L_0+a} dy \, \Phi(y) \, y^{1/2} \tag{4.18}$$

Inserting eqns (4.13) and (4.15) into eqn (4.18) and again neglecting a compared with L_0, we obtain

$$h_f = 4\left[\frac{W(1 - \nu)}{\pi\mu}\right]^{1/2} L_0^{1/2} + \frac{4(1 - \nu)}{\pi\mu}\sigma^* L_0 \tag{4.19}$$

Equation (4.19) can be rewritten as

$$A = L_0 + B^{1/2} L_0^{1/2} \tag{4.20}$$

where

$$A = \frac{\pi\mu\, h_f}{4(1 - \nu)\,\sigma^*} \tag{4.21}$$

Using eqns (4.7), (4.16) and (4.20) the zero-rate fracture energy, G_0, is found to be (plain strain)

$$G_0 = \frac{(1 - \nu)\,(K_0)^2}{2\mu}$$

$$= \frac{4(1 - \nu)(\sigma^*)^2}{\pi\mu}\left[\frac{B}{4} + B^{1/2} L_0^{1/2} + L_0\right]$$

$$= \frac{4(1 - \nu)(\sigma^*)^2}{\pi\mu}\left[\frac{B}{4} + A\right] \tag{4.22}$$

Hence (Brown et al., 1993):

$$G_0 = W + h_f\,\sigma^* \tag{4.23}$$

The zero-rate fracture energy G_0 is therefore larger than the thermodynamic work of adhesion W. This result is nontrivial since one would expect the pull-out contribution to the fracture energy to vanish when the crack propagation rate goes to zero. In fact, as explained in section III, there is a minimum force f^* (eqn (3.4)) required for a fibril to exist, even at zero pull-out rate. As the force on a chain that is being pulled-out remains finite as $V \to 0$, the existence of a zero-rate fracture energy G_0 that is larger than the work of adhesion W is expected.

Equation (4.23) can also be proven by calculating the work done against stresses in the cohesive zone (Rice, 1968; Brown et al., 1993)

$$G_0 = \int_0^{h_f} \sigma \, dh = -\int_0^{L_0+a} \sigma(x)\left(\frac{dh}{dx}\right) dx \tag{4.24}$$

Now, assuming plane strain, we have from eqns (4.3), (4.4), (4.13) and (4.15) :

$$\sigma(x) = \sigma^* \tag{4.25}$$

for $0 < x < L_0 - a$, and

$$\sigma(x) = \sigma^* + \frac{E}{2(1+v)} \left[\frac{W}{4\pi \mu a^2 (1-v)} \right]^{1/2} 2[x - (L_0-a)]^{1/2} \qquad (4.26)$$

$$\frac{dh}{dx} = -4(1-v) \left[\frac{W}{4\pi \mu a^2 (1-v)} \right]^{1/2} [(L_0+a) - x]^{1/2} \qquad (4.27)$$

for $L_0 - a < x < L_0 + a$. Combining eqns (4.24)-(4.27) we obtain

$$G_0 = h_f \sigma^* + \frac{2W}{\pi a^2} \int_{L_0-a}^{L_0+a} dx \, [x - (L_0-a)]^{1/2} [(L_0+a) - x]^{1/2}$$

$$= h_f \sigma^* + W \qquad (4.28)$$

i.e., we recover eqn (4.23).

VI.3. Propagation at Finite Velocity

We now consider the steady state propagation of the crack at a finite velocity V. In order to simplify the discussion, we will assume that $h_f \sigma^* \gg W$. We can then ignore the contribution of the intermolecular forces to the fracture energy (see eqn (4.23)) and eqns (4.8) and (4.11) reduce to

$$\int_0^x dy \, \Phi(y) (x-y)^{-1/2} - \sigma^*/\mu = \lambda \int_x^L dy \, \Phi(y) (y-x)^{-1/2} \quad (0 < x < L) \qquad (4.29)$$

$$\sigma(x=L) = \sigma^* \qquad (4.30)$$

It turns out that the system (4.29)-(4.30) has an exact solution of the form (Fager et al., 1991)

$$\Phi(x) = \mu^{-1} \sigma^* \frac{\cos\pi\varepsilon}{\pi} x^{-[(1/2)+\varepsilon]} (L-x)^\varepsilon \qquad (4.31)$$

with

$$\tan(\pi\varepsilon) = \lambda \qquad (4.32)$$

The length L of the cohesive zone can be determined by using eqn (4.4) and the boundary condition (4.12)

$$4(1-v) \int_0^L dy \, \Phi(y) y^{1/2} = h_f \qquad (4.33)$$

Using eqn (4.31) we get

$$L = \frac{\pi \, \mu \, h_f}{4(1 - v) \, \sigma^*} \frac{1}{\Gamma(1 + \varepsilon) \, \Gamma(1 - \varepsilon)} \frac{1}{\cos\pi\varepsilon} \qquad (4.34)$$

with the limiting behaviors

$$L = \frac{\pi \, \mu \, h_f}{4(1 - v) \, \sigma^*} \qquad\qquad V = 0 \qquad\qquad (4.35)$$

$$= \frac{\pi \, \mu \, h_f}{4(1 - v) \, \sigma^*} \frac{2}{\pi} \frac{V}{V^*} + \dots \qquad V \gg V^* \qquad (4.36)$$

In eqn (4.34), $\Gamma(x)$ is the usual gamma function (see e.g. Abramowitz and Stegun, 1970). The fracture toughness K satisfies (see eqn (4.6))

$$\sqrt{\pi/2} \, \frac{E}{(1 + v)} \int_0^L dy \, \Phi(y) = K \qquad (4.37)$$

Using eqn (4.31) we obtain

$$K = \frac{2 \, \sqrt{2} \, \Gamma(1 + \varepsilon)}{\Gamma(\frac{1}{2} + \varepsilon)} \sigma^* \, L^{1/2} \qquad (4.38)$$

From eqns (4.7), (4.31) and (4.38) the fracture energy G is found to be

$$G = h_f \, \sigma^* \frac{\Gamma(1 + \varepsilon)}{\Gamma(1 - \varepsilon) \, [\Gamma(\frac{1}{2} + \varepsilon)]^2} \frac{\pi}{\cos\pi\varepsilon} \qquad (4.39)$$

with the limiting behaviors

$$G = h_f \, \sigma^* \qquad\qquad V = 0 \qquad\qquad (4.40)$$

$$= h_f \, \sigma^* \frac{\pi}{2} \frac{V}{V^*} + \dots \qquad V \gg V^* \qquad (4.41)$$

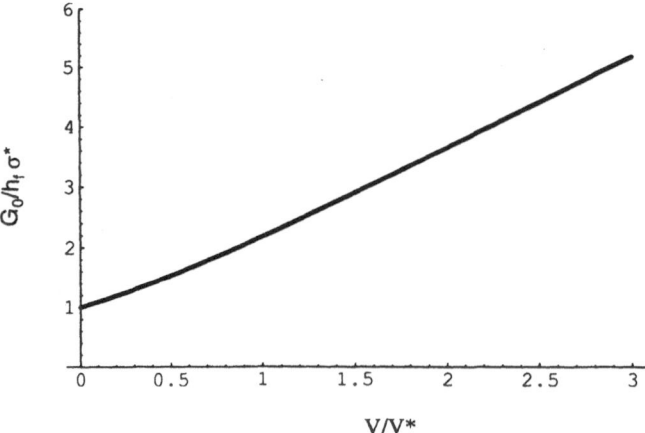

Figure 6. Plot of $G/h_f \, \sigma^*$ versus V/V^*.

Figure 6 represents the behavior of $G/h_f \sigma^*$ as a function of the ratio V/V^*. It is important to notice that the slope of the curve at the origin, $s(0)$, is non zero: $s(0) = (4 \ln 2)/\pi \cong 0.88$, and does not differ greatly from the slope at infinity, $s(\infty) = \pi/2 \cong 1.57$. Thus, the curve does not exhibit a plateau for $V < V^*$ and it might be difficult to determine the value of V^* just by looking for a crossover in the curve behavior (this point has emerged from discussions with L. Léger; see also Xu et al. 1991). An more reliable way to determine V^* would be to look for the velocity at which $G \cong 2.19\, h_f \sigma^*$.

V. DISCUSSION

In the preceding two sections we have presented a simple model for the adhesion between two cross-linked elastomers in the presence of connectors. The model predicts that when the crack velocity tends to zero, the fracture energy takes the simple form (eqn (4.23))

$$G_0 \cong W + (\gamma_A a^2) N \Sigma \qquad (5.1)$$

The model also predicts that the fracture energy increases linearly with the crack velocity when the velocity is well above a critical value V^* (eqns (4.10) and (4.41)).

What is the situation on the experimental side? Ellul and Gent (1984, 1985) have shown that the incorporation of free chains into a cross-linked network increased significantly the fracture energy of interfaces between cross-linked elastomers at finite crack growth rates, but had no effects at very low rates. More recently, Reichert and Brown (1993) placed a known amount of diblock copolymer at the interface between cross-linked polyisoprene and polystyrene. Using a peel test they found that, at the pull-out rate they used, the presence of block copolymer increased the fracture energy by up to a factor 5. The pros and cons of peel tests are reviewed by Brown (1993). An alternative test is the JKR test (after Johnson, Kendall and Roberts, 1971) in which an elastic spherical cap is pushed against a flat plate. The contact area is a function of the applied load, the radius of curvature and the elastic moduli of the cap, and the thermodynamic work of adhesion W between the two materials. If the load is released, the contact area will decrease with time and the measured work of adhesion can be interpreted as the fracture energy G (Brown, 1993).

The effects of chain pull-out on the adhesion of elastomers has been recently investigated by Brown (1993) and by Creton, Brown and Shull (1993) using the JKR technique. A thin layer of polystyrene-polyisoprene diblock copolymer was placed at the interface between a polystyrene coated substrate and a polyisoprene cross-linked lens. Over the whole range of crack speeds investigated (10^{-10} - 10^{-7} m/s), the presence of the copolymer produced a large increase in the fracture energy G. This increase is believed to be due to the pull-out of the polyisoprene chains of the diblock from the bulk cross-linked polyisoprene. At low crack speeds the fracture energy of the interface G was found to increase linearly with velocity from a threshold value G_0. At higher crack speeds, a transition occurred after which the G increased at a much lower rate, which could be attributed to viscoelastic bulk losses. The measured value of G_0 was in good agreement with the predictions (5.1). Furthermore, G_0 increased linearly with the areal density of copolymer present at the interface, Σ, and monotically with the degree of polymerization of the polyisoprene chains, N. This agrees well with the predictions of the model presented in sections III and IV. The observed value of V^* seems, however, to be much lower than the prediction (4.10). In his preliminary study, Brown (1993) suggested that this might be caused by the polyisoprene chains forming multiple stitches. Indeed, as shown by Hong Ji and de Gennes (1993), if each connector crosses the interface many times, G_0 is not altered but V^* is reduced by a factor N from the value obtained in the single-stitch case. According to Creton et al. (1993), this suggestion seems rather unlikely to be correct. Another explanation for the low observed value of V^* has been proposed by Creton, Brown and Shull (1993). It is based on a very recent model of Rubinstein et al. (1993) for the problem of slip between a solid surface with attached grafted chains and a cross-linked elastomer. Rubinstein et al. predicted that at very low velocity the friction is due to a balance between chain stretching and chain relaxation and is considerably larger than the standard Rouse friction. According to Creton et al. (1993), a similar augmentation of the friction should occur in their case, leading to a significant reduction of V^* (see eqn (4.10)).

Quite recently, Marciano, Hervet and Léger (1993) have conducted peel tests on a system made of a thin ribbon of elastomeric polydimethylsiloxane (PDMS) brought into contact with a flat silicon wafer grafted with PDMS chains. The internal structure of the grafted PDMS layer (i.e. the loop and tail distribution) was adjusted by varying the polymer concentration in the reaction bath, f. From peel force measurements at very low velocity (50 Å/s), Marciano et al. estimated G_0. They found that G_0 (as a function of f) exhibited an optimum[†] around $f \approx 30\%$. The occurence of an optimum can be qualitatively understood by noting that at low f the number of connectors is rather small whilst at high f the number of connectors is high but they penetrate into the network with difficulty (Marciano et al., 1993). More theoretical work will be required to understand quantitatively the penetration of the connectors into the network and the corresponding fracture energy (O'Connors and McLeish, 1993).

ACKNOWLEDGMENT

Several parts of this work have been done in collaboration with H. R. Brown and C.-Y. Hui. We would like to thank them for very useful discussions and correspondence. Valuables comments on the manuscript by C. Creton, L. Léger, Y. Marciano and D. Williams are gratefully acknowledged. J. R. Rice draw our attention to the work of Cottrell.

REFERENCES

Abramowitz, M., and Stegun, I.A., 1970, "Handbook of Mathematical Functions", Dover.

Brown, H. R., 1991, *Annu. Rev. Mater. Sci.* 21:463.

Brown, H. R., 1993 *Macromolecules* 26:1666.

Brown, H. R., Hui, C.-Y., and Raphaël, E., 1993, *Macromolecules* submitted.

Creton., C., Brown, H. R., and Shull, K.R., 1993, *Macromolecules* submitted.

Cottrell, A., 1969, *in* "Physics of Strenght and Plasticity," A.S. Argon, ed., MIT Press.

de Gennes, P.-G., 1989a, *J. Phys. France* 50:2551.

de Gennes, P.-G., 1989b,*C. R. Acad. Sci. (Paris) II* 309:1125.

de Gennes, P.-G., 1990, *Canadian J. of Phys.* 68:1049.

Ellul, M.D., and Gent, A.N., 1984, *J. Polym. Sci., Polym. Phys. Ed.* 22:1953.

Ellul, M.D., and Gent, A.N., 1985, *J. Polym. Sci., Polym. Phys. Ed.* 23:1823.

Fager, L.-O., Bassani, J. L., Hui, C.-Y., and Xu, D.-B., 1991, *Int. J. Fracture Mech.* 52:119.

Griffith, A. A., 1920, *Phil. Trans. Roy. Soc. A* 221:163.

Hui, C.-Y., and Raphaël, E., 1993, *Int. J. Fracture Mech.* 61.

Irwin, G.R., 1964, *Appl. Mater. Res.* 3:65.

Ji, H., and de Gennes, P.-G., 1993, *Macromolecules* 26:520.

Johnson, K.L., Kendall, K., and Roberts, A.D., 1971, *Proc. R. Soc. London A* 324:301.

[†] corresponding to an increase of the fracture energy by a factor 4 to 5.

Kanninen, M., and Popelar, C., 1985, "Advanced Fracture Mechanics," Oxford University Press.

Kinloch, A.J., 1987, "Adhesion and Adhesives, Science and Technology," Chapman and Hall.

Lee, L.-H. (ed.), 1991, "Fundamentals of Adhesion," and "Adhesive Bonding," Plenum.

Landau, L.D., and Lifshitz, E. M., 1986, "Theory of Elasticity,", third english edition, Pergamon Press.

Marciano, Y., Hervet, H., and Léger, L., 1993, preprint.

O'Connor, K. P., and McLeish, T.C.B., 1993, preprint; this paper is concerned with the penetration of end-tethered chains into a cross-linked elastomer in the limit of low surface coverages.

Raphaël, E., and de Gennes, P.-G., 1992, *J. Phys. Chem.* 96:4002.

Reichert, W.F., and Brown, H. R., 1993, *Polymer* to be published.

Rice, J. R., 1968, in "Fracture II , An Advanced Treatise," H. Liebowitz, ed., Academic Press.

Rubinstein, M., Ajdari, A., Leibler, L., Brochard-Wyart, F., and de Gennes, P.G., 1993, *C. R. Acad. Sci. (Paris) II* 316:317.

Vakula, V.L., and Pritykin, L.M., 1991, "Polymer Adhesion, Physico-Chemical Principles," Ellis Horwoood.

Wu, S., 1982, "Polymer Interface and Adhesion", Dekker.

Xu, D.-B., Hui, C.-Y., Kramer, E.J., and Creton, C., 1991, *Mech. Mater.* 11:257; this model is aimed mainly at the pull-out process in glassy polymers.

DYNAMICS OF LATE STAGE PHASE SEPARATION IN POLYMER BLENDS

Shimon Reich
Department of Materials and Interfaces
Weizmann Institute of Science
Rehovot, Israel 76100

(In memory of my beloved wife Klara)

INTRODUCTION

The physical properties of multicomponent polymeric materials are determined to a great extent by their phase behavior. For polymer systems forming separate phases, important issues are phase morphology, i.e. size, shape, and connectivity, and the nature of the interfacial adhesion between the phases. These features of morphology are key issues in the formation of microporous structures in membranes, hollow fibers, and may also be used for preparation of high-impact, high toughness composites.

Blends with two phases can be organized into a variety of morphologies. One phase may be dispersed in a matrix of the other, and this can occur in a variety of ways. For example, the dispersed phase may exist as spheres. Variables used to characterize this morphology are the total volume fraction of the dispersed phase, the average size and size distribution of the spheres, etc.. On the other hand the dispersed phase may exist as parallel fibrils as observed in certain fibers formed from polymer blends. Another type of morphology is one in which there is no well defined dispersed phase but, rather, both components are to some degree continuous throughout the structure. Such interpenetrating networks of phases have been observed many times in polymer blend systems. Spinodal decomposition is a mechanism by which such morphologies can be generated.

There has been considerable effort directed at investigating the thermodynamic characteristics of polymer blends.[1-11] And some light scattering experiments[6,7], directed at understanding dynamical aspects of the phase separation processes in these systems.

On the theoretical side, there has been an effort to understand the dynamics of concentration fluctuations in polymer blends, mostly by de Gennes[8], Pincus[9], and Binder[10].

A very sensitive measure of the non-equilibrium structural developments which occur during a phase separation process is provided by the scattering structure factor. In classical spinodal processes one expects to observe a single scattering maximum. The peak position

Soft Order in Physical Systems, Edited by Y. Rabin
and R. Bruinsma, Plenum Press, New York, 1994

provides information related to the time development of concentration correlation length, while the intensity provides information related to the time development of concentration gradients, and the breadth describes the distribution of correlation lengths.

Phase separation processes have been investigated in metals[12-18], glass,[19,20] alloy systems and small molecule fluids[21-24]. Scattering measurements which report the position and intensity of the maximum in the scattering curve have provided the most information about phase separation dynamics. In the Al-Zn metallic alloy system, it was found[17] that the position of the scattering maximum scaled like $t^{0.75}$. No information concerning the growth rate of the scattering intensity was reported. Extensive measurements on phase separation dynamics in small molecule systems which underwent a very shallow temperature quench have been performed by Goldburg[22-24] and Knobler[25] in which the time behavior of both the intensity and position of the peak in the scattering curve were measured. Information related to the breadth of the scattering curve was not reported. It was found from the position of the maximum in the scattering function, that the correlation length ξ could be represented by $\xi \sim t^\beta$ where $\beta = 0.3$ for "early" stages and β approaches 1 for "late" stages in the process. The intensity of the scattering maximum was reported[25] to obey the relationship $I_m \sim t^\gamma$, with γ varying from 1.1 to 2.3, and appeared to depend both on concentration and quench depth. In other cases[23] it was found that in several small molecule systems, the time dependent growth of the maximum scattering intensity could be represented by $I_m \sim t^\gamma$, where $\gamma \simeq 0.6 - 0.9$ during early stages and $\gamma \simeq 3$ during the later stages.

Snyder and Meakin[11] investigated phase separation in blends of polystyrene and poly(vinyl methyl ether) in the spinodal regime, using small-angle light scattering techniques. They found that scaling does occur over a limited time range, however β is not time invariant over the entire range. In addition, they found, $\gamma \simeq 2.1$

Hashimoto et.al.[28-29] investigated spinodal decomposition in critical mixtures of PS and PVME using a time resolved light scattering technique. They suggested three stages to spinodal decomposition. In the early stage, evolution of the concentration fluctuations is predicted well by the linearized theory of Cahn[34] in which the wavelength of the fastest growing fluctuation, $\lambda_m = 2\pi/q_m(t = 0)$, is invariant with time and only the amplitude of the concentration fluctuations increases with time. In the intermediate stage both the wavelength and amplitude of the dominant mode of fluctuations grow with time. In the late stage the amplitude of the concentration fluctuations have reached their maximal value. However, the wavelength and size of the unmixed structure grows with time in a self similar manner. In the late stage they found that $\xi \sim t^\beta$ and $I_m \sim t^\gamma$ with $\gamma = 3.3$. In addition they found that the scattering function $I(q, t)$ obeys a scaling law given by $I(q, t) \sim \xi(t)^3 S(q\xi(t))$ (where $S(q, t)$ is the Fourier transform of the structure factor) with $\xi(t) \sim t^\beta$ where q is the scattering vector and $\xi(t) = q_m^{-1}(t)$ is the correlation length of the unmixed structure at time t. And that the universal scaling function S for a critical mixture, is in good agreement with that predicted by Furukawa[40], $S(x) \sim \frac{x^2}{\gamma/2 + x^{2+\gamma}}$. Where $\gamma = 2d$ for critical mixtures, and $\gamma = d + 1$ for off critical mixtures.

More recently Bates and Wiltzius[72] investigated spinodal decomposition in a symmetric critical mixture of perdeuterated and protonated 1,4 polybutadiene. Here again light scattering was used. They observed four stages in the decomposition. The early stage is accounted for by the Cahn[34] theory. Nonlinear effects cause the breakdown of this description in the intermediate stage. In this stage they found that $\lambda_m(t) \sim t^{\beta_{eff}}$, but in this case β_{eff} is temperature dependent. As the composition fluctuation amplitude approaches the equilibrium value, the process enters a transition stage characterized by decreasing interfacial thickness and increasing $\lambda_m(t)$. Once the interfacial profile equilibrates, the final stage is reached. In this stage a universal structure factor is obtained which scales with $\lambda(t)$.

The virtues of digital image analysis and a wide range of possible applications to polymer systems were discussed by Tanaka et. al.[73,74]. Tanaka and Nishi[75] studied a blend of PS and PVME as it phase separated via spinodal decomposition and observed the process through a phase contrast microscope. They applied digital image processing to monitor $P(c, t)$, the probability that at time t the concentration at a given point is c. This is done by observing the distribution of gray level values in the image, and noting that when using a phase contrast microscope these values are directly proportional to the relative concentrations of the components. They found that the initial distribution is a sharp Gaussian. As the phase separation proceeds this distribution broadens and eventually becomes binodal. In a later paper where Tanaka et.al.[76] apply image analysis to study the late stages of phase separation, they emphasized the motion of the interface between the phases and realized that the morphology in the phase separated pattern changes to minimize the interfacial energy of the system, however they assume that the concentration changes within and between the phases ends at a rather early stage and only coarsening takes place with no change in concentration.

THERMODYNAMICS OF POLYMER BLENDS

The treatment of the stability of polymer blends is based on the concept of phase equilibria. The change in the free energy upon mixing is given by

$$\Delta G_m = \Delta H_m - T \Delta S_m \tag{1}$$

Schematic plots of the free energy of mixing at a given temperature is shown in Fig.1. If the curve has more than one minimum (as do all but the curve corresponding to T_0 in Fig.1.) there is a miscibility gap.

The points of contact of the double tangent with the curve having two minima define the points of the binodal curve, in the temperature-composition diagram. This is the boundary of the area of stable homogenous solutions.

The inflection points of the free-energy curve define points of the spinodal in the temperature-composition curve (Fig.2.). The spinodal represents the boundary between the areas of metastable and unstable solutions. Within the metastable region, between the binodal and the spinodal, the free energy curve is concave, and the solution is stable towards small composition fluctuations, as they will raise the free energy higher than the homogeneous solution. The mixture is unstable towards large fluctuations, such as nuclei, and will phase separate to two phases whose composition is given by the binodal (at the given temperature).

Within the spinodal, the free energy curve is convex, and the mixture is unstable even to small composition fluctuations, as the free energy of even infinitesimally separated solution is lower than the homogenous solution.

The thermodynamic condition for stability towards composition fluctuations in a binary mixture is:

$$\left(\frac{\partial^2 \Delta G_m}{\partial^2 C} \right)_{T,P} \geq 0 \tag{2}$$

where C is the composition parameter.

The critical point is the temperature at which both minima coalesce, and is given by

$$\left(\frac{\partial^2 \Delta G_m}{\partial^2 C} \right)_{T,P} = 0 ; \qquad \left(\frac{\partial \Delta G_m}{\partial C} \right)_{T,P} = 0 \tag{3}$$

simultaneously.

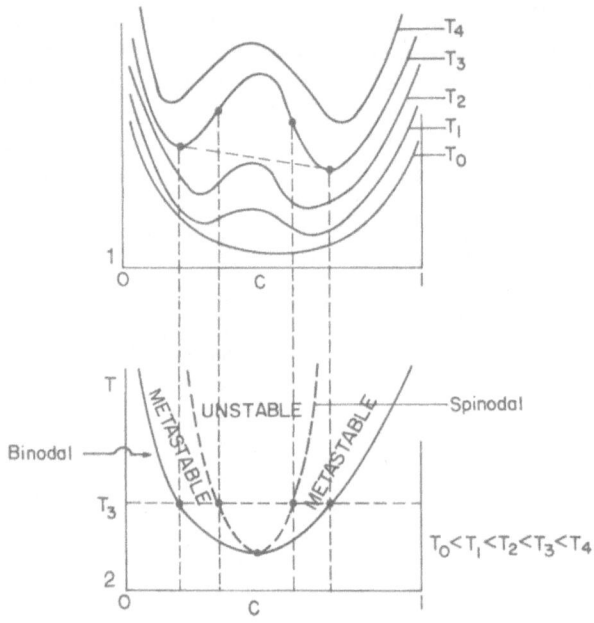

Figure 1
Free Energy of mixing vs. concentration. Several isotherms are shown. At T_0 the mixture is stable while for T_1, T_2, T_3, T_4 the mixture is unstable.

Figure 2
Phase Diagram in the Temperature vs. composition plane. Below the binodal the mixture is stable and only one phase exists. Above the binodal the mixture phase separates. Phase separation occurs via nucleation and growth if the mixture is initially in the metastable region. If it begins in the unstable region phase separation occurs by spinodal decoposition.

For high-polymer blends the binodal and spinodal curves are usually convex, in other words they exhibit a minimum in the temperature-composition plane. This means that phase separation occurs upon heating. This behavior is called lower critical solution temperature (LCST). Typical solutions with low molecular weight solvents and solutes (if they phase separate at all) usually exhibit an upper critical solution temperature (UCST), i.e. the spinodal and binodal curves have a maximum, which means the phase separation occurs upon cooling.

Why LCST?. Mixing polymers is in general not a trivial task. Although many compatible pairs are known, an arbitrary pair of polymers will in general not mix. In order for polymers to mix, there must be an attractive interaction between the monomers of different polymers. For example, assume that in a blend of polystyrene (PS) with poly(vinyl methyl ether) (PVME) the oxygen in the ether is attracted to the hydrogen opposite the benzene ring of polystyrene[41]. This interaction is weak and directional. By directional it is meant that it can "hold" two monomer together if they are oriented properly, for example, if the oxygen is "facing" the hydrogen opposite the benzene ring.

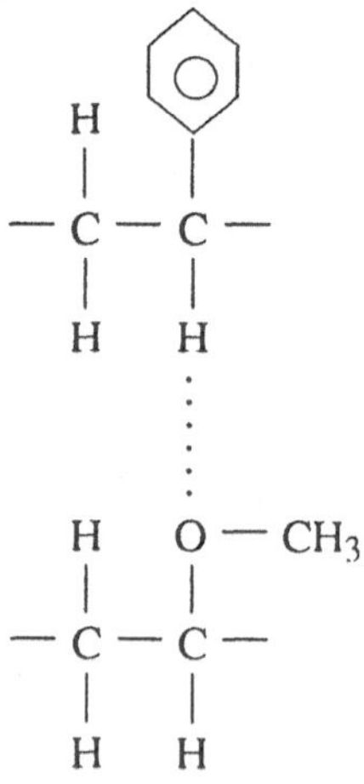

At low temperatures, the thermal agitation gently moves the monomers until their interaction potential pulls them into the desired position. The energy will be minimized when every monomer of one polymer species is "facing" a corresponding "partner" from the other polymer. Thus the polymers mix at low temperatures. These "interlocked" polymers will move together, preserving their relative orientation (monomers "facing" each other).

When the temperature is raised the thermal agitation becomes too strong, and the weak bond between different species breaks. When this happens, the molecules can rotate separately and move separately. This causes a great increase in the entropy. Now the main interaction between monomers is a Van der Waals interaction, therefore the energy is minimized when monomers of like type are together. Therefore a mixed phase separates under these conditions - i.e. upon temperature increase.

THE MECHANISMS OF PHASE SEPARATION

We differentiate two distinct regimes for phase separation, depending on whether the process takes place in the metastable or the unstable region of the Temperature-Composition plane (Fig.2.). These are called "nucleation and growth" and "spinodal decomposition", respectively. These mechanisms are competitive in determining the mode of phase-separation, the dominant effect being dependent on how far from the equilibrium binodal the mixture has been thrusted. Polymer-polymer systems, because of the slow diffusion times, can be brought to well beyond the spinodal even at concentrations far from the critical point.

Dynamics: Nucleation and Growth

In the metastable region only large composition concentration fluctuations lead to phase separation. The domains of new phase must be sufficiently large to compensate for the energetically unfavorable process of the creation of an interface. Domains which are too small are unstable and disappear (are reabsorbed).Therefore nucleation and growth [30,31,33] is expected to proceed by formation of a critical nucleus, about which a phase much richer in one of the mixture components clusters (Fig.3.a). This clustering leads to a depletion region of that component in the vicinity of the nucleus resulting in a concentration gradient. More of that component then diffuses down the concentration gradient and adds to the cluster, resulting in subsequent growth of the nucleus into a spherical droplet whose composition remains invariant during the growth process. As the droplets grow, it may occur that neighboring droplets coalesce. This "coagulation" mechanism may get greatly enhanced if the droplets move around as in the "cluster diffusion and coagulation" mechanism proposed by Binder et.al.[42,43,44]. Of course, the cluster diffusion constant decreases with increasing cluster size.

Dynamics: Spinodal Decomposition

Within the spinodal, the mixture is unstable towards concentration fluctuations of all sizes, including the ever present small fluctuations. Phase separation is expected to proceed by continuous growth of these small amplitude fluctuations. The kinetics of this mechanism, called spinodal decomposition since it occurs within the spinodal were first treated by J.W. Cahn [34] who showed that there is a preferred wavelength of fluctuation which has a maximum growth rate. The phase separation can be viewed schematically as resulting from the growth of this specific fluctuation alone, as described in Fig.3.b. The growth of the fluctuation can be viewed as being caused by diffusion of the components into the area rich in it, that is "uphill diffusion" against its own concentration gradient. The initial phase separated structure emerges throughout the volume of the mixture very rapidly, and during the initial stages of

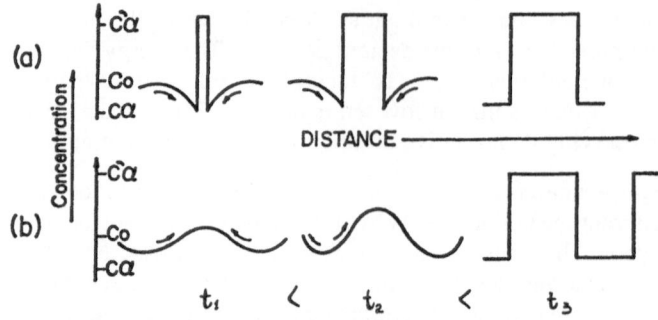

Figure 3.   Modes of phase separation. (a)Nucleation and growth. (b)Spinodal decomposition. In the figure the horizontal axis represents spatial position. The concentration profiles are shown at three times during the evolution. The earliest time is on the left while the latest time is on the right. Note that in nucleation and growth the components move down the concentration gradient whilst in spinodal decomposition the components move against the concentration gradient.

the separation process the spatial structure remains invariant, while the composition of the distinct phases changes continuously with time.

Cahn's kinetic theory is a formulation of a generalized diffusion equation for an inhomogenous system, which accounts for the phenomena described above. The flux of matter is phenomenologically related to the gradient in chemical potential

$$-J_a = M\nabla\mu_a \tag{4}$$

where:

J_a - is the interdiffusional flux of component a
μ - is the chemical potential
M - is a phenomenological constant
for a homogenous system:

$$\mu_a = \frac{\partial f(C)}{\partial C_a} \tag{5}$$

where $f(C)$ is the free energy per unit volume of homogeneous material at concentration C. The Cahn theory, originally derived from a variational formulation of the free energy of an inhomogenous system [35], introduces into the chemical potential in equation (5) a term proportional to the curvature of the concentration profile - the so called "gradient energy" term

$$\mu_a = \frac{\partial f(C)}{\partial C_a} - 2K\nabla^2 C_a \tag{6}$$

This term is introduced when the concentration gradients are such that the average concentration varies appreciably within the range of intermolecular interactions. The greater the range of molecular interaction the larger this effect will be, so K should be related to that range.

Substituting eq. (6) into eq. (4) and using the chain rule we obtain:

$$-J_a = M \frac{\partial^2 f}{\partial^2 C_a'} \nabla C_a' - 2MK \nabla^3 C_a' \tag{7}$$

It should be noted that eq. (7) looks like a "generalized Fick's Law". Fick's Law gives the relation between the flux of matter and the concentration gradient, and can be expressed as:

$$-J_a = D \nabla C_a \tag{8}$$

Comparing eq. (7) with Fick's law (eq. (8)) we see that the diffusion coefficient D in eq. (8) is given by:

$$D = M \left(\frac{\partial^2 f}{\partial^2 C_a} \right) \tag{9}$$

and since M is by definition positive, and within the spinodal

$$\frac{\partial^2 f}{\partial^2 C_a} \leq 0 \tag{10}$$

the diffusion coefficient is negative. This accounts for the so-called "uphill diffusion".

Taking the divergence of the flux in eq. (7) , and recalling that $\partial C / \partial t = -\nabla J$, we find

$$\frac{\partial C_a}{\partial t} = M \frac{\partial^2 f}{\partial^2 C_a} \nabla^2 C_a - 2MK \nabla^4 C_a \tag{11}$$

Equation (11) is the generalized linear diffusion equation relevant for the early stages of spinodal decomposition. A solution of eq. (11) has sinosodial spatial behavior with an amplitude which grows exponentially in time

$$C' - C_0 = exp[R(\bar{q})t] cos(\bar{q} \cdot \bar{r}) \tag{12}$$

$\bar{q}$ is the angular wave, related to the wavelength of the composition wave.

$$\lambda = 2\pi/q \tag{13}$$

$R(\bar{q})$ - is the exponential growth rate
$\bar{r}$ - is the position vector
$R(\bar{q})$ is obtained by back substitution of eq. (12) in eq. (11) and the result is:

$$R(\bar{q}) = -Mq^2 \left(\frac{\partial^2 f}{\partial^2 C} + 2Kq^2 \right) \tag{14}$$

A complete time dependent concentration profile is obtained by summing of eq. (12) over all wavelengths having positive growth rates (i.e. positive $R(\bar{q})$ in eq. (14)). In fact, due to the exponential character of the growth rate, the phase-separation will be dominated by the most rapidly growing wavelength. Differentiating eq. (14) with respect to q and setting to zero leads to

$$\lambda_{max} = 2\pi/q_{max} = 4\pi \left(\frac{1}{K} \frac{\partial^2 f}{\partial^2 C} \right)^{-1/2} \tag{15}$$

79

To summarize, in the Cahn theory there are three main aspects which characterize the early stages of spinodal decomposition.

(a) The composition of the phases changes continuously with time while the spacing of the phase pattern remains nearly constant.

(b) Negative ("uphill") diffusion.

(c) For isotropic materials, the fluctuation waves of λ_{max} have random orientation and phase angles, so that the spinodal structure should be almost uniform, yet random and interconnected.

After bulk thermodynamic equilibrium has been reached via a spinodal mechanism, i.e. the mixture has phase-separated to phases of composition denoted by the equilibrium binodal, a coarsening process begins. At first the size increases linearly with time, and later exponentially. This fast coarsening of the structure can occur because the high level of interconnectivity allows viscous flow of both phases.

Coarsening

In later stages, the coarsening of the structure may proceed via a mechanism proposed by Lifshitz and Slyozov[33]. Here one considers the random evaporation and condensation from and to clusters; these events maintain diffusion fields between the clusters with gradients such that the largest clusters will most likely grow and the smallest shrink, until they are finally dissolved. This process leads to a power law for the average growth of for the linear dimension $\xi(t)$, namely $\xi(t) \sim t^{1/3}$. To see this, consider an A—B binary system. The concentration of B around a cluster of radius R is

$$C = C_0 + \frac{a}{R} \tag{16}$$

where C_0 and a are constants. Using this equation we find that the concentration of B around a cluster of radius $R + \Delta$ is

$$C_1 = C_0 + \frac{a}{R + \Delta} \tag{17}$$

which is smaller than C. Therefore there is a gradient in the concentration of B between these two clusters. If the separation between the clusters is L, the gradient is

$$\frac{dC}{dx} = \frac{C - C_1}{L} = \frac{a\Delta}{R(R + \Delta)L} \tag{18}$$

Assuming in addition that $L \sim R \sim \Delta$ we have that

$$\frac{dC}{dx} = \frac{a\Delta}{R(R + \Delta)L} \sim \frac{a'}{R^2} \tag{19}$$

where a' is another constant. The flux J of B caused by this gradient is

$$J = D\frac{dC}{dx} = D\frac{a'}{R^2} \tag{20}$$

The growth rate of the of the volume($\sim R^3$) of the larger cluster is given by the product of the cluster surface area ($\sim R^2$) and the flux J, so

$$\frac{dR^3}{dt} = R^2 J = Da' = constant \tag{21}$$

From this we conclude

$$R^3(t) \sim t \tag{22}$$

or

$$R(t) \sim q_m{}^{-1} \sim t^{1/3} \tag{23}$$

Siggia[77] investigated the effects of hydrodynamic interactions and surface tension on the coarsening rate. He found that for a sufficiently rarefied precipitate the above condensation-evaporation mechanism is dominant, and he recovered the result $R(t) \sim t^{1/3}$. When the minority phase is continuous, as in a quench at the critical composition, surface tension effects lead to a crossover from $r \sim t^{1/3}$ to $r \sim t$.

Remarks. From the above it is clear that there has been a massive effort made to clarify and understand the phase separation behavior in polymer blends. Theoretical considerations lead to models which predict scaling behavior in various stages of the phase separation and much of the experimental effort is made to observe scaling, obtain numerical values for exponents and verify the relations between exponents. Much of the experimental work uses scattering techniques whose resolution limits the maximal size of objects (domains) which can be observed and therefore limits how much of the phase separation can be observed. The experiments which will be reported on in the following chapters study the phase separation at stages which follow the stages for which light scattering is applicable. Hence these are late stages of phase separation.

Experimental Considerations

The model systems presented in this chapter are thin films ($\sim 7 \mu m$ thick) of polymer blend (polystyrene/poly-vinyl methyl ether in varying weight proportions) supported by a glass microscope slide. The film is rapidly introduced into the miscibility gap where phase separation proceeds under isothermal conditions. Laser light scattering is used to observe the initial stages of the phase separation and to determine whether the phase separation begins as spinodal decomposition (i.e. in the unstable region) or as nucleation and growth (i.e. in the metastable region). Later stages of the phase separation are observed with the aid of an optical microscope and the process is recorded on a video tape.

The film thickness is an important parameter in this experiment. Reich and Cohen[37,38] showed that there is a thickness effect on the phase separation behavior when the film thickness is smaller than $1 \mu m$. On the other hand the film should be thin enough so that the two dimensional images obtained faithfully represent the observed physical system and that the growth behavior undergoes the 3-d to 2-d transition before any images are acquired and processed. It was found that films $\sim 7 \mu m$ thick meet all the above criteria. Thus crossover effects from 3-d to 2-d systems do not contribute to the observed phenomena in the late stages of phase separation in these experiments.

After the process is recorded, a set of video images is selected and digitized. Each digitized image is then processed so as to obtain an image with only two brightness levels Fig.4. In this experiment the image processing is such as to give the "new" phase one brightness level, black, and the remaining phase, white. The resulting images were analyzed using image analysis techniques which can be used to quantify the extent of phase separation and examine the time evolution of the size distribution of droplets, the amount of interface, etc.

Additional information about the phase separation is obtained by imaging a sample using UV light. A sample supported on a quartz slide is quenched into the miscibility gap, (as before), and allowed to phase separate. The process is stopped after a certain time and the resulting morphology is "frozen in" by rapidly cooling the sample to room temperature. The sample is then viewed through a transmission UV microscope equipped with a UV sensitive video camera which is used as a detector. UV light with $\lambda \simeq 264 nm$ is used to irradiate

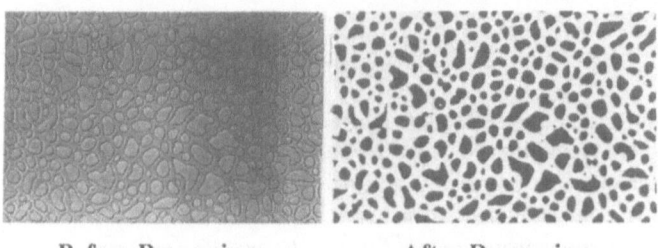

Before Processing   After Processing

Figure 4.   An image before and after digital processing

the sample since at this wavelength the absorption of PS is about ten times greater than the absorption of PVME. An image obtained in this manner contains a "dark" phase which is richer in PS than the "light" phase which is richer in PVME. Using this microscope we can obtain qualitative information about the spatial distribution of concentrations, we can identify the phases, and we can determine if the "new" phase is rich in PS or PVME (Fig.13.)

The effects of the interaction between the polymer film and the substrate were examined by comparing the phase separation behavior when the polymer film is prepared on clean glass (advancing contact angle with water $\leq 7°$) with that of a film cast on a dense monolayer of dimethyl octadecyl chloro silane (advancing contact angle with water = 50°).

Interference microscopy is used to check if the the underlying morphology which develops during phase separation influences the thickness of the film. Indeed, it was found that there is a modulation of the film thickness directly related to the morphology(Fig11.)

OBSERVATIONS OF PHASE SEPARATION

Digital image processing and analysis has not been extensively used to study phase transitions or phase separating systems. In order to effectively use digital image analysis in this kind of study, it is necessary to decide what information contained in the images is relevant to the phenomena under study and then to decide on a method to extract this information from the images. The parameters used in this study to characterize and measure the phase separation process were chosen or defined because their value can yield some information about the phase separation process as a whole, the morphology which evolves or of the phases.

In what follows we shall be describing and discussing various aspects of the phase separation process which manifest themselves in the images observed and the parameters measured. The following is a set of definitions of some terms:

"phase"   During the phase separation process a domain boundary appears, the term "phase" is used to describe the material on either side of the domain boundary. It should be noted that this material may or may not be a phase as is described in a thermodynamic phase diagram, since here the material is undergoing a process which takes it far from thermodynamic equilibrium.

"new phase"   Sometimes, the domain boundary appears as closed loops. We call the material within these loops the "new phase" and distinguish it from the outside material which we call the "background".

"background"   This is the material outside the domains which appear during phase separation. In the initial stages of phase separation, the "background" spans the sample, is connected, and is the matrix from which the "new phase" grows.

"object" A processed image contains only two gray levels: black (0) and white (255). The "new phase" is assigned the black and the "background" is assigned the white. In a processed image the black domains are referred to as "objects".

n - n-density The number of "objects" or "new phase" domains per unit sample area.

A - area density Total area of "objects" or "new phase" per unit sample area.

A_m The area of the "object" with largest area in a given image.

sf - shape factor The shape factor for an "object" is defined as $sf = \frac{4\pi A}{P^2}$ where A and P are respectively the area and perimeter of the object. This parameter has a value of one (1) for a circle and zero (0) for a line, thus this parameter gives an indication how similar to a circle an object is.

$SF = \frac{\sum A_i sf_i}{\sum A_i}$ This is the area averaged shape factor. A_i and sf_i are the area and shape factor respectively of the i-th object, the sum is performed over all objects in a given image. If this value is close to one (1) for a certain image, we know that the "new phase" is composed of round domains, if the value of this parameter is low then the "new phase" has long perimeters and small areas.

correlation length The power spectrum of the gray level distribution in an image is given by the two dimensional Fourier transform image, $|\mathrm{FT}|^2$. The correlation length is obtained from the wave number at which the power spectrum attains its maximum. The "phase-phase" correlation length is obtained from a processed image. Since in a processed image one phase is painted black and the other white, the value obtained for the correlation length is not equivalent to the concentration correlation length obtained in light scattering experiments. Instead, this parameter yields a "typical" domain size.

For each image the following parameters have been measured or calculated: the number density, n, the area density, A, the area of the object with maximal area, A_m, and the area averaged shape factor, SF. We also looked at the "phase-phase" correlation length for each image, as measured from the image Fourier transform. The time evolution of these parameters, for each of the experiments, is discussed in the following sections.

Two representative cases will be discussed. The 1/2 PS/PVME ratio by weight is representative of decomposition in the unstable region (spinodal decomposition) while the 1/9 PS/PVME ratio by weight is representative of decomposition in the metastable region (nucleation and growth). Other mixtures fall into one of these categories and their phase separation behavior is qualitatively similar to that of the representative case described.

Phase Separation in the Thermodynamically Unstable Region

Here is a brief description of the morphological evolution in the polymer blend from the moment it undergoes a rapid temperature jump from room temperature ($\sim25°C$) to $135°C$ or $150°C$ when it is inserted into the preheated hot stage, and observed through the microscope.

Figure 5. shows a sequence of processed images taken during the phase separation. The chronological order of these images is indicated. Note that the scale changes along the sequence of images. The bar at the top left of each image represents $100\mu m$.

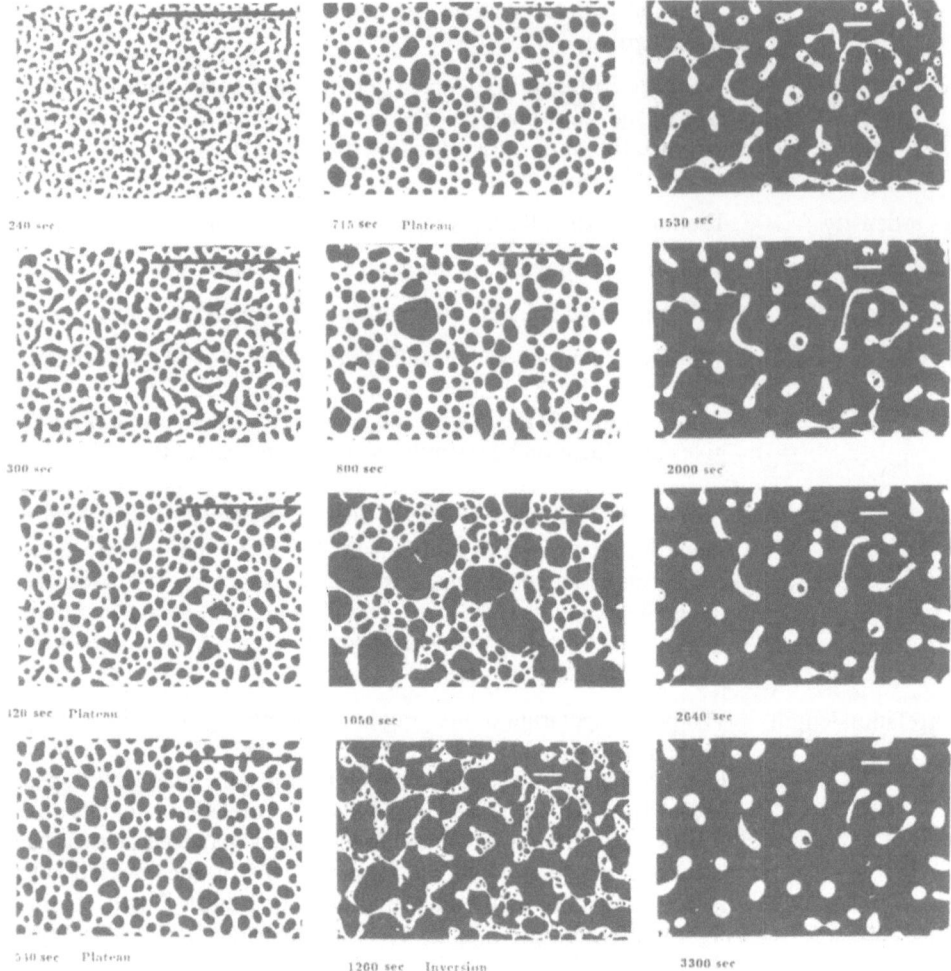

Figure 5. A Sequence of Images taken along the phase separation of a 1/2 PS/PVME blend at 135°C. The time, in seconds, elapsed after the quench is shown beneath each image. At 1260 seconds a "phase inversion" occurred. Black Bar is $100\mu m$.

During the first minute or so there is no observable change in the blend. Then the image becomes fuzzy and darker. After a little while a random distribution of rings develops. There are two phases. We call the phase enclosed by the rings the "new" phase, and the phase outside the rings is the "background" phase. The rings make up the interface. In the processed images, Fig.5., the "new" phase is the black phase. As the phase separation continues, we see the rings grow, and eventually some rings touch and coalesce (Fig5. 240 sec.-420 sec.). Coalescence is slow and the shape of coalesced regions is oblong. As the process continues the oblong regions become more round and compact, and also grow (Fig5. 420 sec.-715 sec.). Some of the oblong regions sometimes coalesce with some other region and create longer domains. As these regions increase in size and take up a larger fraction of the available space, they touch and coalesce again but now there are many places where more than two particles coalesce. This gives rise to complicated and contorted domains containing the "new" phase (Fig.5. 800 sec. - 1050 sec.). Then, rapidly, the "new" and "background" phases seem to form an interpenetrating network (Fig.5. 1260 sec.- Inversion). At this point the interface is practically flat in most places. As the process continues, "background" phase seems to slowly disappear as the "new" phase becomes the majority phase. We have witnessed a phase inversion. The "new" phase continues to grow and coalesce restricting the "background" phase into, at first, oblong pockets which break up and become rounded. These pockets slowly drift, touch and eventually coalesce.(Fig.5. 1530 sec.- 3300 sec.).

The results from the light scattering experiment for this blend indicate that phase separation was initiated via spinodal decomposition.

Fig.6. shows the number density monotonically decreasing throughout the process. This is an indication that we are well into the coarsening regime. No new objects appear, the existing objects appear to grow by two mechanisms, diffusion and coalescence. However, two rates are clearly visible - "before" and "after" the phase inversion.

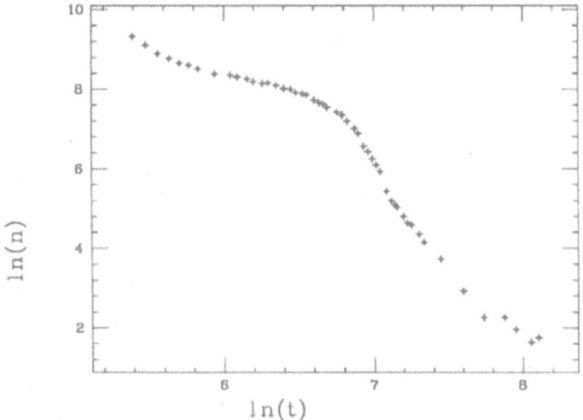

Figure 6.   Variation of n-density as a function of time for a 1/2 blend quenched to 135°C. We note that n is monotonically decreasing.

Fig.7. shows the area density and shape factor as a function of time. The area density, which is expected to give an indication as to "how far" the phase separation has proceeded, shows unexpected time dependence. After an initial growing period, the area density suddenly stops increasing and remains constant for a period of time, (which seems to depend on quench depth), and then grows again rapidly (as it goes through the phase inversion) until it reaches a constant value. When this last constant value is attained we assume that equilibrium concentrations have been attained within the respective domains, and only coalescence remains. We observe that during the time when the area density is in its first plateau, (see Fig. 5. 440, 540 and 715), the area averaged shape factor increases, attains its maximum at the end of the plateau and then drops rapidly and remains low till the end. Apparently with a large shape factor (which implies a small perimeter to area ratio) phase separation cannot proceed as rapidly as when the shape factor is small. Geometrically this is logical since a large interface will allow more material to cross than a small interface (assuming the flux is the same in both cases). However, this would only explain a decrease in the rate of growth of the area density, not a plateau.

In Fig. 8. we observe the time evolution of the area of the largest object in each image, A_m, along with its shape factor. Note that this is a log-log plot that emphasizes the plateau. This is a representative object for the process. In general it will have evolved a

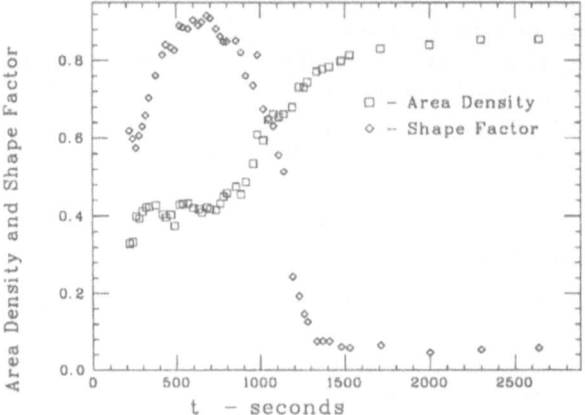

Figure 7.   Time dependence for the area density and shape factor for a 1/2 blend quenched to 135°C. Note that after an initial growth period the area density becomes constant - during this time the shape factor increases to it's maximal value, and then the area density increases again to another "plateau".

little more than the "average" object in each image, but its evolution from image to image is representative, and is only slightly shifted along the time axis, relative to the evolution of the "typical" or "average" object. Here again, we see the same characteristics in the time evolution that were observed in the Area density, except some features are more pronounced. We see an initial increase in the area, then a plateau during which the shape factor increases and reaches a maximal value, and then the area increases rapidly to its final value (the shape factor decreases to a value close to zero).

The "phase-phase" correlation length was computed for the images preceding the phase inversion. This is done by taking the the Fourier Transform (FT) of the image. The image

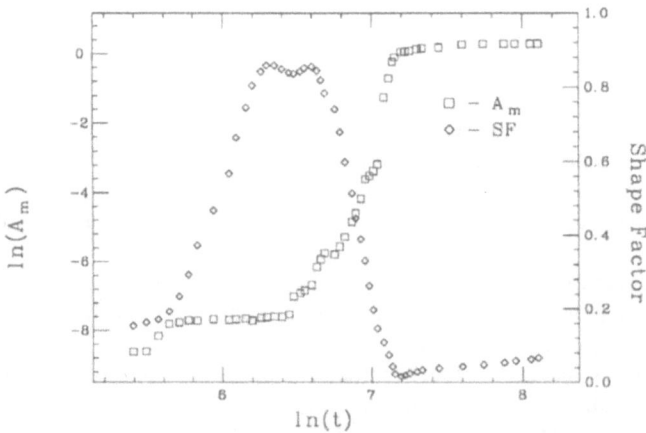

Figure 8.   Time dependence for the area of the largest object for a 1/2 blend quenched to 135°C. Note that after and initial growth period the area becomes constant - during this time the shape factor increases to it's maximal value, and then the area increases again to another "plateau".

obtained from the (FT)^2 has a "scattering ring". This is used to obtain a scattering profile from which we can find the correlation length. In Fig.9. we have an image and its Fourier transform, and with the Fourier transform we have "scattering profile". This profile is obtained by plotting the value of (FT)^2 along a vertical line through the center of the (FT)^2 image. The "phase-phase" correlation length, see Fig.10., also exhibits an initial growth period, followed by a plateau, and another rapid growth period. It is to be expected that the correlation length and area density behave similarly since the objects which evolve are fairly regular. This may not be the case near the phase inversion. It is not possible to measure the correlation length beyond the point seen in Fig.10. because of the limited resolution dictated by the spatial resolution, discretized space, of 512×512.

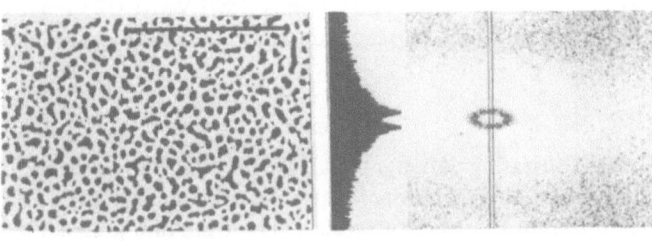

Processed Image   FT of Processed image

Figure 9.   A processed image, and it's FT image. On the side of the FT image we see a "scattering profile" obtained from the intensity of the FT image along a vertical line through it's center. The black bar at the top of the processed image is $100\mu m$.

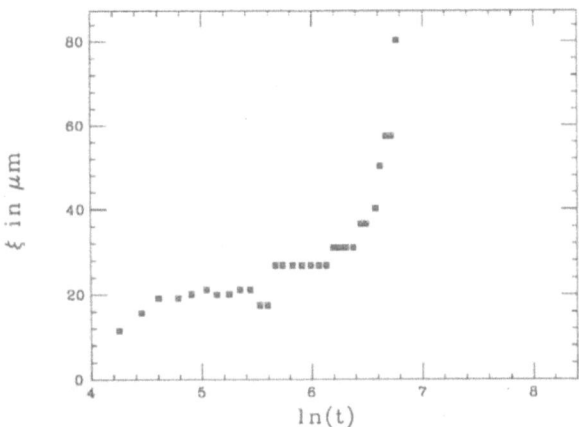

Figure 10.   The "phase-phase" correlation length as a function of time for a 1/2 blend quenched to 135°C. Here too we see a plateau prior to the phase inversion, and the a rapid increase in the correlation.

The behavior described above, (a plateau in the time evolution), was unexpected. It may be conjectured that such behavior is due to a modulation of the thickness of the polymer film. This was checked by interference microscopy. Fig.11. shows an interference image taken from a sample 1200 seconds after the quench to 135 °C. We see, at most, 2 interference fringes. This implies a modulation of about $0.5\mu m$ for $\lambda = 546nm$. This is to be compared with a film thickness of $7\mu m$. This small modulation cannot account for the observed behavior.

Since this is observed in a thin film of polymer blend on a glass slide, one may wonder if this behavior is due to the interaction between the blend and the substrate or if this phenomenon

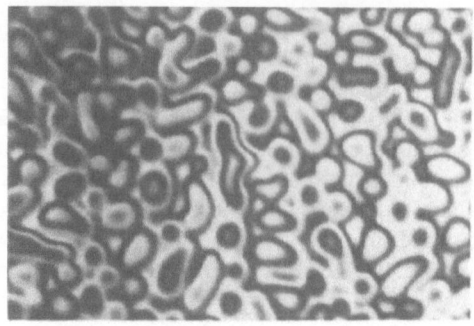

Figure 11.   Interference image after 1200 seconds. PS/PVME 1/2 blend, 135°C. We see at most 2 fringes, which correspond to modulation of approximately $0.5\,\mu m$.

is intrinsic to the phase separating polymer blend. We tested these two possibilities by radically changing the chemical and physical properties of the substrate surface. The experiment was performed on three different substrate surfaces: (1) on a sodium glass slide cleaned with 5% HF solution which produces a hydrophilic surface whose advancing contact angle with water is $\leq 7°$. (2) On a dense monolayer of CH_3-$(CH_2)_{17}$-$Si(CH_3)_2Cl$ which has an advancing contact angle with water of 50°, and (3) on glass cleaned with sulfochromic acid which has an advancing contact angle with water of $\approx 28°$. In all cases the same qualitative behavior is observed. In Fig.12. we see the area density and shape factor for the first twocases mentioned above. From this we conclude that the behavior observed is intrinsic to the polymer film; the interaction with the substrate is of secondary importance.

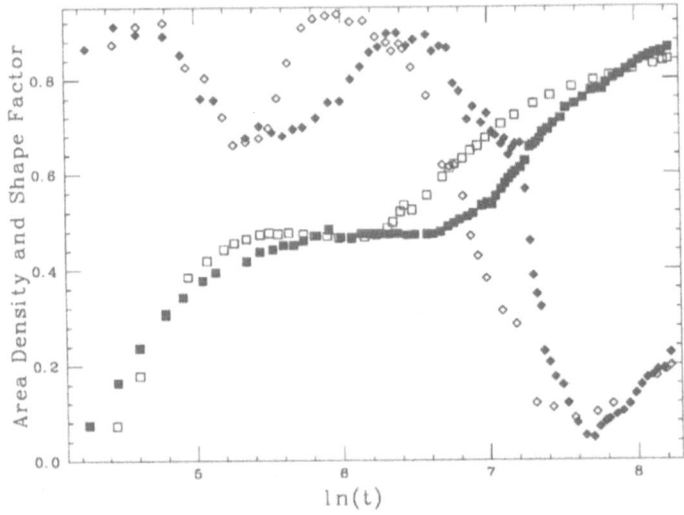

Figure 12.   Area density and shape factor as a function of time: $\square$; $\Diamond$, polymer thin film on glass: $\blacksquare$; $\blacklozenge$, on dimethyl octadecyl chloro silane dense monolayer.

In Fig.13. we have two images obtained with the UV microscope. In addition these images contain a plot of the intensity vs. position along a vertical line (also displayed) on the image. The darker regions on these pictures represent PS rich regions and the lighter regions are richer in PVME. A feature which stands out in these images is that the intensity

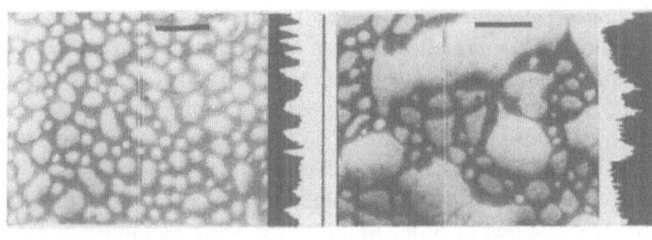

Earlier   Later

Figure 13.   UV images of 1/2 blend at two stages of the phase separation process - both before phase inversion. The images also contain an intensity plot, taken along the vertical line. Note the non uniform intensity. This implies a non uniform distribution of components. The black bar at the top of the images is $100\mu m$.

is not uniform in the brighter (or darker) regions, as is expected from the accepted coarsening theories. Since the brightness in each region is proportional to the relative concentrations of PS and PVME, the non-uniform brightness implies a non-uniform distribution of components within the domains. In addition, it can be seen that some domains are significantly brighter than others. This shows that there may be significant differences in the composition of separate (distinct) domains.

By observing the experiment at different times we confirmed that the "new" phase which appears is indeed the PVME rich phase.

The accepted view with regard to the dynamics of phase separation in the late stages is that there exist a set of scaling relations which govern these dynamics. Assuming that as the coarsening process proceeds there is no change in the concentration of the evolving phase as the clusters or domains grow, essentially the same distribution remains except the characteristic length has increased. Images obtained at different times can be made to look similar by shrinking/expanding one of the images by the appropriate amount. The scaling relations of the correlation length, $\xi(t)$, and the maximum scattered intensity at time t, $I_m(t)$, are

$$\xi(t) \sim t^\phi \tag{24}$$

and

$$I_m \sim t^\theta \tag{25}$$

Since under the conditions described above, the scattering intensity is proportional to the scattering volume, i.e. $I_m \sim \xi^d$, where d is the dimension of the system, we find that

$$\frac{\theta}{\phi} = d \tag{26}$$

These last results represent the idea that there is no change in the concentration of the demixed phases.

We have examined the late stages of phase separation for the system PS/PVME 1 to 2 by weight, and the results clearly indicate that the above ideas completely fail to describe the phase separation behavior.

The number density is strictly decreasing, and a look at the $\ln(n(t))$ vs. $\ln(t)$ plot in Fig.6. shows that if $n(t)$ obeys a scaling law, it is only in the time region preceding and around the phase inversion. At this stage the number density drops off very rapidly, $n(t) \sim t^{-5.5}$ (between $\ln(t)=6.4$ and $\ln(t)=7.4$). The significance of the numerical value of the exponent is not clear, however, a very strong decrease in the number density is to be expected in the region of phase inversion. We also notice that in the time region where $5.6 \leq ln(t) \leq 6.6$ the number density decreases slowly. This indicates that there is very little coagulation during this time interval. The plots of the area density and shape factor vs. time (Fig.7.), maximal area vs. time (Fig.8.) and the correlation length vs. time (Fig.10.) display the interesting feature that the area density, A_m and correlation length appear to be constant while the shape factor attains its maximum and begins to decrease. This means that the correlation length, ξ, does not exhibit dynamic scaling behavior. The interpretation of these facts according to the accepted theories of late stage separation and coarsening[40,77,33] leads to the conclusion that the phase separation process has stopped in order to allow for some geometrical rearrangement. This interpretation is a direct consequence of the assumption that all the material in the "new" phase is at its equilibrium concentration - the concentration at the binodal and there is no redistribution of components between the phases. This is indeed the accepted view about systems undergoing nucleation and growth, and systems in the late stages of phase separation. Since the system under study is far from equilibrium, we drop this assumption and reinterpret the results.

Assuming that the "new" phase does not appear at the equilibrium concentration, or in turn that at this late stage in the process the concentration is not yet the equilibrium concentration but rather some value intermediate to the initial concentration and the final equilibrium concentration*, we reinterpret the results as follows:

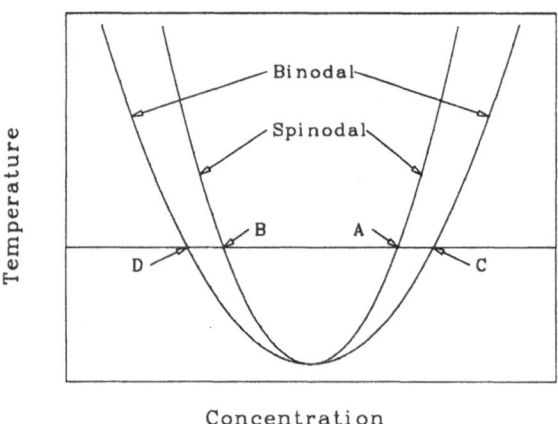

Figure 14.   Schematic phase diagram with Binodal and Spinodal curves.

The flux of a component is given by

$$j = -mC'\nabla\mu \tag{27}$$

where m is the mobility, C' is the concentration, and $\nabla\mu$ is the gradient of the chemical potential. This equation can be written as

$$j = -mC'\frac{\partial\mu}{\partial C'}\nabla C' \tag{28}$$

---

* Indeed, we need not assume that all the "new" regions have the same concentration. It is enough that there be a "large enough" concentration difference between the "inside" and the "outside" of the regions to maintain the interface.

The mobility and the concentration are always positive therefore the flux will be in the direction of the concentration gradient if $\frac{\partial \mu}{\partial C} > 0$ and opposite the concentration gradient if $\frac{\partial \mu}{\partial C} < 0$.

From the light scattering experiment we know that phase separation proceeds via spinodal decomposition, that is, where $\frac{\partial^2 G}{\partial C^2} = \frac{\partial \mu}{\partial C} < 0$. In this region the blend is unstable and the motion of the components is against the concentration gradient, causing even small concentration fluctuations to grow. Obviously, surface tension does not play a role at this stage. The concentration of one of the components, in this case PVME, grows and forms a domain. The concentration in this domain grows until it's concentration reaches the spinodal curve[78] (point A in Fig.14.). At this point an increase in the concentration within the domain would put it in the metastable region (between points A and C in Fig.14.). But in the metastable region $\frac{\partial^2 G}{\partial C^2} = \frac{\partial \mu}{\partial C} > 0$. Therefore the flux must be in the direction of the concentration gradient. Within the domain no such gradient exists, so the concentrations remains fixed. However, the material outside the domain is still in the unstable region, and it will flow up the concentration gradient into the domain. Thus, the domain size increases without changing it's concentration. This will continue until the material outside the domain reaches the concentration on the other branch of the spinodal (point B in Fig.14.). This accounts for the initial increase in the area density. At this point there are two "phases", each at it's respective spinodal concentration. The system is still far from equilibrium and the chemical potential of the components in the "new" phase and the "background" phase is different. The system will try to equate the chemical potentials. However, it cannot do this instantaneously over the entire sample. It will first do this near the interface by transferring components across it. This will cause the region near the interfaces to be richer in the corresponding component. A concentration gradient is created between the region near the interface and the inner parts of the domains. Now components can move with the concentration gradient within the domains. Thus, phase separation continues by the mutual transfer of components between the phases. This causes an increase in the concentration of PVME in the domain without changing it's size - hence the "plateau". In this regime, the surface tension increases, since it is proportional to the concentration difference between the "phases" and it's effects are clearly expressed in the increasing shape factor. This continues until one of the phases (again in this case the PVME rich phase) reaches its binodal concentration (point C in Fig.14.). The concentration of the "background" phase is still between the spinodal and binodal (between points B and D in Fig.14.). When this happens, any increase to the PVME rich phase expresses itself in an increase in size since it's concentration has reached it's equilibrium value. At this point the area density increases and continues to do so until the concentration of the "background" phase reaches the binodal (point D in Fig.14.). As the area of the domains increases, the surface tension per unit length of interface begins to decrease since it is inversely proportional to the size, (radius), of the object. These two effects combined with rapid coagulation make the dynamics near the phase inversion very fast. This is one of the reasons for the the very sharp decay in the number density, n(t), and the rapid increase of the correlation length.

The experiment with the UV microscope was designed to test the assumption that the concentration in the "new" phase may change, and may not initially be at the equilibrium concentration. At this stage we cannot make real time dynamical measurements with this microscope, and the results we obtain are only qualitative in nature. The images produced through the UV microscope clearly indicate that the concentration of the PVME rich phase (which is the "new" phase) is not always uniform. There are concentration differences within domains and different domains may have different concentrations within certain minimal and maximal values, (Fig.13.). These results lend support to the assertion that the concentration within the domains observed during the "resting period" of the phase separation is not the equilibrium concentration. In light of the above an interesting observation comes to mind. The "new" phase which is initially observed appears at the spinodal concentration and during the stagnation period (plateau in the area density) passes through the metastable region. Then, in principle, it should be possible for this "new" metastable phase to undergo nucleation and growth. This is indeed observable in cases when the stagnation period is long enough. For example, see Fig.15., in PS/PVME 1/2 ratio by weight at 130°C. Sometimes, before the phase inversion, the "new" phase does not remain homogeneous and an internal structure appears. This structure evolves into small droplets surrounded by a homogeneous phase, these droplets slowly coalesce. The area covered by these droplets within the "new" phase is relatively small. This appears to be nucleation within the metastable material constituting the "new" phase.

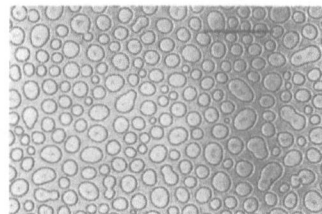

Figure 15.   During phase separation initiated as spinodal decomposition of PS/PVME 1/2 at 130°C, some of the "new" domains undergo nucleation and growth. In the figure the large domains contain small drops which resulted from nucleation and growth. The black bar at the top of the images is $100\mu m$.

Phase Separation in the Metastable Region

PS/PVME blends of 1/9 ratio by weight heated to 140°C, 150°C, 160°C display behavior which is characteristic of phase separaion in the metastable region, see Fig.16. At first the image becomes dark, eventually small round domains appear, which coalesce slowly creating slightly larger domains which are separated by large distances. Ostwald ripening, that is growth of large domains at the expense of small ones, can be seen. As this occurs, the smallest domains dissolve (disappear) into the "background" phase, while the larger domains grow.

At all times the volume fraction of the "new" phase is very small relative to the "background" phase. In Fig.17. we see a plot of ln(n) vs. ln(t). The number density is monotonically decreasing. This implies that we are in the coarsening stage. Here the number density decreases by two mechanisms. In one mechanism two domains touch and unite. In the other, the smallest domains dissolve and disappear as the larger domains grow.

The area density remains constant throughout the process at $A = 0.09$. The shape factor also remains constant with a value near 1. See Fig.18.. This indicates that the redistribution of components occurred before digital image analysis was applied.

In this example the accepted theories adequately describe the phenomena observed. The system is quenched into the metastable region. Because the system is metastable small concentration fluctuations are damped out, and it takes a long time for large enough fluctuations to appear. This is why the system remains homogenous for such a long time ($\sim$ 23 minutes). In the late stages, which is what we observe, both phases are at their final concentrations

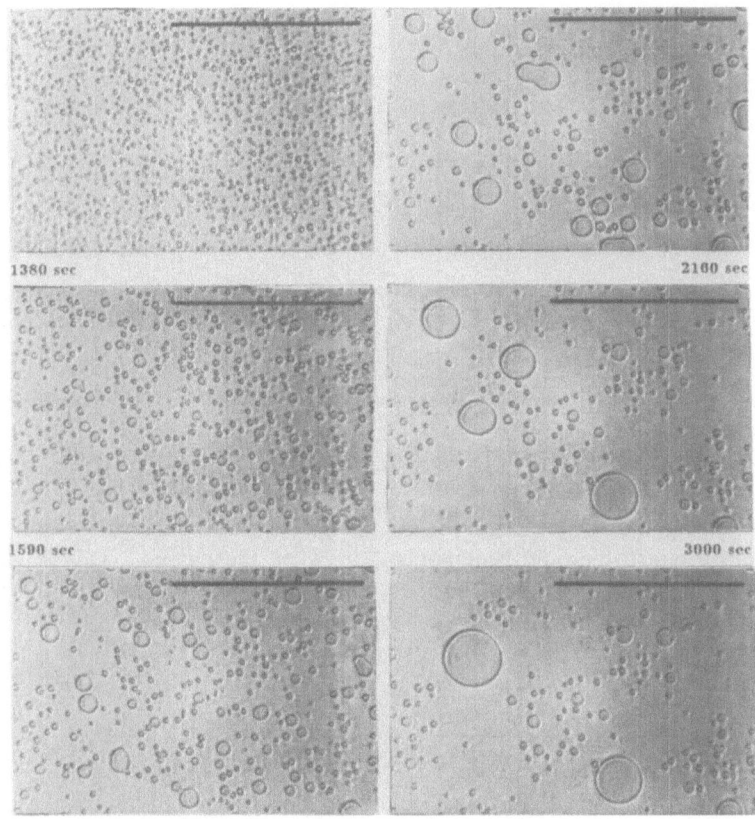

Figure 16. A sequence of images taken along the phase separation of a 1/9 blend. Bar is $100\mu m$. Note the time for the first picture is very large, $\sim$ 23 min. The bar in the images is $100\mu m$.

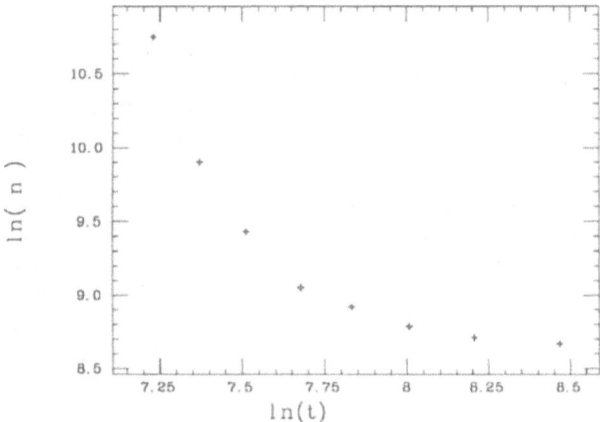

Figure 17. Time evolution of the number density, for a 1/9 blend quenched to $140°$C.

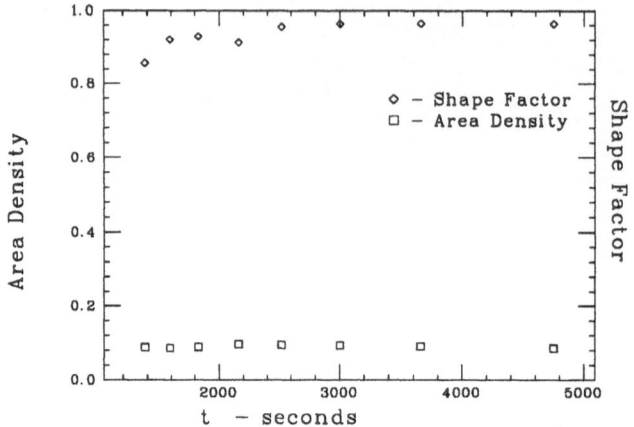

Figure 18.   Time behavior of area density and shape factor for 1/9 blend quenched to 140°C. Note the area density and shape factor remain constant throughout the process.

and there is no redistribution of components between the phases. This is why the area density remains constant throughout the process. The approach to equilibrium is achieved by unification of colliding domains and by Ostwald ripening.

SUMMARY

Late stages of phase separation in PS/PVME thin films were observed with a light microscope and analyzed using digital image analysis. This was done both for phase separations initiated via spinodal decomposition and nucleation and growth mechanisms. Additional information on the late stages of phase separation was obtained by observing the phase separated sample through a UV transmission microscope, sensitive to the relative concentrations of PS and PVME.

The effects of the interaction between the polymer film and substrate on the phase separation behavior was studied by casting films on surfaces which are chemically different and have different physical characteristics. The results indicate that the polymer-substrate interaction is of secondary importance. The phase separation behavior is intrinsic to the polymer film and is qualitatively the same on the different substrates.

The results from the image analysis for various blends were surprising and are attributed to previously unknown aspects of spinodal decomposition. When looking at the area density, that is the relative area occupied by one phase, as a function of time, an initial growth is observed, then no growth for a period of time, then more growth until the final value is attained. The first period where there is no growth in the area density cannot be explained by accepted coarsening mechanisms. The initial growth stage is interpreted as growth of a "phase" at the spinodal concentration, this growth stops when both phases attain their respective spinodal concentrations. The fixed area density is interpreted as an exchange of components between the "phases" causing a change in concentration between the phases without a change in volume. This continues until one of the phases reaches the binodal concentration and then continues to grow. Growth stops when both phases attain their respective binodal concentrations. The above description is schematically summarized in Fig.19.

It is generally accepted that there is a set of scaling relations which govern the dynamics

$$-j = mC\nabla\mu = mC(\partial\mu/\partial C)\nabla C$$

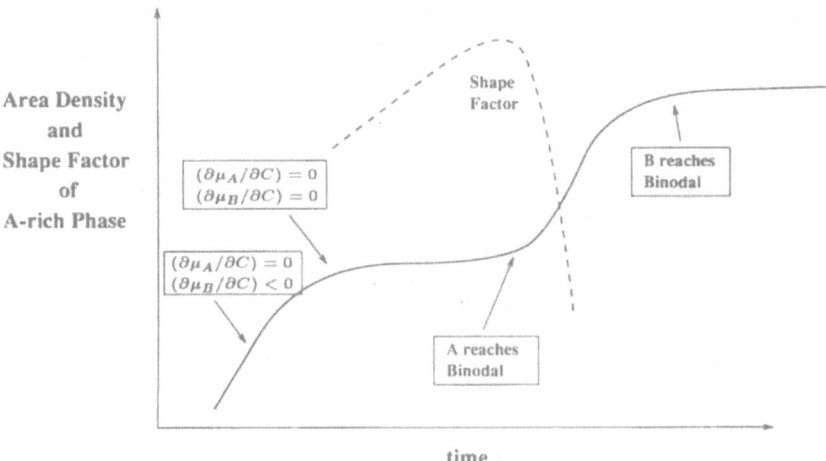

Figure 19. Schematic representation of the phase separation behavior in a blend undergoing spinodal decomposition. The first increase in the area density represents growth of A-rich "phase" at the spinodal concentration while the concentration in the background "phase" changes. The plateau in the area density begins when both "phases" are at their respective spinodal concentrations. During the plateau there is redistribution of components between between the "phases" and their concentrations approach their respective binodal concentrations. The area averaged shape factor increases during the plateau attaining a maximum at the end of the plateau. The plateau ends and the area density increases when the A-rich phase attains its binodal concentration. The area density continues to increase and reaches its final value when the other phase reaches it's equilibrium concentration.

in the late stages of phase separation. These scaling relations are a result of the assumption that there is no change in the concentrations of the evolving phases during the late stages of the phase separation process. It was found that the late stages of phase separation of the PS/PVME system cannot be described by simple scaling laws. Surface tension between the phases and the redistribution of components are dominant factors in the dynamics of the morphological changes observed throughout the process even after the phase inversion point.

The experimental work presented in this chapter was performed by Dahn Katzen as a part of his Ph.D. program at the Weizmann Institute of Science. This work profited from many suggestions and useful discussions with Shlomo Alexander and Yachin Cohen.

REFERENCES

1. L.P McMaster, *Macromolecules* **6**, 760 (1973).

2. R. Koningsveld, L.A.Kleintjes, H.M. Schoffleers, *Pure Appl. Chem.* **39**, 1 (1974).

3.   G.D. Patterson, A. Robard, *Macromolecules* **11**, 690 (1978).

4.   G.D. Patterson, *Polym. Eng. Sci.* **22**, 64 (1982).

5.   T. Nishi, T.T. Wang, T.K. Kwei, *Macromolecules* **8**, 227 (1975).

6.   H.L. Snyder, S. Reich, P. Meakin, *Macromolecules* **16**, 757 (1983).

7.   H.L. Snyder, P. Meakin, *Journal Chemical Physics* **79**, 5588 (1983).

8.   P.G. de Gennes, *Journal Chemical Physics* **72**, 4756 (1980).

9.   P. Pincus, *Journal Chemical Physics* **75**, 1996 (1981).

10. K. Binder, *Journal Chemical Physics* **79**, 6387 (1983).

11. H.L. Snyder, P. Meakin, *J. Polym. Sci. Polym. Sym.*, **73**, 217 (1985).

12. K.B. Rundman, J.E, Hilliard *Acta Metall.* **15**, 1025(1967).

13. T.L. Bartel, K.B. Rundman, *Metall. Trans.* **A6**, 1887(1975).

14. V. Gerold, W. Merz, *Scr. Met.* **1**, 33 (1967).

15. D.T. Lewandowski, K.B.Rundman, *Metall. Trans.* **A6**, 1895(1975).

16. S. Agarwal, H. Herman, *Scr. Met.* **7**, 503 (1973).

17. G. Laslaz, P. Guyot, G. Kostorz, *J. De Physique* **C7**, 406(1977).

18. S.P. Singhal, H. Herman, *J. Appl. Crystallogr.* **11**, 572(1978).

19. J. Schroeder, C.J. Montrose, P.B. Macedo, *J. Chemical Physics* **63**, 2907(1975).

20. G.R. Srinivasan, R. Colella, P.B. Macedo, V. Volterra, *Phys. Chem. Glasses* **B14**, 90(1973).

21. W.I. Goldburg in *Scattering Techniques and Applications to Supramolecular and Nonequilibrium Systems* ed. by Sow-Hsin Chen, B. Chu, and R. Nosul 383 (1981).

22. W.I. Goldburg and J.S. Huang, in *Fluctuations, Instabilities and Phase Transitions* ed. T. Riste, 87 (19753).

23. A.J. Schwartz, J.S. Avery, W.I.Goldburg, *J. Chemical Physics* **62**, 1874(1975).

24. Y.C. Chou, W.I.Goldburg, *Phys. Rev.* **A20**, 2105 (1979).

25. N.C. Wong, C.M. Knobler, *J. Chemical Physics* **69**, 725 (1978).

26. H.E. Cook, *Acta Metall.* **18**, 297 (1970).

27. J.W. Cahn, *Trans. AIME.* **242**, 166 (1968).

28. T. Hashimoto, M. Itakura, H. Hasegawa, *J. Chemical Physics* **85**, 6118(1986).

29. T. Hashimoto, M. Itakura, N. Shimidzu, *J. Chemical Physics* **85**, 6773(1986).

30. Landau and Lifshitz, "Statistical Physics", 533, Pergamon Press.

31. E.M Lifshitz and L.P. Pitaevski, "Physical Kinetics", 427, Pergamon Press.

32. S. Reich, *Physics Letters*, **114A**,No. 2,90 (1986)

33. I.M. Lifshitz, V. V. Slyozov, *J. Phys. Chem. Solids*, **19**,35 (1961).

34. J.W. Cahn *J Chem. Phys.*, **42**, 93 (1965).

35. J.W. Cahn, J.E. Hilliard, *J Chem. Phys.*, **28**, 258 (1958).

36. T. K. Kwei,T. Nishi,R.F. Roberts, *Macromolecules* **7**, 667 (1974).

37. Y. Cohen, Masters Thesis, (1980)

38. S. Reich,Y. Cohen, *J. Polymer Sci. Pol. Phys. Ed.* **19**, 1255 (1981).

39. F.Family, P.Meakin, *Phys. Rev. Let.***61**, 428 (1988)

40. H.Furukawa, *Physica* **123A**, 497 (1984)

41. F.Cangelosi, M.T.Shaw, *Polymer Compatibility and Incompatibility* 107, ed. Karel Šolc (1982)

42. K. Binder, D. Stauffer,*Phys. Rev. Let.***33**, 1006 (1974)

43. K. Binder, *Phys. Rev.***15B**, 4425 (1977)

44. K. Binder, M. H. Kalos,*J.Stat. Phys.***22**, 363 (1980)

45. L. D. Landau, V.G. Levich, *Acta Physicochimica URSS* **17**, 42 (1942).

46. M. Gordon, J.S. Taylor, *J. App. Chemistry***2**, 493 (1952).

47. T. K. Kwei, H. L. Frisch, *Macromolecules***11**,1267 (1978).

48. K. Naito, G. E. Jonson, D. L. Allara, T. K. Kwei, *Macromolecules* **11**,1260 (1978).

49. M. Bank, J. Leffingwell, C. Theis, *J. Poly. Sci. A2* **10**,1097 (1972).

50. M. Bank, J. Leffingwell, C. Theis, *Macromolecules* **4**, 43 (1971).

51. D.C. Wahrmund, R.E. Bernstein, J.W. Barlow, D.R. Paul, *Polym. Eng. Sci.***18**, 677 (1978).

52. R.E. Bernstein, D.R. Paul,J.W. Barlow,*Polym. Eng. Sci.* **18**, 683 (1978).

53. R.E. Bernstein, D.C. Wahrmund, J.W. Barlow, D.R. Paul, *Polym. Eng. Sci.***18**, 1220 (1978).

54. T.R. Nassar, D.R. Paul, J.W. Barlow, *J. Appl. Polym. Sci.***23**, 85 (1979).

55. D.C. Wahrmund, D.R. Paul, J.W. Barlow, *J. Appl. Polym. Sci.***22**, 2155 (1978).

56. R.Mohn, D.R. Paul, J.W. Barlow, C.A. Cruz *J. Appl. Polym. Sci.***23**, 575 (1979).

57.   C.A. Cruz, D.R. Paul, J.W. Barlow, *J. Appl. Polym. Sci.*23, 589 (1979).

58.   J.W. Schurer, A. de Boer, G. Challa, *Polymer* 16, 201 (1975).

59.   E. Roerdink, G. Challa, *Polymer* 19, 173 (1978).

60.   R.Gelles, C.W. Frank, *Macromolecules*15, 1486 (1982).

61.   R.Gelles, C.W. Frank, *Macromolecules*16, 1448 (1983).

62.   T. Hashimoto, J. Kumaki, H. Kawai, *Macromolecules,*16, 641 (1983).

63.   T. Hashimoto, K. Sasaki, H. Kawai, *Macromolecules,*17, 2812 (1984).

64.   C.C. Han, M. Okada, Y. Muroga, F.L. McCrackin, B.J. Bauer, Q.Tran-Cong, *Polym. Eng. Sci.,*26, 3 (1986).

65.   S. Nojima, K. Tsutsumi, T. Nose, *Polym. J.,*14, 225 (1982).

66.   S. Nojima, Y. Ohyama, M. Yamaguchi, T. Nose, *Polym. J.,*14, 907 (1982).

67.   M. Okada, C.C. Han, *J. Chem. Phys.,*85, 5317 (1986).

68.   T. P. Russell, G. Hadziiannou, K. Warburton, *Macromolecules,*18, 78 (1985).

69.   K. Sasaki, T. Hashimoto, *Macromolecules,*17, 2818 (1984).

70.   I.G. Voigt-Martin, K.H. Leister, R. Rosenau, R. Koningsveld *J. Polymer Sci. Pol. Phys. Ed.,*24, 723 (1986).

71.   M. Takenaka, T. Izumitani, T. Hashimoto, *Macromolecules,*20, 154 (1987).

72.   F. Bates, P. Wiltzius, *J. Chem. Phys.*, 91, 3258 (1989).

73.   H. Tanaka, T. Hayashi, T. Nishi,*J. Appl. Phys.*, 59, 653 (1986).

74.   H. Tanaka, T. Hayashi, T. Nishi,*J. Appl. Phys.*, 59, 3627 (1986).

75.   H. Tanaka, T. Nishi,*Phys. Rev. Let.*, 59, 692 (1987).

76.   H. Tanaka, T. Hayashi, T. Nishi,*J. Appl. Phys.*, 65, 4480 (1989).

77.   E.D. Siggia,*Phys. Rev. A.*, 20, 595 (1979).

78.   S. Alexander, private communication (1992)

SWELLING AND UNIAXIAL EXTENSION OF POLYMER GELS AS SEEN BY SMALL ANGLE NEUTRON SCATTERING

J. Bastide [1], F. Boué [2], E. Mendes[3], A. Hakiki [1], A. Ramzi [2], J. Herz [1]

[1] Institut Charles Sadron-CRM, CNRS-Université Louis Pasteur
  6, rue Boussingault, F-67083 Strasbourg Cedex France
[2] Laboratoire Léon Brillouin (CEA-CNRS)
  Centre d'études, Saclay, F-91191 Gif-sur-Yvette France
[3] Laboratoire de Physicochimie Structurale et Macromoléculaire
  ESPCI, 10, rue Vauquelin, F-75251-Paris Cedex France

INTRODUCTION

One of the surprising features of rubber elasticity is that a very simple theory[1], describing a polymer network as a collection of independent "entropic springs" (the elementary "meshes" of the net, pinned by their extremities), deformed affinely in the macroscopic strain, is enough to give the right order of magnitude for its elastic modulus. However, as shown by S. Alexander[2], some fundamental questions, which are at the same time very complex and very rich, arise when approaching rubbers and gels without making a priori such a drastic simplification[3]. These considerations are not purely abstract ones. The extreme difficulties which are encountered when trying to reach a slightly higher level of precision in the modeling originates very likely in an excessive reduction of the problem: it is, for instance, difficult to describe self-consistently (in a relatively precise manner) the elastic properties of a rubber and its swelling degree when put in presence of some appropriate diluents (it is then called a "gel"). But how to know at which level it is worth introducing some additional complexity in the theories?

WHY NEUTRON SCATTERING EXPERIMENTS ON GELS

A possibility among others is to perform experiments meant not only to check predictions of the theories but also to test some aspects of the representation underlying them. Neutron scattering is one of the techniques which can be employed to this end. For instance, a controversed point is the characteristic length scale of affine deformation (the deformation being less than affine on distances smaller than this length scale). Such a question is relevant also when the deformation of the network is a "swelling" in a solvent of the precursor chains. In such a gel state, a neutron scattering "contrast" can be created between the polymer and the diluent by letting the system absorb some deuterated solvent. Under these conditions, one expects, in analogy to the case of "semi-dilute solutions" (solutions of long strongly overlapping chains which behave as networks at high frequencies[4]) to measure the variation, with the concentration of polymer in the gel, of a correlation length ξ, representing the typical length scale above which the "monomer units" of the chains are distributed randomly in space.

Soft Order in Physical Systems, Edited by Y. Rabin
and R. Bruinsma, Plenum Press, New York, 1994

TYPICAL SCATTERING BEHAVIOUR OF GELS

Before doing the experiment, we hoped to get a clue to the deformation process by positioning the experimental law of variation of ξ with respect to two limits: (i) the "soft gel" limit for which $\xi \sim \phi^{-0.77}$, as in semi-dilute solutions. Such a behaviour would correspond to a generalization of the c* theorem by de Gennes[4] and would be possible only if the swelling was accounted for by a strongly non-affine unfolding[5] of the network on length scales comparable with the size of the elementary chains. (ii) the "strong gel" limit for which $\xi \sim \phi^{-1/3}$ ($\xi \sim L \sim V^{1/3}$, L and V being respectively the lateral dimension and the volume of the sample). This limit would mean a deformation affine in the macroscopic strain up to the scale of ξ (i.e. a rather constrained network). But for all gels we have studied, and for which it was possible to determine a correlation length, we found values which were too large and which were varying too much, even with respect to the "soft gel" limit. Therefore, it was not easy to interpret the data in the framework of a simple blob picture[5-7]. Moreover, the absolute value of the scattering intensity of the gels was much larger than expected from the available thermodynamical description of gels[8]. In other words, the fluctuations of polymer concentration in gels appeared to be surprisingly large. What is then the origin of the phenomenon? Note that a static neutron scattering experiment is not able to tell whether the spatial fluctuations which give rise to the scattering are dynamical (they form and unform then as a result of brownian motion) or, on the other hand, are nearly frozen in. There is now a theory, by Y. Rabin and R. Bruinsma[9] which suggests that the large scattering by gels might be driven by thermal fluctuations. But, since it was well known that polymer networks are often rather irregular[10], we tried to explore another way: we focused on the question of the network imperfections. L. Leibler, J. Prost and one of us proposed[11] that the "quenched disorder", one of the well known sources of complexity of the rubber elasticity problem, might manifest itself as"heterogeneities" and lead to enhanced scattering.

RANDOMLY CROSSINKED SEMI-DILUTE SOLUTIONS: MODEL[11] AND EXPERIMENTS[12]

In order to focus on a simple situation, the case of well formed gels, prepared by statistical crosslinking of semi-dilute solutions in good solvent, was more specifically addressed[11]. From the assumed presence of a preexisting "lattice" of interchain contact points in the solution, it was possible to propose a "two levels" model. According to this description, such statistical gels should contain regions "more crosslinked" ("harder") than the average, having the "shape" of polydisperse branched clusters. Under certain conditions, the "clusters" (with eventually a local polymer concentration bigger than in the rest of the gel), were expected to spread over large length scales and, however, remain "invisible" for a scattering experiment, because having been formed at random (a random growth implies a screening of the correlations). It was argued that the swelling of such a gel should, however, be accompanied by a sort of "unscreening" of the clusters and, thus, should lead to a strong increase of the scattered intensity for lower values of the scattering vector amplitude q. Accordingly, the expansion of the network should be achieved in a very inhomogeneous way (the presence of the clusters leading to strong distortions) and therefore the correlation length was expected to scale like ϕ^{-s} (ϕ being the polymer concentration inside the gel), with s always larger than for semi-dilute solutions (and s=5/3 in a particular case). A new series of experiments was then undertaken and the experimental behaviours were always found to be in agreement with the model predictions. See a typical result in Fig. 1. More quantitatively, the experimental values of the correlation length seemed to scale (in a limited range of ϕ) like ϕ^{-s}, with s always larger than 1 and sometimes very close to 5/3.

Uniaxially stretched gels This model was then generalized to the case of the stretching[11]. Classically, one expects, in an elongated network, a decrease of the correlations in the direction of extension (because correlations are supposed to exist only along the same elementary strand of the network and because these strands are supposed to be stretched to some extent). As a result, the scattering intensity should decrease for **q** parallel to the stretching axis and some anisotropic iso-intensity lines should show up, with a long axis perpendicular to the direction of stretching. This new approach predicts an entirely different behaviour, provided that, before deformation, some large clusters (spreading on distances bigger than the elementary strand) are still partly masked by a screening effect. In analogy with the situation of the swelling, one may then imagine that the stretching leads to an unscreening of some intra-cluster correlations for **q** parallel to the elongation: both the scattering intensity and the correlation length should therefore increase strongly. Conversely, for **q** perpendicular, the effect of the elongation may be compared to a deswelling and both the intensity and the correlation length should decrease. At the same time, the iso-intensity lines were predicted to exhibit, for the lower q values, double winged shapes (like the lobes observed by Inoue et al[13] and the "butterfly patterns" described by R.Oeser[14]) with a long axis parallel to the stretching axis. In practice, experiments were performed on the same gels as those studied under swelling and behaviours in good agreement with the model predictions were effectively observed: some "butterflies" were found in the iso-intensity lines (see Fig. 2) and strong variations of the correlation length in the parallel direction were obtained (in one case, the variation was bigger than the extension ratio to the cube).

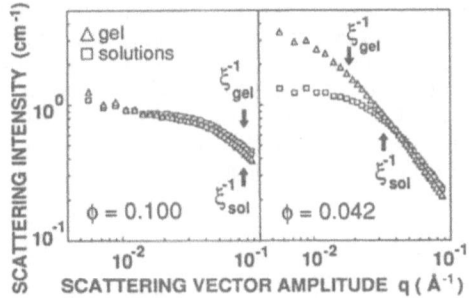

Figure1. Scattered intensity from a statistical gel (a semi-dilute solution of polymer volume fraction $\phi = 0.1$, transformed into a "gel" by random reaction of tie-points between the chains; concentration of crosslinking molecules $c_x = 0.8\%$, in moles per mole of "monomer unit"). The neutron scattering contrast between the network and the solvent is achieved by swelling the gels with deuterated toluene. The signal from the gel is compared to that of a semi-dilute solution of the same polymer concentration, here for two swelling degrees only: <u>left</u>, nearly that of preparation; <u>right</u>: maximum swelling approximately

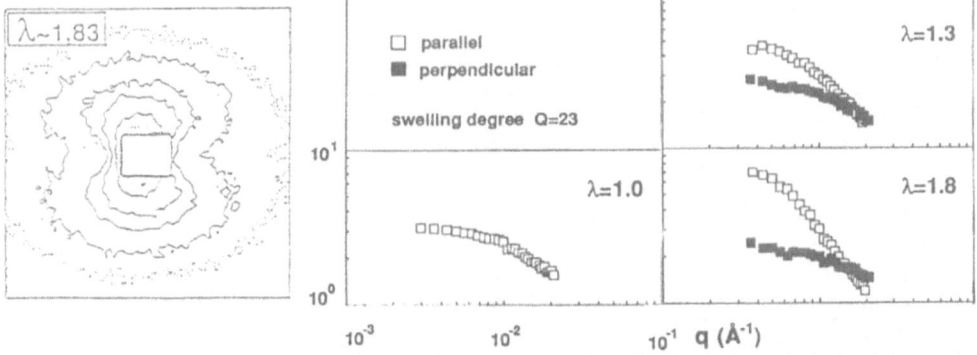

Figure 2. <u>right</u>: Scattered intensities (in cm^{-1}) as a function of the scattering vector amplitude q (in Å^{-1}), for **q** parallel.(□) and **q** perpendicular (■) to the stretching axis, for the same gel as in Fig. 1, at maximum swelling in deuterated toluene ($\phi \approx 0.042$) and different elongation ratios λ.<u>left</u>: several iso-intensity lines for the same gel, elongated by a factor $\lambda = 1.83$. The stretching axis is vertical.

WHICH PICTURE FOR GELS IN GENERAL?

After this set of experiments on randomly crosslinked semi-dilute solutions, we have studied (and we are still studying) new series of gels prepared by various methods performed either in the presence of solvent or in the bulk[15,16]. Comparable enhancements of scattering intensity upon increase of swelling degree are found in most cases. Under uniaxial stretching, the appearance of "butterfly patterns" seems to be rather frequent. Then, one may wonder why these phenomenona are so general. On one hand, since they denote an amplification of fluctuations in space of the polymer concentration, it is true that they may be compared, to some extent, to a decomposition induced by the strain (an expansion or an extension of the network). According to Y. Rabin and R. Bruinsma[9], such an analogy makes sense and the fluctuations which are amplified are effectively the thermal ones. Another possibilty is the generalization[17-19] of the model[11] recalled above: the effects which are observed would originate essentially in the presence of some quenched disorder. Some large scale inhomogeneities, say "hills and valleys" of crosslinking density, would exist in the networks. The typical size of the hills might be larger than the correlation length in the state of preparation of the gels (because the hills and valleys are thought to be formed by a random process). However, the presence of these hills and valleys would lead to strong distortions of the gel, on molecular scales, upon swelling or uniaxial deformation. In a scattering experiment, such "defects" would be made more "visible" when the system is deformed, because of a sort of unscreening process (the hills being deformed less than the valleys). We hope that some quasi-elastic scattering experiments will be able to discriminate between these noticeably different ways of interpreting the data.

REFERENCES

1. see for example: P.J. Flory, Proc. Roy. Soc. 351:351 (1976)
2. S. Alexander, J. de Physique (Paris), 45:1939 (1984), and in "Physics of Finely Divided Matter", N. Boccara and M. Daoud Eds., Springer Proc. in Physics 5, p. 162 (1985)
3. R. T.Deam, S.F. Edwards, Philos. Trans. Roy. Soc. London, Ser A,280:317 (1976)
4. P.G. de Gennes, "Scaling Concepts in Polymer Physics", Cornell University Press
5. Candau, J. Bastide, M. Delsanti, Adv. Polym. Sci., 44:27 (1982)
6. J. Bastide, F. Boué, M. Buzier, in "Molecular Basis of Polymer Networks", A. Baumgärtner and C. Picot Eds., p. 48, Springer Verlag (1989)
7. E. Geissler, A. M. Hecht, S. Mallam, F. Horkay, M. Zrinyi, Makromol. Chem., Macromol. Symp, 40:101-108 (1990) and references there in.
8. T. Tanaka, L.O. Hocker, G.B. Benedek, J. Chem. Phys. 59:5151 (1973) see also A. Onuki, in "Space-Time Organization in Macromolecular fluids", p. 54, F. Tanaka, T. Ohta and M. Doi Eds., Springer Verlag (1989)
9. Y. Rabin, R. Bruinsma, Eur. Phys. Lett. 20:79 (1992)
10. see, e.g., N. Weiss, A. Silberberg, British. Polym. J. 9:144 (1977) and S.J. Candau et al. J. Chem. Phys. 17:83 (1979)
11. J. Bastide, L. Leibler, Macromolecules 21:2647 (1988), J. Bastide, L. Leibler, J. Prost, Macromolecules 23:1821 (1990)
12. E. Mendes, P. Lindner, M. Buzier, F. Boué, J. Bastide, Phys. Rev. Lett. 66:1595 (1991) and J. Bastide, E. Mendes, F. Boué, M. Buzier, P. Lindner, Makromol. Chem., Macromol. Symp, 40:81-99 (1990)
13. T. Inoue, M. Moritani, T. Hashimoto, H. Kaway, Macromolecules, 4:500 (1971)
14. R. Oeser, C. Picot, J. Herz, in "Polymer Motions in Dense Systems", D. Richter and T. Springer Eds., p.104, Springer (1988) see also e.g.J.Bastide, M. Buzier, F. Boué, same issue p.112
15. F. Zielinski, M. Buzier, C. Lartigue, J. Bastide, F. Boué, Prog.Col.Polym Sci. 90:115 (1992), F. Zielinski, Thesis Université Pierre et Marie Curie Paris VI (1991)
16. A. Hakiki, Thesis, Université Louis Pasteur, Strasbourg, France (1991).
17. A. Onuki, J. Phys. II, France 2:45-61 (1992)
18. S. Panyukov, JETP Lett. (1992)
19. J. Bastide, F. Boué, R. Oeser, E. Mendes, F. Zielinski, M. Buzier, C. Lartigue, P. Lindner, Mat. Res. Soc. Symp. Proc. 248:313-324 (1992)

SILICA-POLY(DIMETHYLSILOXANE) MIXTURES: AN NMR APPROACH TO THE INVESTIGATION OF CHAIN ADSORPTION STATISTICS

Roger Ebengou and Jean-Pierre Cohen-Addad

Laboratoire de Spectrométrie Physique associé au CNRS, Université Joseph Fourier, Grenoble I, B.P. 87 - 38402 St Martin d'Hères, France

INTRODUCTION

This work deals with the statistical description of poly(dimethylsiloxane) chains adsorbed upon the surface of silica particles within silica-siloxane mixtures. In these systems, the incorporation of fumed silica into the PDMS polymer melt is obtained by mechanical mixing. Subsequently, the adsorption which occurs through the formation of hydrogen bonds between the oxygen atoms on the polymer chains and the silanol groups located on the silica's surface is a solvent free or melt adsorption process. Consequently, the law of adsorption observed from these systems is specific to such a process.

It has already been shown both theoretically and experimentally[1], that the law of adsorption of end-methylated PDMS chains obeys the formula:

$$Q_r^1 = \chi_a \overline{M}_n^{1/2} \tag{1}$$

where Q_r^1 denotes the amount of polymer left bound to 1g of silica after the completion of the adsorption process and $\overline{M}_n$ is the number average molecular weight of PDMS chains.

The treatment that led to the formula (1) relies on two basic findings. Firstly, from the investigation of the surface properties of fumed silica,[2] one can assume that no roughness exists on to the silica surface over an average area which is of the order of magnitude of the area covered by a polymer chain. Secondly, some experimental features[1] suggested that, at equilibrium, the silica's surface is saturated and the PDMS chains remain gaussian in the mixtures as in the pure melt. Therefore, the adsorption of a single chain was pictured as the cross between a three dimensional random-walk (namely the polymer chain) and an adsorbing plane. The average number of contact points, or fixed monomeric units, of one PDMS chain upon the silica surface $\langle n_c \rangle$ was found to be proportional to the square root of the number N of monomeric units.

$$\langle n_c \rangle = \varepsilon_a \sqrt{N} \tag{2}$$

Here ε_a is a numerical constant which takes into account the chain stiffness. The adsorption law observed for these systems and given by equation (1) was shown to be the direct consequence of this particular statistics of PDMS chain adsorption.

We have recently shown work[3] that this law of adsorption is still observed for end-hydroxylated PDMS chains. However, the amounts of polymer left bound to silica particles were larger compared to those measured for end-methylated PDMS chains. This demonstrated the fact that the adsorption was most efficient in the case when PDMS chain ends were hydroxylated.

In this work we approach the statistics of adsorption of both end-methylated and end-hydroxylated PDMS chains by an independent way. We use the magnetic relaxation properties of protons attached to the polymer chain to estimate the relative number of monomeric units which are fixed on the silica surface. We show that the molecular weight dependence of this quantity is in accordance with the gaussian statistics of PDMS chains adsorption.

I. EXPERIMENTAL

I.1. Samples

The silica-PDMS samples were kindly supplied by the Rhône Poulenc, Co. The surface area of fumed silica (Aerosil 150) determined by BET methods was $150 m^2 g^{-1}$. The initial silica concentration were $c^i_{Si}=0.17$ (w/w) and $c^i_{Si}=0.17$(w/w) for end-methylated PDMS samples and $c^i_{Si}=0.29$ (w/w) for end-hydroxylated PDMS samples. The samples were observed after the completion of the adsorption process.

I.2. Extraction of non-adsorbed chains

The non-adsorbed polymer chains were removed from the mixtures by maintaining each sample in a large excess of methylcyclohexane for two days. Subsequently, each sample was dryed and the amount of polymer left adsorbed onto the silica surface was determined by microanalysis of the silicon-carbon ratio.

I.3. NMR Measurements

NMR measurements were performed using a Bruker pulsed spectrometer operating at 60 MHz. The protons transverse magnetic relaxation function was analysed from spin echoes which were formed by applying Carr-Purcell pulse sequences to the spin system.

II. RELATIVE NUMBER OF FIXED MONOMERIC UNITS

II.1. Principle of NMR observation

Typically the proton magnetic relaxation function measured after the removal of the free chains exhibits a two components behavior. Therefore, the whole relaxation function can be written as:

$$M_x(t) = \tau_B M_x^B(t) + (1 - \tau_B)M_x^I(t) \qquad (3)$$

wher $M_x^B(t)$ is the fast decay component observable at short time scales and attributed to protons on fixed or bound monomeric units. $M_x^I(t)$ is the slow decay component observable at large time scales and attributed to protons on loops and tails. τ_B is the relative number of fixed monomeric units. According to equation (2) this number is given by:

$$\tau_B = \gamma_v \langle n_c \rangle / N \qquad or \qquad \tau_B = \gamma_v \varepsilon_a \sqrt{M_m} / \sqrt{M_n} \qquad (4)$$

where γ_v is a numerical factor which accounts for the number of monomeric units which are very close neighbours of adsorbed monomers along one chain; these neighbours may be observed from NMR as fixed although they undergo slow motions. M_m is the molecular weight of the monomeric units.

II.2 High silica concentrations

Figure 1 shows, for an initial silica concentration $c^i_{Si}=0.29$ (w/w), the relative number of fixed monomeric units τ_B for both end-methylated and end-hydroxylated samples.

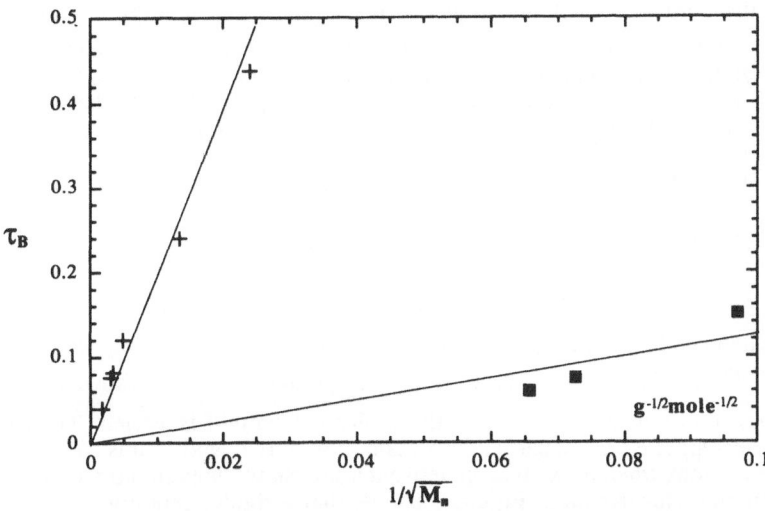

Figure 1. Relative number of bound monomeric units τ_B as a function of the inverse of the square root of the molecular weight: end-methylated PDMS chains (+), end-hydroxylated PDMS chains (■)

τ_B obeys equation (4) denoting a gaussian statistics of PDMS chains adsorption. The lower value of τ_B observed for end-hydroxylated PDMS chains is not surprising if one assumes the silica surface to be saturated at equilibrium as it has been shown in previous works[3,4]. Since the specific amount of polymer adsorbed onto silica surface is higher in the case when the PDMS chain ends are hydroxylated, the mineral surface available per chain is lowered in the case of end-hydroxylated PDMS chains and the nubmer of contact points per chain is also lowered.

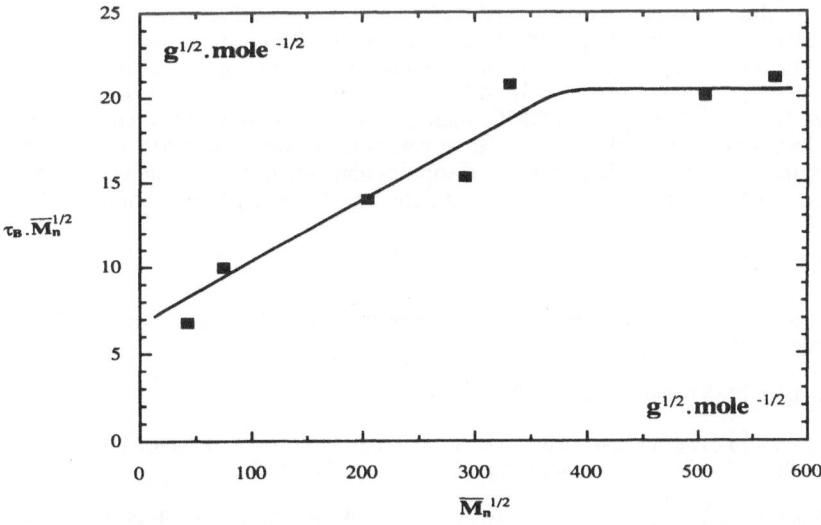

Figure 2. Product $\tau_B \sqrt{\overline{M_n}}$ for end-methylated samples corresponding to $c_{Si}^i = 0.17$ (w/w) as a function of the square root of the molecular weight

From the comparison of the slopes of the straight line of Figure 1 and the law of adsorption[1,3], we estimate the values $\gamma_v = 3.9$ for end-methylated PDMS chains and $\gamma_v = 5$ for end-hydrooylated PDMS chains. Accordingly, the number of close neighbours which are perceived from NMR as fixed is higher for end-hydroxylated PDMS chains than for end-methylated ones. This result, which gives evidence of a higher amount of topological constraints exerted on one chain in the case of end-hydroxylated PDMS chains, is probably due to the formation of double hydrogen bonds at chain ends when they are hydroxylated.

II.3. Low silica concentration

In the case of end-methylated PDMS chains, deviations from the law $Q_r^l = \chi_a \overline{M}_n^{1/2}$ were observed when $c_{Si}^i = 0.17$ (w/w)[1]. Deviation from the law $\tau_B \propto 1/\sqrt{\overline{M}_n}$ was also observed from the NMR measurements of τ_B for the same samples. Figure 2 shows that the product $\tau_B \sqrt{\overline{M}_n}$ is not constant anymore when the molecular weight is varied. To explain these deviations we suggested that a partial desorption of weakly bound chains could occur for low silica concentrations when the sample underwent a solvent washing in order to extract the non-adsorbed chains. Nevertheless, Figure 2 shows that a tighter binding of polymer chains occurs when the number of contact points per chain is sufficiently large ($\sqrt{N} \geq 50$) to avoid the desorption process induced by the solvent washing. In this molecular weight range we recover the $\tau_B \sqrt{\overline{M}_n}$ = constant behavior.

CONCLUSION

The law of adsorption observed from silica-PDMS mixtures is specific to the solvent free or melt saturated adsorption that occurs within these systems. The number of monomeric units of one chain fixed onto the silica surface is proportional to the square root of the number of monomers. As a direct consequence of these particular statistics and the saturation of the silica's surface at equilibrium, the specific amount of adsorbed polymer Q_r^l is proportional to the square root of the molecular weight.

This statistics of chain adsorption can be observed using two different and independant ways. The first one, which is macroscopic consists of the direct measurement of the specific amount of adsorbed polymer after extracting the non-adsorbed chains. The second one, which is microscopic consists of the estimate of the relative number of bound monomeric units from relaxation NMR measurements. We show that the two independent ways for observing the law of adsorption are in accordance with each other.

The results observed for end-methylated PDMS chains and for end-hydroxylated PDMS chains show that the hydroxylated chain ends plays an important role in the adsorption process. For the same chain lengths, the amount of adsorbed polymer is increased by the presence of hydroxylated chain ends. Accordingly, the NMR study of the interfacial layer shows that for end-hydroxylated PDMS chains, the relative number of fixed monomeric units is lower than for end-methylated ones. However, in the presence of hydroxylated chain ends, NMR results suggest a higher number of topological constraints exerted on PDMS chains, probably because of the possibility of the formation of double hydrogen bonds at chain ends.

REFERENCES

1. J.P. Cohen-Addad, Silica-siloxane mixtures. Structure of the adsorbed layer: chain length dependence, Poymer, 30:1820 (1989)

2. D.W. Schaefer, MRS bulletin, XIII:22 (1986)

3. J.P. Cohen-Addad and R. Ebengou, Silica-siloxane mixtures. Investigation into the adsorption properties of end-methylated and end-hydroxylated chains, Polymer, 33:379 (1992)

4. J.P. Cohen-Addad, Ph. Huchot, Ph. Jost and A. Pouchelon, Hydroxyl or methyl terminated poly(dimethylsiloxane chains: kinetics of adsorption on silica in mechanical mixtures, Polymer, 30:143 (1989)

LIQUID CRYSTALLINE POLYMERS IN NEMATIC SOLVENTS: EXTENSION AND CONFINEMENT

A. Halperin[1] and D.R.M. Williams[2]

[1]Department of Materials
University of California
Santa Barbara, CA 93106

[2]Matière Condensée
Collège de France
75231 Paris Cedex 05

INTRODUCTION

Solution behavior occupies a central position in polymer science[1]. Here we briefly review some recent results for th solution properties of main chain thermotropic liquid crystalline polymers (LCPs) dissolved in liquid crystalline solvents. We concentrate on the distortional properties of such LCPs.

Thermotropic LCP consist of mesogenic monomers joined by flexible spacer chains. Two extreme architectures are encountered[2]: Main chain LCP, in which the mesogenic monomers are incorporated into the backbone of the polymer, and side chain LCP, where the mesogens are attached to the backbone by short pendant chains. Our discussion concerns the behavior of main chain LCP solutions where the coupling between the LCP configurations and the nematic medium is strongest. The main chain LCP experience an anisotropic molecular field due to the nematic order. This field affects *both* the mesogenic monomers and the flexible spacer chains. It has important effects on the configurations of the LCP. The theoretical models[3,4], in a solution as well as in a melt, picture the LCP as a persistent chain of length L. The chemical structure is smeared out and the polymer is envisioned as a uniform line characterized by an elastic bending constant or rigidity, ϵ, and a nematic coupling constant, a_n. The elastic free energy density of the chain is thus $(\epsilon/2)(d\theta/ds)^2$ while the nematic free energy density is $(3a_nS/2)\sin^2\theta$ where θ is the local angle between a tangent to the chain and the nematic director $\mathbf{n}$, s is the arc length position along the chain and S is the nematic order parameter. The important length scale in the isotropic case is the persistence length, $\zeta \approx \epsilon/kT$. This defines an "elastic blob" for which the elastic energy, ϵ/ζ, is comparable to kT. For long chains $L \gg \zeta$, the chain configuration is that of an isotropic random walk of L/ζ steps of length ζ. Accordingly, the chain size is roughly $\sqrt{L\zeta}$. Two new length scales appear in a nematic medium: The deflection length, $\lambda \approx (\epsilon/a_nS)^{1/2}$, and the nematic length, $l \approx kT/a_nS$. The elastic energy, ϵ/λ, is comparable to the nematic energy $a_nS\lambda$ for segments of length λ. l defines a "nematic blob" for which the nematic energy, a_nSl, is comparable to kT. The configurations of LCP in nematic media result from the superposition of two components. One corresponds to the undulations of the elastic line. These may be thought of as the meandering of the chain trajectory within a virtual conical tube due to the nematic field. The chain is deflected whenever it encounters the "wall" of the capillary. The decay of the angular correlations in this regime is specified by λ. λ replaces ζ which characterized the angular correlations in the isotropic medium. Hairpin defects are the second component. These are abrupt reversals in the trajectory of the chain. Entropy favors hairpins since they enable otherwise inaccessible configurations. However, the bend region of the hairpins is oriented perpendicular to the nematic field and is thus energetically penalized. Each hairpin is thus associated with an energy $U_h \approx (a_nS\epsilon)^{1/2}$ of order ϵ/λ. Since the hairpins are marginally stable and thus long lived excitations, one may view them as qasiparticles. Their number in equilibrium, n_o, is obtained by minimizing the free energy of an LCP viewed as a one dimensional gas of n hairpins, $F = nU_h + nkT\ln(nl/L)$, thus leading to $n_o = (L/l)\exp(-U_h/kT)$. Each of the two components determines one of the two equilibrium dimensions of the LCP. The lateral undulations determine the coil span perpendicular to $\mathbf{n}$, $R_{\perp_o}$. $R_{\perp_o}$ may be attributed to a two dimensional random walk of L/l steps of length l so that $R_{\perp_o}^2 \approx Ll$. The hairpins give rise to a one dimensional random walk in the direction of $\mathbf{n}$. It consists of n_o steps of length L/n_o and thus $R_{\parallel_o} \approx L^2/n_o$. Altogether, the LCP are ellipsoidal objects with major axis $R_{\parallel_o} \gg R_{\perp_o}$.

EXTENSION BEHAVIOUR: THE ISING ELASTICITY

The familiar, flexible polymer chains are simply modeled as isotropic, three dimensional random walks[1]. The monomer size, a, is the step length and the polymerization degree, N, is identified with the number of steps. The probability density function of the end to end distance is $\mathcal{P}(\mathbf{R}) \sim \exp(-R^2/R_o^2)$ where $R_o^2 \approx Na^2$ characterizes the average end to end distance. The elastic stretching penalty is thus $F_{el}(R)/kT \approx -\ln \mathcal{P}(R) \approx R^2/R_o^2$. The presence of configurational random walk components suggests a similar elastic behavior for LCP. In particular, one expects $F_{el}(R_{\parallel}) \approx R_{\parallel}^2/R_{\parallel_o}^2$ and $F_{el}(R_{\perp}) \approx R_{\perp}^2/R_{\perp_o}^2$. Only one of these expectations, regarding $R_{\perp}$, is fulfilled. A qualitatively different situation is encountered when $R_{\parallel}$ is extended[5,6,7]. In this case the individual steps are due to the presence of n hairpins. However, these topological defects can be removed by stretching the chain. Since the number of steps is not conserved, the familiar Gaussian free energy is inappropriate. A suitable description is suggested by the subdivision of the LCP by the hairpins: It is possible to distinguish $+$ segments, where the motion along the chain trajectory coincides with the spatial displacement, and $-$ segments where the two motions are opposite. Furthermore, $+$ segments are favored by the applied tension, f, while $-$ segments are penalized. This suggests a correspondence with the magnetization of a one dimensional Ising chain by an external field. $+(-)$ segments correspond to domains of up(down) spins and hairpins are thus analogous to domain boundaries. The tension plays the role of the external field while the magnetization is analogous to the end to end distance, $R_{\parallel}$. Finally, the lattice constant is identified with the nematic length, l. In the case of $n \gg 1$, and outside the very strong stretching regime, (where both n and the $-$ domains are small), an analytic solution is easily obtained. The number of hairpins is $n = n_o(1 - R_{\parallel}^2/L^2)^{1/2}$ while $F_{el}(R_{\parallel})/kT = n_o - n$. For weak extensions, $R_{\parallel} \ll L$, when the "saturation" factor $(1 - R_{\parallel}^2/L^2)^{1/2} \approx 1$, a Gaussian behavior is recovered and $F_{el}(R_{\parallel})/kT \approx R_{\parallel}^2/R_{\parallel_o}^2$. However, this $F_{el}(R_{\parallel})$ is rather weak compared to $F_{el}/kT \approx R^2/R_o^2$ of an isotropic chain because $R_o \ll R_{\parallel_o}$. In the opposite limit, of strong stretching, hairpins are annihilated and the behaviour is on longer Gaussian.

Once the stretching elasticity of the LCP is understood, it is possible to consider their swelling behavior in good nematic solvents[5,6]. Monomer-monomer interactions are strongly repressed in short, hairpin free LCP. Their importance grows as the chain length and the number of hairpins increases. Accordingly, long LCP in good nematic solvents are expected to swell beyond their dimensions in a θ solvent or a melt: $R_{\perp_o} \approx N^{1/2}(a/l)^{1/2}$ and $R_{\parallel_o} \approx Na/n_o^{1/2} \approx N^{1/2}(a/l)^{1/2} \exp(U_h/2kT)$. A simple modification of the Flory theory of swelling describes this effect. For isolated LCP the swelling results in weak extension of $R_{\parallel}$ and $F_{el}(R_{\parallel})/kT \approx R_{\parallel}^2/R_{\parallel_o}^2$. Altogether, the free energy per chain, F_{chain}, is comprised of three terms. An interaction term, $F_{int}/kT \approx v_n N^2/R_{\parallel}R_{\perp}^2$, and two elastic terms, $F_{el}(R_{\parallel})$ and $F_{el}(R_{\perp})$. Here v_n is an excluded volume parameter. The equilibrium condition, $\partial F_{chain}/\partial R_{\parallel} = \partial F_{chain}/\partial R_{\perp} = 0$, yields $R_{\parallel} = (l/a)^{4/5} \exp(2U_h/5kT)R_F$ and $R_{\perp} = (l/a)^{7/10} \exp(-U_h/10kT)R_F$ where

$R_F = v_n^{1/5} N^{3/5} a^{2/5}$ is the Flory radius of an isotropic chain characterized by v_n. While both $R_\parallel$ and $R_\perp$ exhibit the familiar Flory $N^{3/5}$ scaling behavior, their temperature, T, dependence is very distinctive. The two radii are predicted to show an exponential T dependence, much stronger than that found in flexible chains in isotropic solvents. Furthermore, $R_\parallel$ is predicted to shrink with increasing temperature while $R_\perp$ is predicted to grow.

THE ANCHORING AND TILTING TRANSITIONS

The swelling behavior discussed above may be observed in solutions oriented by a magnetic or an electric field[8]. The anisotropic LCP are then aligned with their major axis parallel to the director, **n**, thus allowing the measurement of $R_\parallel$ and $R_\perp$. These two traits, the anisotropy and the orientability, give rise to novel interfacial behavior of the LCP solutions. In particular, the tilt of the major axis of the LCP with respect to the surface becomes important as a route to lower the free energy of the system[6,9]. Furthermore, the tilt occurs as a second order phase transition. These effects involve an extra ingredient, anchoring, that is the alignment of the nematic by an appropriately treated surface[10,11]. The surface anchoring aligns the nematic solvent and, with it, the LCP. This allows us to orient the LCP with respect to the surface and, independently, increase their free energy by confinement or imposed chain-chain overlap. In certain cases the free energy of the LCP is lowered by tilt. However, this results in a distortion of the nematic medium and is thus penalized. The interplay between the free energy of the LCP and the distortion penalty of the nematic give rise to the tilting phase transition. In the following we mostly focus on this effect as it occurs in "brushes"[12,9] of LCP i.e., flat layers of densely grafted LCP.

The LCP are cigar-like objects. When the surface anchoring is homeotropic, **n** is perpendicular to the surface and so is the major axis of the LCP. In this case one expects continuous swelling of the brush as the grafting density is increased beyond the overlap threshold. A qualitatively different scenario is predicted for a brush grafted onto a surface imposing homogeneous alignment such that **n** is parallel to the interface. Chain-chain interactions gain importance as the grafting density is increased. The development of tilt is then beneficial as a way to increase the layer thickness thus lowering the monomer concentration and the interaction free energy. A simple analysis of this effect is enabled by appropriate modification of the Alexander model of polymer brushes[13,14]. The two original assumptions are retained: (i) The chains are considered to be uniformly stretched with their ends straddling a single surface at height H. (ii) The concentration profile is assumed to be step like. This determines the the interaction free energy per chain $F_{int}/kT \approx v_n N^2/\sigma H$, with σ the grafting density. Because tilt is possible one must allow for stretching of both $R_\parallel$ and $R_\perp$. As a result it is necessary to incorporate $F_{el}(R_\parallel)$ as well as $F_{el}(R_\perp)$ into F_{chain}. Finally, it is necessary to account for the distortion of the nematic field because of the LCP tilt . The last term is obtained from the continuum theory of nematics[10] by assuming a height dependent tilt angle of the form $\theta(z) = Q \sin(\pi z/2H)$, where Q serves as an order parameter. The associated distortion energy, within the one constant approximation, is $F_{dis} = (\sigma K/2) \int_0^H (d\theta/dz)^2 dz = (\pi^2/16)(\sigma K/H)Q^2$. In the spirit of the Alexander model we adopt a global description[12,9] of the LCP assuming uniform tilt with an angle of $\bar{\theta} = \frac{1}{H} \int_0^H \theta(z)dz$. Ignoring numerical prefactors, $\bar{\theta}$ and H are related via $H = R_\parallel \sin \bar{\theta} + R_\perp \cos \bar{\theta}$. With these relationships in mind we minimize $F_{chain} = F_{dis} + F_{int} + F_{el}(R_\parallel) + F_{el}(R_\perp)$ with respect to H and $R_\parallel$ thus obtaining H, $R_\parallel$ and $R_\perp$ as functions of Q. In turn, this enables us to obtain the Landau free energy per chain in powers of Q. Two dimensionless parameters are involved $\mu = R_{\perp_o}/R_{\parallel_o}$ and $\alpha = (K\sigma/2kTH_\parallel)(R_{\perp_o}/H_{\parallel_o})^2$. The coefficient of the Q^2 term changes sign at $\alpha = \alpha_c = (32/\pi^4)(2\mu^{-2} - 1)$. The change in sign of the second coefficient is the signature of a second order tilting phase transition at $\alpha = \alpha_c$. As viewed from the bulk, the onset of $Q \neq 0$ is an anchoring phase transition since it involves a change in the boundary conditions imposed on $\mathbf{n}$. Such phase transitions have been observed experimentally in different systems but are poorly understood at present.

The anchoring transition considered above is driven by repulsive interactions as controlled by the grafting density. Related transitions may be driven by geometrical confinement[6,7,12], which can induce a tilting transition. The transition is reminiscent of the Frederiks transition[10] in single component nematics, but is driven by the elasticity of the LCP rather than by an external field.

PERSPECTIVE

Thus far, little research effort has been directed at solutions of main chain LCP in nematic solvents. Yet, as we have seen, this field affords rich opportunities for theoretical as well as for experimental research. Furthermore, these systems have a potential for practical applications in the design of liquid crystalline devices. A tilting transition may be induced by a combination of an external field and a polymeric driving force[6,12,9]. When the polymeric contribution is subcritical, the external field can be made arbitrarily small. This effect may allow for low voltage drives of practical importance in the design of portable devices incorporating liquid crystalline displays.

REFERENCES

1. (a) P.-G. deGennes, *Scaling Concepts in Polymer Physics*, Cornell University Press, Ithaca, (1979).(b) J. des Cloizeaux and G. Jannink, *Polymers in Solution*, Clarendon Press, Oxford, (1990).

2. A. Ciferri(Ed), *Liquid Crystallinity in Polymers*, VCH Publishers Cambridge (1991).

3. T. Odijk, *Macromolecules*, **19**, 2313 (1986).

4. (a) P.G de Gennes in *Polymer Liquid Crystals* Ed. A. Ciferri, W.R. Krigbaum and R. Meyer, Academic Press, NY, 1982. (b) J.M.F. Gunn and M. Warner, *Phys. Rev. Lett.*, **58**, 393 (1987). (c) M. Warner, J.M.F. Gunn and A.B. Baumgartner, *J. Phys. A: Math. Gen.*, **18**, 3007 (1985).

5. A. Halperin and D.R.M. Williams *Europhys. Lett.*, **20**, 601 (1992).

6. D.R.M. Williams and A. Halperin *Macromolecules* submitted.

7. D.R.M. Williams and A. Halperin *Europhys. Lett.*, **19**, 693 (1992).

8. (a) F.Volino, M.M. Gauthier, A.M. Giroud-Godquin and R.B. Blumstein, *Macromolecules*, **18**, 2620 (1985). (b) F.Volino and R.B. Blumstein, *Mol. Cryst. Liq. Cryst.*, **113**, 147 (1984). (c) J.F. D'Allest et al *Phys. Rev. Lett.*, **61**, 2562 (1988).

9. A. Halperin and D.R.M. Williams *Europhys. Lett.*, **21**, 575 (1993).

10. P.G. deGennes, *The Physics of Liquid Crystals*, Clarendon, Oxford, (1974).

11. B. Jerom, *Rep. Prog. Phys.* **54**, 391 (1991).

12. D.R.M. Williams and A. Halperin *Macromolecules* in press.

13. S. Alexander, *J. Phys. (France)* **38**, 977 (1977).

14. (a) A. Halperin, M. Tirrell and T. Lodge, *Adv. Poly. Sci*, **100**, 33 (1992). (b) S.T. Milner, *Science*, **251** 905 (1991).

THE REVEALING OF HETEROGENEITIES BY FREE LINEAR
CHAINS IN A   NETWORK

Jyotsana Lal[1], Jacques Bastide[2], Rama Bansil[3] and Francois Boué[4]

[1]College de France, Matière Condensée, F- 75231, Paris Cedex 05

[2]ICS, F-67083, Strasbourg Cedex

[3]Department of Physics, Boston University, Boston MA-02215, U.S.A.

[4]LLB, CE- Saclay, F-91191, Gif-Sur-Yvette, Cedex

Using Small-Angle Neutron Scattering (SANS) we first, studied Methylmethacrylate (MMA) gels with no free chains of polystyrene (PSD) inside them. Gels show increased scattering compared to the solutions, contrary to the predictions of the classical theories. Second, linear PSD chains (their $c<c*$) were introduced in the system: again increased scattering is observed from chains in the gel compared to the chains in an equivalent semi-dilute solution. These results imply that when the translational entropy of free PSD chains is weak (especially for the larger ones) they would tend to mix less efficiently with the gel than with the solution. At low q, the scattering from the PSD chains increases with molecular weight and progressively tends to a master curve varying as $q^{-2.5}$. We discuss an explanation of this effect in terms of the percolation model of heterogeneities. The larger molecular weight PSD chains define the contour of the large percolation clusters in the gel.

RESULTS AND DISCUSSION

Gels With No PS Chains Some samples were made by polymerisation of MMA in deuterated toluene with no free chains of PS inside (details are given in reference 1). The scattering from such samples is that from the gel itself. Figure 1 shows the scattering from gels of different crosslinking ratio on a log S(q) versus log q plot. On the same plot we show scattering from a semi-dilute solution of PMMA chains (Mw=463,000) of the same concentration as the MMA gels. At the high q values all the data seem to fall on a straight line of slope 1.6. At low values of q there is always a levelling of intensity when decreasing q, but the value of S(q=0) increases and the value of q at which intensity levels off decreases with increasing crosslink ratio. One observes that the signal from the gels is higher than that from the semi-dilute solution.

This higher scattering at low q for gels than for solution is not in agreement with the chemical theories or the c^* theorem. It could be related to a heterogeneous structure, for which we propose a fractal, percolation like structure. A percolation model[2] has been used to describe the clusters of regions more crosslinked than the average above the gel point in a randomly cross-linked gel, where lattice points are occupied by blobs. The average size and the spread of the size of clusters of frozen blob increase when the percolation threshold is approached with increasing crosslink densities. For a semi -dilute solution of long chains, many more crosslinks must be introduced to reach the percolation threshold of frozen blobs than to pass the gel point of the chains. There is a progressive unscreening of these clusters on overswelling (on swelling from the preparation state) and stretching in a uniaxial direction because the tightly crosslinked regions are perturbed less than the soft interstitial regions. This establishes a contrast between the two kinds of regions in space as the system is diluted and it leads to greater scattering at low q values. In the preparation state however, the clusters are interpenetrated and the scattering is the same as in a solution.

Experimentally identical S(q) are observed for solution and a gel of not too high crosslinking density in case of randomly crosslinked gels. However, using the same chemistry but slightly higher crosslink ratio makes the scattering from the gel slightly higher giving the impression that the scattering progressively tends to $q^{-1.6}$ law[3]. A proposed explanation is that when the gels are more tightly crosslinked, one can expect that some regions of very high crosslink density will start to expel the solvent. There the heterogeneities will be revealed even in the preparation state. In any case where the scattering is larger than the solution, it would be decomposed into two q ranges[2,4]. For $q > 1/\xi_{sol}$: $S(q) \approx q^{-(5/3)}$ here 5/3 is the excluded volume exponent (self avoiding random walk) of the chains of the frozen blobs. For $1/\xi_{gel} < q < 1/\xi_{sol}$, we observe $S(q) \approx q^{-D_f^S(3-\tau)} \approx q^{-(8/5)}$ where ξ_{gel} being the correlation length of the gel in a pure

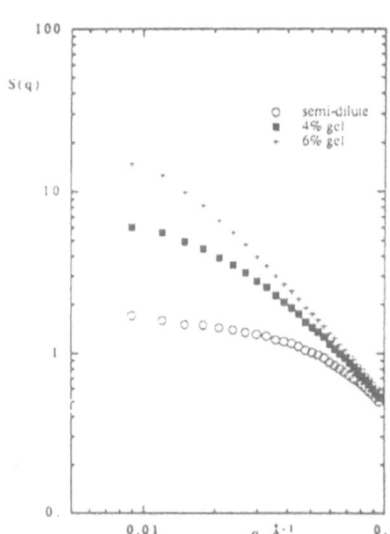

Figure 1. Log S(q) (units cm^{-1}) versus log q of MMA gels in deuterated toluene and semi-dilute PMMA solution of the same polymer concentration as the gel.

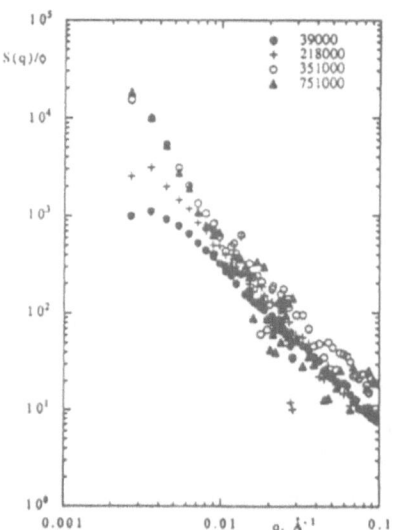

Figure 2. Log S(q)/Φ (units cm^{-1}) versus log q of free PSD chains of increasing molecular weight inside MMA gel of crosslinking density of 4.0% at 22° C.

solvent and ξ_{sol} being in the semi-dilute solution at the same concentration. D^s_f is the exponent of swollen percolation clusters and τ is the exponent which characterizes the distribution of cluster sizes in the percolation model.

The MMA gels in our experiment are observed in the preparation state. These are prepared not by crosslinking of long polymer chains but, by polymerization of monomeric and polyfunctional units. This procedure might make them more heterogeneous.

Gels With PS Chains In this case the solvent is a mixture of hydrogenated and deuterated toluene, which contrast matches the MMA chains of the network. Therefore the scattering is due to only the deuterated polystyrene (PSD) chains.

Figure 2 presents the SANS data $\log S(q)/\Phi$ versus $\log q$ for PSD chains of increasing molecular weight inside of MMA gels with 4% crosslinking ratio at 22° C. At large q, the data are noisy, especially for PSD chains of high molecular weights because their concentration is low, but they approximately falls on a single curve of slope 1.6. At low q, as the molecular weight of the linear PSD chain increases, the scattering intensity at low q also increases although, the concentration is quite low, $S(q->0)$ is high. The value of q at which the intensity levels off is lower for $M_w=218,000$ than for 39,000. The curves of two largest molecular weight PSD chains ($M_w = 351,000$ and $751,000$) show no levelling off in intensity in the available q range. Their scattering follows a slope of 2.5 at low q. The same trends in the scattering from PSD chains in MMA gels were observed for two independent series of gels with crosslinking ratio ranging between 4% and 6%, but at two different temperatures 22° C and 45° C. As for 5.5% crosslinked samples at 45° C, the scattering for PSD chains of $M_w=218,000$ is larger than the one for 4% at the same temperature and falls approximately on the curves of $M_w=351,000$ and 751,000.

Scattering from PSD chains in the gel can also be compared to scattering from such chains in an equivalent semi-dilute solution. At high q side again the data falls on a single curve of slope 1.6. At similar molecular weights, the scattering observed here is more than ten times lower than from free PSD chains inside MMA gels[1]. Thus the scattered intensity from labelled chains in a gel is much larger than a semi-dilute solution of the same polymer concentration. Such a behavior means that the distribution of free chains is not as uniform in the gel as in the semi-dilute solution.

Building a network around free chains might classically (Flory-Huggins) predict to lead to phase separation for the following reasons: first, the entropy of mixing of the free chains maybe insufficient to balance the elasticity of the network. Second, if the chains are of different species than the network or the crosslinker, the χ parameter is usually positive. A slope of 2.5 could be due to uncomplete phase separation but should vary for different M_w chains depending on the degree of separation which is at variance with our results, where we do not observe a change of slope at small q, with the increase in size of the large polymer chains.

We are thus tempted by the following picture: due to slight expulsion from the network, the PSD chains trace its heterogeneous structure by preferentially going in less crosslinked regions i.e. regions of lower elastic free energy. This means that the local concentration of free labelled chains will create a "contrast" that reveal or label the clusters of high crosslinking densities in the network. We can postulate, as before, that such regions can be represented as percolation clusters, the labelling by PSD chains would be similar to that by an ordinary deuterated solvent. Thus, in the (limited) range of q such as: $1/\xi_{gel} < q < 1/\xi_{sol}$ one would get: $S(q) \approx q^{-(8/5)}$. As said before, the "effective" exponent 8/5 can be related to the fractal dimension of the clusters and to their polydispersity or distribution[4] i.e. $8/5 = D^s_f.(3-\tau)$ where $D^s_f = 2$ is the fractal dimension of the larger clusters in the (limited) domain of distance scales (ranging from ξ_{sol} to ξ_{gel}) in which the clusters are swollen by excluded volume and $(3-\tau) = 0.8$.

For $q < 1/\xi_{gel}$, one should then observe an Ornstein-Zernicke law: But we do not expect long polymer chains which are free to move inside a network to behave exactly as an ordinary solvent. A small molecule can penetrate freely in all the holes of clusters of any size. Long polymer chains will probably not. Presumably, they will not go into holes smaller than their average radius of gyration R_g: it would cost too much energy. It is easier for them to escape into the relatively larger holes of the larger clusters. In the picture of percolation clusters these holes contain also smaller clusters, but, very likely the chains will cross clusters smaller than R_g as in an ideal mixture rather than being effectively constrained by them. These smaller clusters will remain "invisible" in the scattering. Only clusters of sizes comparable to the R_g are revealed by scattering, thus there are no more polydispersity effects and the $(3-\tau)$ term should also be removed. Note that, in this range of scales, the bigger clusters are not swollen (the excluded volume interactions remain screened off). Their fractal dimension is that of the bulk D_f^b instead of the swollen one D_f^s. Therefore, in the case when $R_g \gg \xi_{gel}$, the labelled chains should trace the bulk fractal dimension D_f^b of the larger clusters only. Since $D_f^b = 5/2$, one expects in the range of q such as: $1/\xi_{lc} (\approx 1/R_g) < q < 1/\xi_{gel}$

$$S(q) \approx q^{-5/2} \qquad\qquad (1)$$

At low q, the intensity should start to saturate and follow an Ornstein-Zernicke law for $q < 1/\xi_{lc}$.

As seen above, the experimental variation ($q^{-2.5}$) that we observe in the case of the larger chains is approximately in agreement with this dependence. In the case of smaller chains, R_g becomes comparable with ξ_{gel}. This is the first reason why one should observe a dependence weaker than the one given by (1): only a crossover regime can be seen. Such an argument is valid in the case of the chains with $M_w = 39,000$ where the observed slopes remains 1.6. In addition, the "segregation" between the more and the less crosslinked regions should be less pronounced for the shorter chains (since the entropy of mixing of the chains is larger). This may explain why, in the case of the 218,000 chains, the $q^{-2.5}$ dependence is attained only in the case of the larger crosslinking density .

Finally, we remark that assuming such a weak segregation process, i.e. a modulation of the concentration of the free chains driven by the local crosslinking density, each chain would stay in a non-confined conformation. The statistics should indeed be gaussian on length scales larger than ξ_{sol}, as in semi dilute solution. Conversely, on length scales smaller than ξ_{sol} ($q > 1/\xi_{sol}$) the local statistics of the PSD chains should remain in an excluded volume state: $S(q) \approx q^{-5/3}$. This is in agreement, with the change of slope we observe for all the systems of chains trapped in gels, for the larger q values.

REFERENCES

1 J. Lal, J. Bastide, R. Bansil and F. Boué, Macromolecules, 26: 6092 (1993).
2. J. Bastide and L. Leibler, Macromolecules , 21: 2647 (1988).
3 F. H. Zielinski, M. Buzier, C. Lartigue, J. Bastide, and F. Boué,
   Prog. Coll. & Pol. Sci. 104, (1992).
4 M. Daoud, and L. Leibler, Macromolecules, 21: 1497 (1988).

NON-DEBYE SCREENING IN POLYELECTROLYTE SOLUTIONS

Kurt Kremer,[1] Mark J. Stevens,[1,*] Philip A. Pincus[2]

[1]Institut für Festkörperforschung
Forschungszentrum Jülich
D-52425 Jülich
Federal Republic of Germany
[2]Materials and Physics Departments
University of California
Santa Barbara, CA 93106
USA

We dedicate this paper to Shlomo Alexander, who for many years has strongly encouraged simulations on polyionic solutions.

INTRODUCTION

Charged polymers or polyelectrolytes, still remain only poorly understood[1-3] in contrast to the well developed theory of neutral polymer solutions[4]. This is despite their great importance in biology and modern technology. Experimentally macroscopic properties such as the osmotic pressure[2] and the bulk viscosity[5] are well known, while a detailed understanding of the conformational properties of the chains is still lacking. Theoretically the difficulties are mainly related to the long range nature of the Coulomb forces, which up to now can only be handled in a rather approximate manner. Usually the electrostatic pair interactions are treated within the Debye-Hückel approximation. Starting from this ansatz two independent physical pictures were developed by deGennes et al[6] and Odijk, Skolnick and Fixman[7,8]. A more detailed discussion of the success and failure of these approachs will be given elsewhere[9]. Here in the present short note we confine ourselves to some striking effects of the charge fluctuations on

---

*Present Address: Corporate Research Science Laboratories, Exxon Research and Engineering Co., Annandale, NJ 08801, USA.

Soft Order in Physical Systems, Edited by Y. Rabin and R. Bruinsma, Plenum Press, New York, 1994

the conformational properties of the chains by actually taking the nonbonded charges explicitly into account. Since this is an extremely difficult theoretical problem, we carry out computer simulations. In the next section we first shortly describe the simulation model and the parameters. Then we present some results for the dilute regime and discuss the influence on the chain overlap concentration, ρ^*. Finally a short and rather preliminary attempt is made to account for the effect of the discrete charges using a Flory type argument.

MODEL AND METHOD

We employ the freely-jointed bead-spring model of a polymer as it has been successfully used for simulations of neutral polymers[10,11]. Each polymer chain consists of N_b monomers of mass m connected by a nonlinear bond potential. The bond potential is given by

$$U_{bond}(r) = -\tfrac{1}{2}kR_0^2 \ln(1 - r^2/R_0^2),$$ (1)

with $k = 7\epsilon/\sigma^2$ used for the spring constant, and with $R_0 = 2\sigma$ used for the maximum extent[11]. Here, as throughout this paper Lennard-Jones units are used. Excluded volume between the monomers is included via a repulsive Lennard-Jones (RLJ) potential with the cutoff, r_c, at $2^{1/6}\sigma$:

$$U_{LJ}(r) = \begin{cases} 4\epsilon \left[\left(\frac{\sigma}{r}\right)^{12} - \left(\frac{\sigma}{r}\right)^6 - \left(\frac{\sigma}{r_c}\right)^{12} + \left(\frac{\sigma}{r_c}\right)^6 \right]; & r \le r_c \\ 0; & r > r_c \end{cases}$$ (2)

The counterions are given a repulsive core by using the same RLJ potential. The polyelectrolyte simulations were done with 16, 32 and 64 bead chains, and in some cases we extended the runs to $N_b = 128$ beads to test chain length dependence. The number of chains was either 8 or 16. The average bond length, $\langle b \rangle$, is 1.1σ. Since the system is neutral and no salt is added, the number of counterions equals the number of monomers. The largest systems contained 2048 charged particles.

Coulomb interactions pose special difficulties for simulations as well as theory[12] and are the limiting factor for the present investigation. The long range interaction requires an Ewald summation to include interactions of the periodic images. In our case we used the Wigner-Setz cell of a bcc crystal as simulation cell. This allowed us to evaluate the Coulomb interactions within a spherical approximation to the full Ewald sum as given by Adams and Dubey[12]. This approximation is better by at least an order of magnitude for the energy calculation than the bare minimum image evaluation which is the best used to date for polyelectrolytes[13]. Our approach is feasable as long as the Coulomb interactions are not much stronger than thermal energies. This holds for the present systems. The Coulomb strength is parameterized by the Bjerrum length, $\lambda_B = e^2/\varepsilon k_B T$ which is here taken to be 1σ, which corresponds to many experimental and theoretical studies[1,3,6]. This is to be compared to the average bond length $l = 1.1\sigma$. For this value of λ_B the Coulomb pair interactions are generally less than $k_B T$.

The simulations were done at constant temperature, $T = 1.2\epsilon$, using the Langevin thermostat with damping constant $\Gamma = 1\tau^{-1}$, and timestep 0.015τ, where $\tau = \sigma(m/\epsilon)^{1/2}$[10]. The length of the simulation is such that the chains move at least 10 times their contour length.

VERY DILUTE REGIME

Compared to neutral chains, polyelectrolytes are strongly stretched. Most theories predict an almost rodlike behavior at very dilute concentrations[6,14,15,8,16-19]. Charges

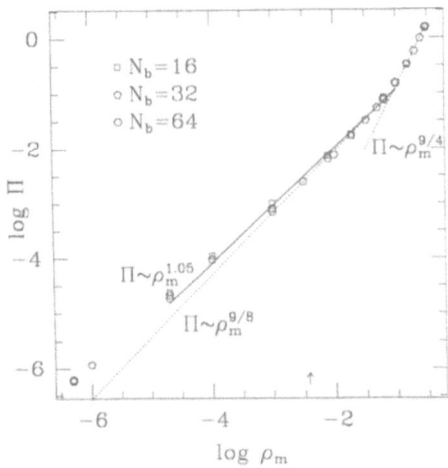

FIG. 1. Ratio $\langle R^2 \rangle / \langle R_G^2 \rangle$ versus monomer density ρ_m for chain length $N = 16, 32, 64$. The included lines are guides to the eye.

are treated as being continously smeared out, not allowing for fluctuations in the charge density, which might arise from their discrete nature. In agreement with recent experiments[20,21] we find, that the chains are not at all rodlike. As a function of density the ratio $r = \langle R^2 \rangle / \langle R_G^2 \rangle$ of the mean squared end to end distance and the mean squared radius of gyration remains below $r = 12$, the rigid rod limit. The asymptotic limit for r is only approached by very long chains, which are far beyond the simulation capabilities and also beyond current experimental possibilities.

There are different effects, which cause the chains to fluctuate locally and thus result in a general reduction of R and r. Entropic effects[9] cause small scale kinks along the chain. This effect is independent of the way the counterions are treated. Here we are looking for the phenomena resulting from the discrete nature of the charges, which is beyond any theoretical model to date. We found that r starts to decrease from the $\rho = 0$ asymptotic plateau value around $\rho_m \simeq 1 \cdot 10^{-4} \sigma^{-3}$, $2 \cdot 10^{-5} \sigma^{-3}$ and $1 \cdot 10^{-6} \sigma^{-3}$ for $N_b = 16$, 32, and 64 respectively. These densities are much too low compared to any of the available theoretical models. One reasonable value of the saturation density, $\rho_m^{(s)}$, is the density at which there is one counterion per polymer volume. For densities below this value, the counterion screening of the intrachain interactions is negligible and one expects r to be constant[9]. It is important to note, that these densities are not at all related to the classical chain overlap concentration ρ_m^*, which are much higher, or the concentration c_1 in Odijk's theory[7], where the persistence length equals the contour length.

Fig. 1 shows the ratio r for 3 different chain lengths. It should be noted that the plateau value for r even for the largest chains considered stays well below the limiting rod value 12. First let us look at the conformations of chains at densities slightly above the onset of the plateau regime. Fig. 2 shows some characteristic conformations of chains of length $N = 32$ at $\rho_m = 10^{-4} \sigma^{-3}$. The conformations are shown in a way that the longest axis of inertia lies horizontally while the second largest lies vertically. For lower densities only small changes occur, as to be expected from Fig. 1. In the lower left corner of Fig. 2 one characteristic conformation at a much higher density of $\rho_m = 0.3 \sigma^{-3}$ in the dense solution regime is shown. At that density the chain conformations approach the uncharged limit. The eight sample conformations should be compared to the rigid or weakly bending rod limit on one side and to the coiled

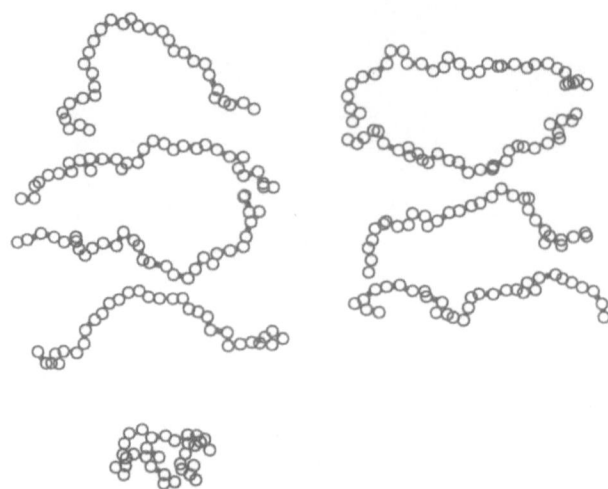

FIG. 2. Typical conformations for eight chains of length $N = 32$ at density $\rho_m = 10^{-4}\sigma^{-3}$. The 9th conformation in the lower left corner gives a chain of $N = 32$ at a density $\rho_m = 0.3\sigma^{-3}$, the dense solution limit.

regime on the other side. It is obvious that they are very stretched compared to the dense limit, however they are certainly also very different than the bending rod limit. The small length scale kink fluctuations can be rationalized by realizing that their entropy and their kink energy are of the same order of magnitude. This implies that there will be a finite density of kinks. This however only can account for the "accordion modes" along a "weakly bending rod". The stronger large scale shape modifications arise from the fluctuations in the local charge density of the counterions. As one can see from a simple spherical solution of the Poisson Boltzmann equation, there is an increased charge density near the center of gravity of the chains. This however is not sufficient to produce the observed effect, especially if one takes into account, that there is no significant change in the conformations at even lower densities than the crossover density, as observed in Fig. 1. Fig. 3 shows the qualitative origin of the occurrence of rather strongly bent or compressed conformations. Again we plot two conformations (from Fig. 2) however explicitly including the the counterions which are located within the volume of the chain. In all cases we observe these strong shape fluctuations to be directly related to fluctuations in the local counterion charge density. It should be emphasized again that because of the smeared out charge density any Debye-Hückel picture cannot account for this effect.

The effect of such charge fluctuations can be discussed in a rather qualitative and preliminary fashion by estimating the end-to-end distance within a simple Flory picture. Here we just give the argument in a very idealized way. We leave out both, numerical prefactors and logarithmic corrections. The effect of the increase of the local charge density is most easily taken into account by considering chains with simply a reduced number of charges. Disregarding any log corrections and additional prefactors we may write for the free energy $F/k_B T$ of a chain

$$\frac{F}{k_B T} = \frac{R^2}{N} + \frac{Z^2}{R} \tag{3}$$

Z being the charge of the chain. Minimizing the free energy trivially yields $R^3 = Z^2 N$. Introducing the effective charge $Z = fN$ by assuming that each counterion within the volume of the chain reduces the charge by one unit. yields

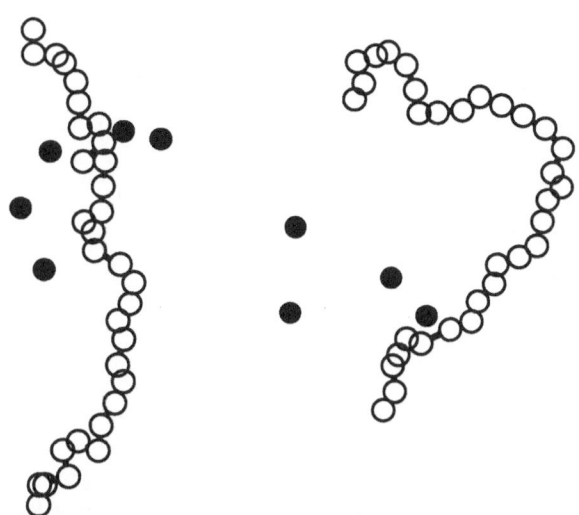

FIG. 3. Two conformations from Fig. 2 but now with the counterions, which are located within the volume of the chain included. Note that the average counterion density $\rho_m{}^{-3}$ is much smaller than the local density observed here.

$$R = f^{2/3} N \qquad (4)$$

with $0 \leq f \leq 1$. This is to be related to the number of counterions n_c within the chain volume. Using $n_c = \rho_m R^3$ one gets $n_c = \rho_m f^2 N^3 = N(f-1)$. Debye screening should become relevant at the density where the Debye length κ^{-1} equals R. With $\kappa^{-2} = 4\pi \lambda_B \rho_m$, which we may use since we are just in the weakly interacting regime with $\kappa^{-1}/L \approx 0.9$. With the above arguments this yields $\kappa R = 1$ at $\rho_m \approx 0.3 N^{-2}$. This is surprisingly close to the simulation results for all considered chain lengths and far above the onset of decay in r. This supports the fluctuation picture put forward above.

DILUTE-SEMIDILUTE TRANSITION

As can be suspected from the above line of arguments the chains will be significantly contracted before they actually start to overlap. We were never able to find a regime, where the observed persistence length is of the order of the strand strand distance. The above shown line of arguments actually would lead to individually collapsed chains, since ρ_m^* automatically means $f = 0$. This would indicate that stretched polyelectrolytes only exist in the dilute phase. This of course cannot hold for the present good solvent systems, however there are indications for this in cases where a poor solvent is present[22]. From the discussion of the previous section we can then expect a significant shift in ρ_m^*. The classical picture which views the chains as rodlike would estimate $\rho_m^* \approx 0.6 N 6/\pi L^3$, L being the contour length of the chains and the factor 0.6 approximately accounts for the random packing of the charged spheres. For the present systems this would yield $\rho_m^* = 0.0041\sigma^{-3}, 0.0009\sigma^{-3}, 0.00022\sigma^{-3}$ for $N = 16, 32, 64$. To compare this, Fig. 4 estimates the crossover density with the same equation, however using the actually measured R values. This gives $\rho_m^* \approx 0.20\sigma^{-3}, 0.05\sigma^{-3}, 0.017\sigma^{-3}$. While there are not enough different chain lengths in order to decide on a power law there is a typical factor of about 200 between the two sets of data. This is also supported by the calculation of the density dependence of the first peak in the chain-chain structure function $S(q)$.

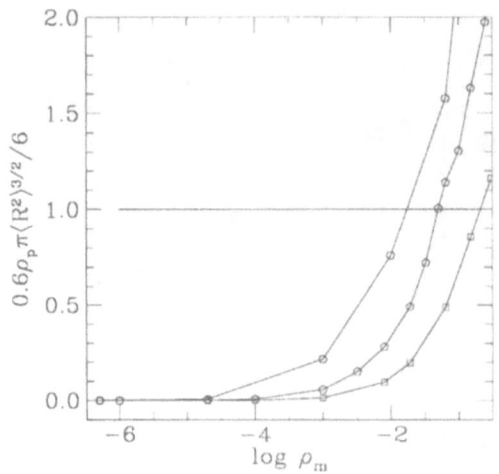

FIG. 4. Plot of $0.6\rho_m$ times the inverse chain volume. $V = \pi R^3/6$ versus density to estimate directly the chain overlapp concentaration.

CONCLUSION

This short summary of recent computer simulation results on the conformational properties of polyelectrolyte solutions showed that the classical pictures of screening and conformation changes for charged polymers need to be revised. At low concentrations the density fluctuations of the discrete counterions cause significant shape fluctuations, which go far beyond the short kink excitation driven by entropy. For increasing density with consequently increasing ionic strength, the chains dramatically shrink, before the classical overlapp concentration ρ_m^* is reached.

ACKNOWLEDGEMENT

We would like to acknowledge a large grant of computer time from the Höchstleistungsrechenzentrum, Germany within the Disordered Polymers Project. A NATO Travel Grant is greatfully acknowledged.

REFERENCES

[1] Hayter, J.; Janninck, G.; Brochard-Wyart, F.; and de Gennes, P.: *J. Physique Lett.* **1980**, *41*, 451.

[2] Wang, L.; and Bloomfield, V.; *Macromolecules* **1990**, *23*, 804.

[3] Witten, T.; and Pincus, P.; *Europhys. Lett.* **1987**, *3*, 315.

[4] de Gennes, P.; *Scaling Concepts in Polymer Physics*: Cornell University Press: Ithaca, NY, 1979.

[5] Kim, M.; and Pfeiffer, D.; *Euro. Phys. Lett.* **1988**, *5*, 321.

[6] de Gennes, P.; Pincus, P.; and Velasco, R.; *J. Physique* **1976**, *37*. 1461.

[7] Odijk, T.; *Macromolecules* **1979**, *12*, 688.

[8] Skolnick, J.; and Fixman, M.; *Macromolecules* **1977**, *10*, 944.

[9] Stevens, M.; and Kremer, K.; (unpublished).

[10] Kremer, K.; and Grest, G.; *J. Chem. Phys.* **1990**, *92*, 5057.

[11] Dünweg, B.; and Kremer, K.; *Phys. Rev. Lett* **1982**, *66*, 2996.

[12] Adams, D.; and Dubey, G.; *J. Comp. Phys.* **1987**, *72*. 156.

[13] Christos G. A.; Carnie, S. L. *J. Chem. Phys.* **1989**, *91*, 439. Reed, C.; Reed, W *J. Chem. Phys.* **1990**, *92*, 6916. Brender, C; Danino, M *J. Chem. Phys.* **1992**, *97*, 2119. Barrat, J. L.; Boyer, D. *J. Phys. II France* **1993**, *3*, 343. See also references therein.

[14] Kuhn, W.; Kunzle, D.; and Katchalsky, A.; *Helv. Chim. Acta* **1948**, *31*, 1994.

[15] Hermans, J.; and Overbeek, J.; *Recents Travaux Chimiques* **1968**. *67*, 761.

[16] Odijk, T.; *J. Polym. Scil, Polym. Phys. Ed.* **1977**, *15*, 477.

[17] Odijk, T.; *Polymer* **1978**, *19*, 989.

[18] Fixman, M.; *J. Chem. Phys.* **1982**, *76*, 6346.

[19] Bret, M. L.; *J. Chem. Phys.* **1982**, *76*, 6242.

[20] Degiorgio, V.; Mantegazza, F.; and Piazza, R.; *Europhys. Lett.* **1991**, *15*, 75.

[21] Schmidt, M.; *Macromolecules* **1991**, *24*, 5361.

[22] Krämer, U.; and Hoffmann, H.; *Macromolecules* **1991**, *24*, 256.

ENTROPY OF KNOTS AND
STATISTICS OF ENTANGLED RANDOM WALKS

Sergei Nechaev[1]†, Alexander Grosberg[2]‡

[1]Lab. Physique Mathématique Université Montpellier II
Pl. E.Bataillon, Case 50, 34095 Montpellier Cedex 05, France

[2]Department of Physics Massachusetts Instiute of Technology
Cambridge, MA 02139, USA

FORMULATION OF THE MAIN PROBLEM

Our main aim concerns the determination of the entropy of the knot embedded in 3D space (or, in other words, the determination of available volume in the phase space for the path with fixed topological state).

Let us characterize the knot by the topological invariant, G_0, which depends only on the topological state of the knot and does not depend on the shape of it.

Entropy, S, of some knot, G_0, equals to

$$S(G_0) = \ln Z(G_0)$$

and the partition, $Z(G_0)$, function reads

$$Z(G_0) = \int \delta\left(G\{\Gamma\} - G_0\right) D\{\Gamma\} \tag{1}$$

where the path integral (1) means that the summation over all chain conformations Γ and δ-function should be understood as an indicator: it vanishes for conformation Γ, if the topology of Γ, $G\{\Gamma\}$, differs from G_0, and does not depend on Γ otherwise, i.e. it cuts off the conformations with given topology. Thus, our problem deals with the calculation of the set of trajectories with given fixed topology.

TOPOLOGICAL INVARIANTS AND LIMIT BEHAVIOR
OF THE RANDOM WALK IN THE LATTICE OF OBSTACLES

We consider the random walk on the infinite plane with a regular lattice of removed points. The elementary cell of the lattice has the form of equalsided triangle. These points play the role of topological constraints. Let us put the following question [1]: What is the probability, $W_0(N)$ to find the random walk of length N to be unentangled (i.e. contractable to the point) in the lattice of obstacles?

Soft Order in Physical Systems, Edited by Y. Rabin
and R. Bruinsma, Plenum Press, New York, 1994

CONSTRUCTION OF TOPOLOGICAL INVARIANTS

1. Using the conformal approach for this model [1] we found the so-called covering space, $w(z)$, for the plane with the lattice of obstacles. This space is free of obstacles in any internal domain and contains the obstacles only at the boundary. This covering space coincides with the so-called modular group [1, 2] and can be realized in the upper half-plane $\mathrm{Im}\, w > 0$ of the plane w as a discrete subgroup of group of motion of Lobachevsky plane defined via relation:

$$w \to \frac{aw + b}{cw + d}, \quad ad - bc = 1$$

Thus we can reduce our problem to the investigation of the random walk on the Lobachevsky plane, where the non-Euclidean distance between ends is the topological invariant.

More formally we have to solve the diffusion equation

$$\frac{\partial}{\partial N} W(w, N) = \frac{a^2}{4} \frac{1}{|z'(w)|^2} \Delta W(w, N)$$

where $z'(w) \equiv dz/dw$ and $z(w)$ is the conformal transformation to the covering space mentioned above.

2. The topological structure of the modular group coincides with the Cayley tree – free group with two generators. Therefore we can consider in the discrete case the random walk on the Cayley tree, where the distance from the origin in the topological invariant.

Both approaches (1) and (2) give the similar asymptotics for the probability to find the N-step closed random walk to be unentangled in the lattice of obstacles on the plane:

$$W_0(N) \sim \frac{1}{N} e^{-N/N_0}$$

ALGEBRAIC INVARIANTS AND ENTROPY OF LATTICE KNOTS

We consider $2D$ square lattice and suppose that it represents the projection to the plane of the knot embedded in $3D$-space. The crossings on projection are lattice vertices, their total number is N. We distinguish two kinds of crossings, b_i, corresponding to different configurations of knot: a) $b_i = +1$ $-|-$ and b) $b_i = -1$ $\overline{}|\overline{}$.

There are 2^N different realizations of the lattice. The natural physical question here is [3]: What is the part $P_0(N)$ of unknotted (i.e. topologically isomorphic to trivial ring) paths on the lattice among 2^N possible ones?

FIXED TOPOLOGICAL STATE OF THE KNOT CAN BE REGARDED AS A QUENCHED DISORDER FOR ORDINARY VERTEX (IN PARTICULAR, POTTS) MODEL

The fixation of the knot topology means freezing of all spins b_i, i.e. quenching of some disorder pattern of these spins on the lattice.

For the identification of the topological state of the knot we use the Kauffman algebraic invariant $K(A)$, which is the Laurent polynomial in A variable. We have shown the Kauffman invariant to be equal to the partition function of some special **disordered Potts model** [3, 4]. The number of equivalent states, q, and the nearest neighbor interaction constant, J_{kl}, are defined as follows:

$$q = \left(A^2 + A^{-2}\right)^2, \qquad J_{kl} = 4b_i \ln A$$

The Potts spins are defined in the centers of plaquettes; k and l are the numbers of plaquettes and i is the vertex lying between k and l.

METHODS OF SOLUTION AND RESULTS

We consider the variables b_i to be independent and randomly distributed with the the density

$$Q\{b_i\} = \frac{1}{2}\left[\delta(b_i + 1) + \delta(b_i - 1)\right]$$

Thus the path integration in (1) is replaced by the simple averaging of $\delta(\ldots)$ over te set of discrete independent variables b_i.

We have constructed the upper estimation for the probability P_0 to find the trivial knot. Since Kauffman invariant is not complete some nontrivial knots with $K(A) = 1$ may exist ($K(A) \equiv 1$ for the trivial knot). Then P_0 is less than the probability to find a knot with $K(A) = 1$ **for all** A-values; which, in turn, is less than the probability to find a knot with invariant $K(A)$ being equal to 1 **for some fixed** value $A = A^*$:

$$P_0(N) < \int \delta\left(\ln K_{\{b\}}^2(A^*)\right) Q\{b\} D\{b\} \qquad (2)$$

Using the integral representation of usual scalar δ-function we can rewrite (2) as follows

$$P_0(N) < \frac{1}{\pi} \int_{-\infty}^{+\infty} \left\langle \left[K_{\{b\}}(A^*)\right]^{2iy} \right\rangle dy \qquad (3)$$

where $< \ldots >$ means the averaging of the type $\int \ldots Q\{b\}D\{b\}$. Thus our problem is reduced to the averaging of moments of the partition function over the realization of lattice disorder – the well known problem in the theory of spin glasses and other systems with **quenched** disorder. As usual such an averaging can be performed for an integer exponent like $\langle Z^n \rangle$ and then analytically continued to non-integer complex values $n = iy$.

The Potts variables q and J are the functions of A. Thus, changing the A-value we are moving along the certain curve on (q, J)-plane of Potts variables. We have chosen A^* at this curve as crossing point with the line of the phase transition in the Potts model from the paramagnet to the spin-glass state.

The Kauffman invariant in terms of Potts partition function reads

$$\left\langle K^{2n} \right\rangle = [2\cosh(2\beta)]^{-2(N+1)} \sum_{\vec{\sigma}} \prod_{kl} \exp\left\{\ln\cosh\left[\beta \sum_{\alpha=1}^{2n} (4\delta(\vec{\sigma}_k^\alpha, \vec{\sigma}_l^\alpha) - 1)\right]\right\} \qquad (4)$$

where $\beta = \ln A - i\pi/2 > 0$ (A is complex).

The calculation of (3) in the thermodynamic limit near the spin-glass transition point (at $\beta = \beta^* \approx 0.35$) gives the answer for the part of unknotted paths [3]

$$P_0(N) < \exp(-N/N_0); \qquad N_0 \approx 2.6$$

where N is the number of lattice vertices, i.e. the number of crossings on the knot 2D projection.

CONCLUISION

The connection between problem of determination of the knot entropy and statistics of entangled random walks is schematically shown in the Table 1. We argue that both these topological questions could be considered from one general point of view of random walk on noncommutative groups. (Some preliminary remarks concerning this connection one can find in [2].)

Table 1.

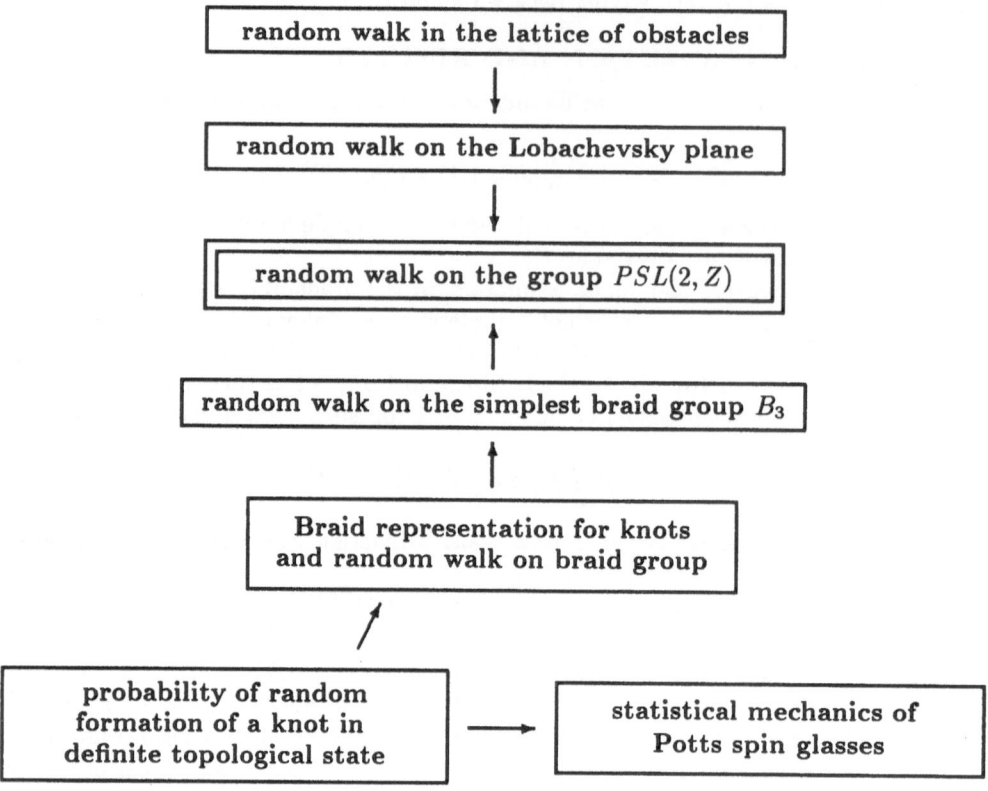

permanent addresses:
† Landau Institute for Theoretical Physics, 117940, Moscow, Russia
‡ Institute of Chemical Physics, 117977, Moscow, Russia

REFERENCES

[1] S.K.Nechaev, "Topological properties of 2D polymer chain in the lattice of obstacles", *J.Phys.A: Math. Gen.*, **21**:3659 (1988).

[2] S.K.Nechaev, Ya.G.Sinai, "Limiting-type theorem for conditional distributions of products of independent unimodular 2×2 matrices", *Bol. Soc. Bras. Math.*, **21**:121 (1991).

[3] A.Grosberg, S.Nechaev "Algebraic invariants of knots and disordered Potts model", *J.Phys.A.: Math. Gen.*, **25**:4659 (1992).

[4] A.Grosberg, S.Nechaev "Averaged Kauffman invariant and quasi-knot concept for linear polymers", *Europhys. Lett.* **20**:613 (1992).

POLYMERS IN A RANDOM ENVIRONMENT AND MOLECULAR QUASI-SPECIES

Luca Peliti[*]

Institut Curie – Section de Physique et Chimie
Laboratoire Curie – 11, rue Pierre et Marie Curie
F-75231 Paris Cedex 05 (France)

THE QUASI-SPECIES MODEL

The quasi-species model was introduced in 1971 by Manfred Eigen[1] to describe evolving populations of self-reproducing (RNA-like) molecules. It lies at the basis of the theory of the origin of biological organization, and in particular of the hypercycle theory, developed by Eigen and P. Schuster[2]. It may be cast in the following form. Consider a population of self-reproducing molecules, whose structure may be described by a collection of N binary variables, $S_i = \pm 1$, $i = 1, 2, ..., N$. Then the fraction $x_{\mathbf{S}}$ of molecules of structure $\mathbf{S} = (S_1, S_2, ..., S_N)$ obeys the following evolution equation:

$$x_{\mathbf{S}}(t+1) = \frac{1}{Z(t)} \sum_{\mathbf{S}'} Q_{\mathbf{SS}'} A(\mathbf{S}') x_{\mathbf{S}'}(t) \tag{1}$$

where $A(\mathbf{S})$ is the average number of offsprings that a molecule of structure $\mathbf{S}$ produces at the next generation (if one assumes infinite environmental carrying capacity) and $Q_{\mathbf{SS}'}$ is the conditional probability that the reproduction of a molecule of structure $\mathbf{S}'$ effectively produces a molecule of structure $\mathbf{S}$, and therefore represents the effects of mutations. A convenient expression for the matrix Q is given by

$$Q_{\mathbf{SS}'} = q^{d(\mathbf{S},\mathbf{S}')}(1-q)^{N-d(\mathbf{S},\mathbf{S}')}, \tag{2}$$

---

[*] Boursier Henri de Rothschild. Permanent address: Dipartimento di Scienze Fisiche and Unità INFM, Università "Federico II", Mostra d'Oltremare, Pad. 19, I-80125 Napoli (Italy). Associato INFN, Sezione di Napoli.

where $0 < q < 1$ is the probability of having one mutation per unit and per generation, and

$$d(\mathbf{S},\mathbf{S}') = \frac{1}{2}\sum_{i=1}^{N}\left(1 - S_i S_i'\right) \tag{3}$$

is the number of different units in the structures $\mathbf{S}$ and $\mathbf{S}'$ respectively. The factor $Z(t) = \Sigma_{\mathbf{S}}\,A(\mathbf{S})\,x_{\mathbf{S}}$ ensures the normalization of $x_{\mathbf{S}}(t)$ at any generation. In order to derive equation (1) one assumes that "generations" of the self-reproducing molecules are non-overlapping and that the number of molecules in the population is sufficiently large to neglect fluctuations in the $x_{\mathbf{S}}(t)$.

It was shown by I. Leuthäusser[3] that the quasi-species model corresponds to a problem in *equilibrium* statistical mechanics. Indeed, the matrix

$$T_{\mathbf{SS}'} = Q_{\mathbf{SS}'}A(\mathbf{S}') \tag{4}$$

can be considered as the transfer matrix for an Ising-like system of Hamiltonian

$$-\beta H[\mathbf{S}(t)] = \sum_{t}\sum_{i=1}^{N}\beta\, S_i(t)S_i(t+1) + \sum_{t}\ln\!\left[A(\mathbf{S}(t))\right], \tag{5}$$

where the "inverse temperature" β is given by

$$\beta = -\tfrac{1}{2}\ln\!\left[q/(1-q)\right], \tag{6}$$

and an irrelevant additive constant has been understood. It is important to keep in mind that the quantities one naturally computes within this approach are the *unrenormalized* fractions $y_{\mathbf{S}}(t)$, which satisfy the evolution equation

$$y_{\mathbf{S}}(t+1) = \sum_{\mathbf{S}'}T_{\mathbf{SS}'}\,y_{\mathbf{S}'}(t). \tag{7}$$

The relation between the y's and the x's is obviously given by

$$x_{\mathbf{S}}(t) = y_{\mathbf{S}}(t)\Big/\sum_{\mathbf{S}'}y_{\mathbf{S}'}(t). \tag{8}$$

This analogy was recently exploited by P. Tarazona[4] to solve the quasi-species model for several choices of the "fitness landscape" $A(\mathbf{S})$. Since the properties which affect the reproducing efficiency of RNA molecules in experiments[5] are rather complex[6], it is costumary to represent $A(\mathbf{S})$ by a random function, such as a spin-glass Hamiltonian. The simplest choice is to assume that $A(\mathbf{S})$ takes on independent, identically distributed, random values for each different "genotype" $\mathbf{S}$. This corresponds to identify the "fitness landscape" with the Random Energy Model of spin-glasses introduced by B. Derrida[7].

POLYMERS IN A RANDOM ENVIRONMENT

We now consider the explicit expression of $y_S(T)$ arising from equations (4),(5) and (7):

$$y_S(T) = \sum_{S(T-1)} \sum_{S(T-2)} \cdots \sum_{S(1)} \exp\left\{\sum_{t=0}^{T-1}\left[\sum_{i=1}^{N} \beta S_i(t)S_i(t+1) + \ln[A(S(t))]\right]\right\}. \tag{9}$$

We have implicitly assumed that $S(0) = S_0$ (a given genotype) and $S(t) = S$. Equation (9) represents the partition function of a polymer, whose monomers are identified by the time label t and are placed on a site of the N-dimensional hypercube $\{S\}$. A random, time-independent potential $V(S) = -\beta^{-1}\ln[A(S)]$ is assigned to each site of the hypercube. No excluded-volume interaction is assumed. This model may be approached by replica techniques[8]. However, one may consider the simplified problem in which β is sufficiently large to allow for only one "mutation" (at most) per generation, and we identify the "genotypes" S with the integers (and S_0 with the origin). In this case, a polymer of length T can explore all sites S with $|S| \leq T$, and will tend to stay on the site corresponding to the minimum value V^* of V. If V is a Gaussian variable, this value increases like $(\ln T)^{1/2}$ (in modulus) and therefore dominates more and more the sum (9) in the limit $T \to \infty$. As a consequence, in this limit, the extension of the polymer is proportional to T and a finite fraction of it is collapsed on the site corresponding to the minimum value of V. The situation becomes even worse if we model the "genotype space" $\{S\}$ by a Cayley tree, since in this case the number of explored genotypes increases exponentially with T, and the minimum value becomes accordingly larger in modulus.

These conclusions are at variance with simulations of finite evolving populations[6,9] and with the reasonable expectation that the population will tend to dwell on an adaptation optimum for a comparatively long time, "hopping" to a better optimum only when a better adapted mutant arises. The origin of this disagreement may be traced on the neglect of fluctuations in the x's, i.e., in the hypothesis of infinite population size. However, it is possible to use the quasi-species approach to yield information on this "hopping" process. Let us consider the weight of a path linking a relative minimum S_1 of V to a deeper minimum S_2:

$$W(S_1, S_2; T) = \sum_{\{S(t)\}} \delta_{S(0),S_1} \delta_{S(T),S_2} \exp[-\beta H[S(t)]]. \tag{10}$$

This weight must be compared with the one of a path staying at S_1 for T generations. The ratio of the two weights is initially exponentially small in the distance between S_1 and S_2, but increases exponentially with T, due to the larger value of $|V(S_2)|$. We can thus identify the "hopping" time as the time in which the ratio of the two weights becomes of the order of the inverse population size, so that the probability that at least one S_2 mutant has arisen becomes finite. This time decreases, albeit slowly, with population size, and vanishes in the infinite population limit. The study of finite population corrections is therefore primordial in the interpretation of results of the quasi-species model.

ACKNOWLEDGMENTS

It is a pleasure to thank M. Mézard for illuminating discussions. I also thank G. Weisbuch and M. Sellitto for helpful suggestions, and Y. Kantor and B. Fourcade for their interest in this work.

REFERENCES

1. M. Eigen, *Naturwissenschaften* **58**, 465 (1971).
2. M. Eigen, P. Schuster, *The Hypercycle: A Principle of Natural Self-Organization* (Berlin: Springer-Verlag, 1979).
3. I. Leuthäusser, *J. Stat. Phys.* **48**, 343 (1987).
4. P. Tarazona, *Phys. Rev.* Phys. Rev. **A45**, 6038 (1992).
5. C. K. Biebricher, *Chem. Scr.* **26B**, 51 (1986).
6. W. Fontana, W. Schnabl, P. Schuster, *Phys. Rev.* **A40**, 3301 (1989).
7. B. Derrida, *Phys. Rev.* **B24**, 2613 (1981).
8. M. Sellitto, Thesis, University of Naples "Federico II" (1993).
9. G. Weisbuch, *C. R. A. S. Paris*, Série III, **298**, 375 (1984).

WETTING FROM MIXTURES OF FLEXIBLE CHAINS

U. Steiner[1], J. Klein[1], E. Eiser[1], A. Budkowski[1,2] and L.Fetters[3]

[1]Dept. of Materials and Interfaces, Weizmann Institute, Rehovot, Israel
[2]on leave from Jagellonian University, Cracow, Poland
[3]Exxon Research and Engineering Corporation, Annandale, NJ 08801, USA

1. INTRODUCTION

The behavior of liquids on surfaces is a matter of everyday experience. The physics of this behavior is determined by the strength and nature of the surface interactions. These act only over microscopic distances, but determine the formation of common macroscopic features such as drops or thin films on a surface.

The shape of drops (in equilibrium with their vapor) on a surface was first described by Young[1], relating the macroscopic dihedral contact angle θ to the balance of surface tensions γ:

$$\cos(\theta) = (\gamma_{vs} - \gamma_{ls})/\gamma_{vl} \qquad (1)$$

with γ_{vs} the vapor-solid, γ_{ls} the liquid-solid and γ_{vl} the vapor-liquid interfacial tensions respectively. The difference of γ_{vs}-γ_{ls} governs the form of the liquid on the surface: if $\gamma_{ls} > \gamma_{vs}$, the liquid phase covers the substrate only partially and the contact angle θ is finite. If γ_{vs}-γ_{ls}=γ_{vl}, θ=0 and the substrate is covered by a continuous wetting film, which may attain macroscopic dimensions. Upon changing the balance of surface tensions i.e. by changing the temperature, a transition from the droplet phase, called partial wetting to the phase featuring a continuous film - complete wetting - is observed.

While the situation of a liquid in coexistence with its vapor phase is related to systems familiar from everyday life, the same considerations apply to a coexisting binary liquid mixture in contact with a third phase (vapor or solid), as first shown in a experiment by Moldover and Cahn[2]. For a binary liquid mixture replace the indices l, v in eq. 1 with α and β, the coexisting phases of the liquid-A/liquid-B mixture. Cahn[13] showed in an elegant argument, that a wetting transition always has to exist as a bulk critical point is approached along the coexistence curve.

While there is an extensive literature on experimental and theoretical aspects of wetting transitions from simple liquids along the gas-liquid as well as the binary coexistence curve[3], experimental studies on wetting from macromolecular mixtures appeared only very recently[4, 5]. Binary polymer mixtures offer distinct advantages in the study of fluid-thermodynamics in general and wetting transitions in particular. The bulk miscibility of two polymers can be 'tuned' almost continuously by changing the molecular weights of the polymers, as well as their chemical composition[6]. Further advantages of using polymers is their relatively large size (radius of gyration), which makes direct observation of wetting profiles possible, and their very low mobilities (due to entanglements, which also suppress convective effects) which allows us to follow the build-up of wetting layers with time. Most significantly, the tiny interfacial energies in phase separated systems even far below the critical point make even relatively small

Soft Order in Physical Systems, Edited by Y. Rabin
and R. Bruinsma, Plenum Press, New York, 1994

differences in surface interactions dominant[5], and one expects wetting to occur more readily for polymer mixtures compared to their small molecule analogs[7].

Here we report the experimental observation of complete wetting from binary olefinic polymer mixtures. We show that wetting can be obtained along two different trajectories in phase space, from the one phase region or by surface directed phase separation as the system is quenched into the two phase region. In the last section, we discuss the dynamics of build-up of wetting layers.

2. EXPERIMENTAL

The polymers used are statistical copolymers of ethyl-ethylene and ethylene monomers, $(A_x B_{1-x})_N$, where the monomers are arranged randomly along the polymer chain. The relative amount x of ethyl-ethylene monomers, $A = -(C_2H_3(C_2H_5))-$, can be varied to any value $0 \leq x \leq 1$. Some of the polymers used in this study were partially deuterated (to an extent e), a prerequisite of our analytic technique as described below. The miscibility of a mixture of such copolymers depends very sensitively on the difference in x, as well as on the difference in deuteration[8]. The materials used are indicated at the beginning of each section. It is noteworthy that mixtures of different x values (and extent of deuteration, e) can exhibit similar critical temperatures. As the material with the higher ethyl-ethylene content (i.e. higher x value) is preferentially adsorbed at the free surface, this allows studies of systems with different effective surface interactions, while bulk miscibility is nearly unchanged.

Samples were prepared by spin coating a film from toluene on a polished silicon wafer. When bilayer samples were required, a second film was spin-coated onto mica and then float-mounted onto the first film. The samples were annealed in vacuum and subsequently quenched and stored at a temperature below the glass transition temperature (of ca. -40°C).

The composition-depth profile $\phi(z)$ of the samples following different annealing times was determined using $^2He(^3He,^4He)^1H$ nuclear reaction analysis (NRA). NRA is described in detail elsewhere[9]. In brief: a 900keV ^3He beam impinges on the polymer sample and the nuclear reaction $^3He+^2H \rightarrow ^1H+^4He+18.35MeV$ takes place. The energy of the outgoing α-particles contains information on the deuterium distribution in the sample: the incident ^3He particles lose energy due to inelastic electronic processes as they penetrate the sample, which results in a reduced energy of the α-particle as compared to a reaction which has taken place at the surface. The α-particles lose additional energy on the way to the detector and the overall energy loss can be related to the depth in which the reaction has taken place.

3. WETTING FROM THE ONE-PHASE REGION

To study wetting at a surface by approaching the coexistence curve from the one-phase region, the concentration of the phase adjacent to the wetting layer has to be increased to reach its coexisting value so that the wetting layer may form. To achieve this, we employ a "self-regulating" geometry: d88 (x=0.88, e=0.37, N=1610) material is placed on the substrate and covered with a h78 (x=0.78, e=0, N=1290) film. Initially, the d88-h78 interface will broaden to an interfacial width comparable to the bulk correlation length and the two materials will partially interdiffuse until the two layers attain the composition of the coexisting phases ϕ_1 and ϕ_2. If the surface interaction to either surfaces is weak, such an experiment can be used to experimentally determine the coexistence curve for the binary

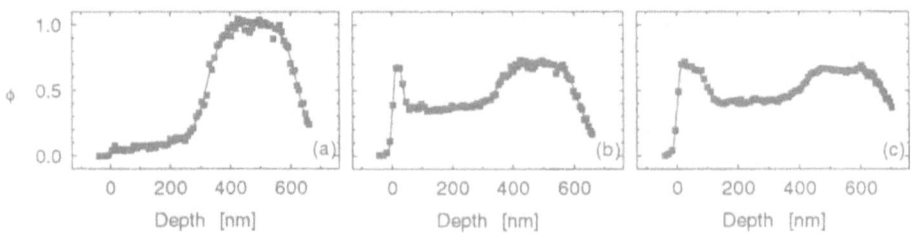

Figure 1. $\phi(z)$ for h78/d88 bilayers following annealing of t=0min (a), t=30min (b) and t=3days (c) at 110°C.

mixture[5,8]. As d88 is preferentially adsorbed at the free surface, a d88 surface peak forms, as soon as there is a finite d88 concentration (ϕ) at the surface. As the d88 in the surface near region approaches its value on the coexistence curve (fig.1(a)), the surface peak grows and for $\phi \approx \phi_1$, the surface layer attains a thickness which is larger than the other length scales in the system (molecule size, correlation length). This behavior is reproduced in fig. 1, which shows the h78/d88 bilayer as prepared, after 30 minutes, and 3 days at 110°C (16°C below the critical point of this mixture $T_c=126°C$). The layer-thickness continues to increase until the entire d88-rich phase is incorporated into the wetting layer, as expected in the case of complete wetting.

4. WETTING FROM THE TWO PHASE REGIME

In another series of experiments thick monolayers of d86(x=0.86, e=0.4, N=1520)/ h75(x=0.75, e=0, N=1267) mixtures with a d86 content varying from 5% to 40% were mounted on Si wafers and annealed. Depth profiles of samples of three initial compositions ϕ_{ini} = 20%, 30% and 40% (of the d86), following annealing for 1 day at 150°C, are shown in Fig 2. For final plateau concentrations $\phi_\infty < \phi_1$, thermodynamic equilibrium is characterized by a peak with maximum d66 concentration $\phi_{max} < \phi_2$ and with a width smaller than the spatial extent of the polymer molecules (as in fig. 2(a)). By analyzing the surface excess (the area under the peak) as a function of ϕ_∞, the surface interaction parameters for this mixture can be obtained[7].

For initial concentrations within the 2-phase region at the annealing temperature, $\phi_{ini} > \phi_1$, the sample will undergo spinodal decomposition. Due to the surface activity of d86, the subsequent growth of the spinodal domains takes place in a surface-directed manner, and for long times the composition profile is characterized by a wetting layer in contact with a phase of composition ϕ_1. Similar experiments in a different binary pair have been reported by Bruder and Brenn[4].

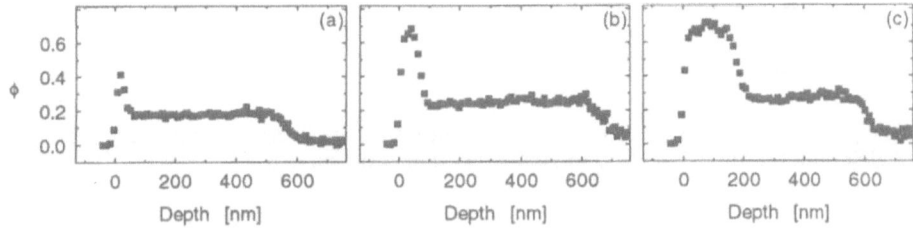

Figure 2. $\phi(z)$ for h75/d86 mixtures of overall d86 composition: (a) 20%, (b) 30% and (c) 40%, after 1 day at 150°C

5. DYNAMICS OF WETTING

In this section, we return to the results described in section 3. As indicated in fig. 1, we are able to follow the build-up of the wetting layer with time on a time scale from 30min to 1 week. The width of the wetting layer, defined by the distance ℓ between the points of inflection of $\phi(z)$ at the air surface and the ϕ_1/ϕ_2 interface, where the wetting layer terminates, is plotted in fig. 3a vs. the logarithm of time. In this representation a straight line appears to best describe our data.

While there is ample theoretical work on the equilibrium properties of wetting, there are few publications on the build-up of wetting layers with time. A recent idea to explain the logarithmic increase of the wetting layer was suggested by Jones[11], following ref.[12]. The following derivation reproduces these results. In this model the assumption is made that the incorporation of the surface active material into the wetting layer is diffusion limited. The sketch in fig. 3b illustrates this assumption: the surface excess of the wetting layer $z^*=A_1$ has to equal the area of the diffusion well A_2 from which the surface active component is incorporated into the layer.

The surface excess z^* can also be expressed in terms of the thickness of the wetting layer $\ell \sim z^*$. The functional dependence of ℓ on ϕ_d, the concentration in the depleted zone adjacent to the growing wetting layer, may be obtained if one assumes the latter to be in (instantaneous) equilibrium with ϕ_d. It has been calculated by Cahn[13] (and by others, e.g refs.[7,12]), for the case of $\phi_d \approx \phi_1$ ($\phi_d < \phi_1$), as $\ell = -const. \log(1-\phi_d/\phi_1)$. If we approximate the depleted area by a square well of width $(Dt)^{1/2}$, then $z^* = (\phi_1 - \phi_d)(Dt)^{1/2}$. Eliminating ϕ_d we find

$$\ell(t) \sim \log(t) \qquad (2)$$

While this result is in agreement with our experimental result, it does not accurately describe our experimental system. This is because the value of the diffusion coefficient D is high[10] so that the spatial extent of the depleted well $(Dt)^{1/2}$ would be expected to exceed - even for short times - the width of the available layer of composition ϕ_1 in our experiments (fig. 1). An accurate description of the build-up with time of the wetting layer for our system has rather to take into account the finite dimensions of the overall film thickness, and the mechanism by which material flows from the "reservoir" (this is the phase of composition ϕ_2 adjacent to the silicon substrate) into the wetting layer.

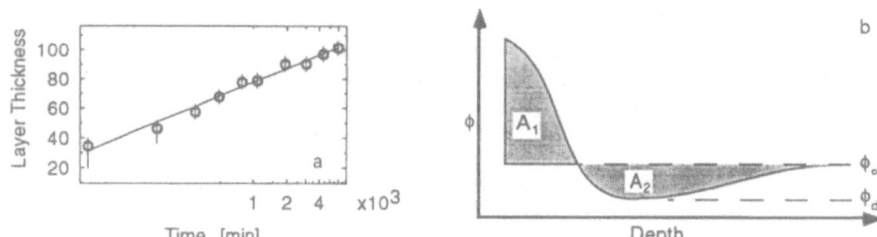

Figure 3. (a) variation of layer thickness with time for the profiles of fig. 1. (b) schematic illustration of diffusion limited growth of a wetting layer.

ACKNOWLEDGMENTS

We thank K. Binder, R. Jones and R. Lipowsky for useful discussions, and the US-Israel BSF and German-Israel Foundation for financial support.

REFERENCES

1. T. Young. *Phil. Trans. R. Soc. London* 95:65 (1805).
2. M. Moldover and J.W. Cahn. *Science* 207:1073 (1980).
3. S. Dietrich, in "Phase Transitions and Critical Phenomena", C. Domb and J. Lebowitz, eds., Academic Press, London (1988).
4. F. Bruder and R. Brenn. *Phys. Rev. Lett.* 69:624 (1992).
5. U. Steiner, J. Klein, E. Eiser, A. Budkowski andL.J. Fetters. *Science* 258:1126 (1992).
6. P.G. de Gennes. "Scaling Concepts in Polymer Physics", Cornell University Press, London (1979).
7. I. Schmidt and K. Binder. *J. Physique* 46:1631 (1985).
8. W.W. Graessley, R. Krishnamoorti, N.P. Balsara, L.J. Fetters, D.J. Lohse, D.N. Schulz andJ.A. Sissano. *Macromolecules* 26:1137 (1193); E.Eiser, U. Steiner, A. Budkowski, J. Klein, L.J. Fetters, R. Krishnamoorti, *ACS Polymer Preprints*, August 1993 - in press.
9. U.K. Chaturvedi, U. Steiner, O. Zak, G. Krausch, G. Schatz andJ. Klein. *Appl. Phys. Lett.* 56:1228 (1990).
10. A. Losch, R. Salomonovic, U.Steiner and J. Klein, unpublished data.
11. R.A.L. Jones, personal communication. See also R.A.L. Jones and E.J. Kramer, *Phil. Mag. B* 62:129 (1990)
12. R. Lipowsky and D.A. Huse. *Phys. Rev. Lett.* 57:353 (1986).
13. J.W. Cahn. *J. Chem. Phys.* 66:3667 (1977).

TWINS IN DIAMOND FILMS

Dan Shechtman

Department of Materials Engineering, Technion, Haifa, Israel
Currently a guest scientist at NIST and NRL, USA

INTRODUCTION

Diamond film technology has made fast progress in the past few years thanks to the unique properties of diamond such as high heat conductivity and hardness at room temperature. Important new discoveries contributed to this progress. The many processes known to produce diamond films yield products of varying quality depending on the deposition parameters. The structural defects which form in these diamond films during deposition, determine to a large extent the degree to which properties like thermal conductivity and fracture toughness differ from those of a perfect single crystal diamond. It is therefore important to define the various defects and to determine their effect on the properties of the films.

The structural defects which were previously identified in diamond films include stacking faults, twins of several orders and grain boundaries. Most of these defects are similar to the ones found in silicon and germanium [1-11]. In this article we define the defect structure of diamond films, discuss the way in which it forms during the film's growth process and evaluate its significance to the properties of the diamond film. High resolution transmission electron microscopy (HRTEM) was the main tool used in the study. Coincidence Site Lattice (CSL) [12] notations will be used in the article for the definition of the various twins.

DIAMOND TWINS

A. General

Diamond films were made by plasma assisted chemical vapor deposition. Our experiments revealed and defined four types of twins in the diamond films. These include the $\Sigma=3$, $\Sigma=9$, $\Sigma=27$ and $\Sigma=81$ boundaries. We speculate upon a $\Sigma=243$ twin boundary but because of the very small probability for its formation, we did not find it in our films. We shall now outline the main crystallographic characteristics of these twins.

Soft Order in Physical Systems, Edited by Y. Rabin
and R. Bruinsma, Plenum Press, New York, 1994

B. Σ=3 Twin Boundaries

These twins are found in abundance throughout the lattice of the diamond crystal with a higher density in its periphery. A detailed analysis of this type of twin was given in another article [1]. Σ=3 twins are apparently the most effective way the lattice copes with accumulating strain which may result from the incorporation of impurities in the lattice.

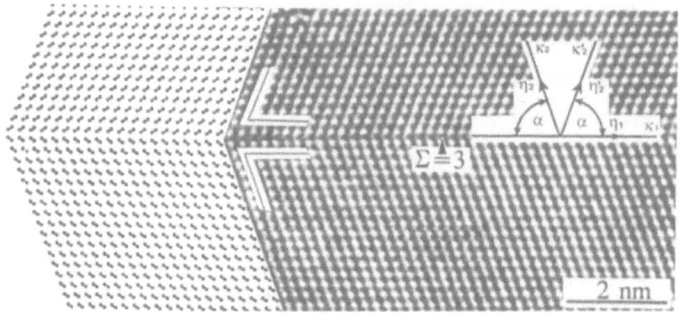

FIG. 1. The Σ=3 twin boundary

The Habit plane of most of the Σ=3 twin boundaries is {111}, in which case they are coherent. A simulation of such a boundary as seen in the [110] direction is shown in figure 1 along with a high resolution electron micrograph. The twining elements of this Σ=3 type twin are (see drawing on the micrograph, figure 1):

K_1 - (11$\underline{1}$), K_2 - ($\bar{1}$1$\underline{1}$), η_1 - [$\bar{1}$12]   all in the matrix coordinate system, η_2 - [1$\bar{1}$2], K'_2 - (1$\bar{1}$1), η'_2 - [1$\bar{1}$2]   in the twin coordinates.

While coherent Σ=3 twin boundaries occur through a simple twining process, non coherent Σ=3 and higher order boundaries form as two twin related parts of the crystal grow toward one another.

C. Σ=9 Twin Boundaries

A Σ=9 boundary is created when two Σ=3 boundaries intersect or, in other words, when two parts of the crystal which are Σ=9 twin related touch each other during the growth process. An illustration of two intersecting Σ=3 boundaries and the resulting Σ=9 boundary is given in figure 2 together with a high resolution image. A Σ=9 boundary is coherent in some cases and then it coincides with the {221} twin plane.

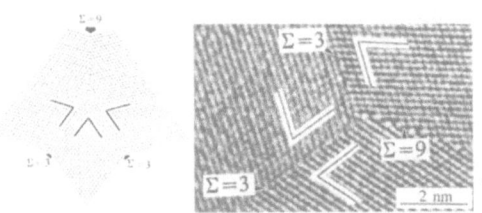

FIG. 2. The Σ=9 twin boundary

D. Σ=27 Twin Boundaries

The formation of a Σ=27 twin boundary is illustrated in figure 3. Here three Σ=3 boundaries meet at a point to form a Σ=27 twin boundary. This is a rather rare case; usually it is the intersection of a Σ=9 and Σ=3 boundaries that form this type of boundary. An example of such a case is shown in the high resolution micrograph (figure 3). Unlike the usually coherent $\Sigma=3$ twin boundaries that form along the {111} twin plane and some of the $\Sigma=9$ boundaries which tend to form along the {221} twin plane, the $\Sigma=27$ boundary wiggles its way across the crystal without any special relationship to its {115} twin plane.

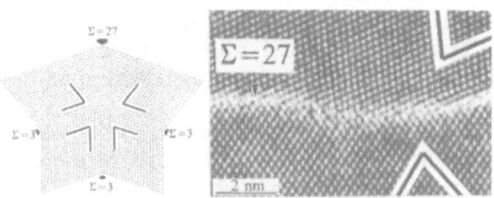

FIG. 3. The $\Sigma=27$ twin boundary

E. The $\Sigma=81$ twin boundary

A $\Sigma=81$ forms when four $\Sigma=3$ boundaries meet at a point or by an intersection of $\Sigma=3$ boundaries and a higher order one. We shall now deal with some characteristics of these $\Sigma=81$ boundaries.

An illustration of the formation of a $\Sigma=81$ boundary by four intersecting $\Sigma=3$ boundaries is shown in figure 4. A twin quintuplet thus formed is rather common in CVD diamond crystals and its importance to the growth of the crystal was discussed elsewhere [1,3]. A high resolution micrograph of one such case is also shown in figure 4. Because the angle between two {111} planes in the diamond structure is 70.529° a mismatch angle of 7.356° forms between section A and B as demonstrated in the model (figure 4). The $\Sigma=81$ twinning angle is therefore 77.885°. This angle is the one needed to rotate twin A around the common <110> axis so as to coincide with twin B.

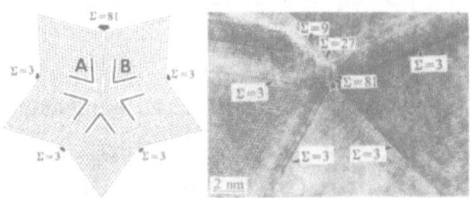

FIG. 4. The $\Sigma=81$ twin boundary as part of a twin quintuplet

The length in this {110} section of the $\Sigma=81$ boundary extends only about 1.5 nm and then intersects with a $\Sigma=3$ boundary to form a $\Sigma=27$ boundary. All the $\Sigma=81$ observed in our study had a relatively short cross section when observed along the <110> zone axis. In the crystal space, this boundary has a shape of a ribbon which extends from the origin of the twin quintuplet to the surface of the crystal along the <110> direction.

Not all twin quintuplets form a $\Sigma=81$ boundary. In some cases the misfit angle is split into several parts, usually two, to form regular grain boundaries which extend from the center of the twin quintuplet. This was detailed elsewhere [3].

A second morphology which leads to a formation of a $\Sigma=81$ boundary was discussed in another article [1].

CONCLUSIONS

We have studied and defined four basic twin boundaries that form during the growth of diamond film from the gas phase. The basic $\Sigma=3$ boundary which occurs apparently due to strain build-up in the growing diamond crystal is usually coherent but can form as an incoherent boundary when two parts of the crystal which have a $\Sigma=3$ relationship grow toward each other. $\Sigma=9$ and $\Sigma=27$ boundaries form in a similar way.

$\Sigma=81$ twin boundaries form in two morphologies. The first is associated with a twin quintuplet and thus forms when four $\Sigma=3$ boundaries emerge from a quintuplet point center. The second morphology results from an intersection of $\Sigma=3$ and higher order boundaries.

The importance of high order twins is linked to the way they form. Because these twins form like grain boundaries, when two parts of the crystal grow toward each other, they are likely to trap impurity atoms which would not attach to a free growing surface otherwise. In addition, our observations indicate that amorphous regions, several atomic distances in size, occur along these high order boundaries. Although the amorphous regions occupy only a small volume fraction, they are nevertheless important phonon scatterers and thus reduce heat conductivity. The heat conductivity of diamond films is therefore only indirectly correlated with the density of $\Sigma=3$ twins, as the interaction among these twins results in the formation of higher order twins which, in turn, impede heat transfer through the lattice.

High order twin boundaries are likely to affect also the fracture toughness of diamond films. Because the structure is columnar and contains practically only one grain per any given line across the thickness, a defect that runs along the grain provides a planar, non deflected path for cracks. Such defects are high order twins which constitute a weak, ribbon-like surfaces in the crystal and a likely path for propagating cracks.

ACKNOWLEDGEMENTS

This study was supported in part by the Office of Naval Research. The assistance of the Louis Edelstein Center of the Technion is gratefully a acknowledged. We wish to thank also Ed Farabaugh of NIST for specimen preparation and John Hutchison of the university of Oxford for his help with electron microscopy.

REFERENCES

1. D. Shechtman, J.L.Hutchison, L.H.Robins, E.N. Farabaugh and A. Feldman, J. Mater. Res., vol 8, No 3, Mar 1993
2. B.E. Williams and J.T. Glass, J. Mater. Res. 4, 373(1989).
3. D. Shechtman, J.L. Hutchison and A. Feldman, in Diamond Optics V, edited by A. Feldman and S. Holly, Proc SPIE 1759 (SPIE, Berlingham, WA, 1992)p. 81.
4. G-H.M. Ma, Y.H. Lee and J.T. Glass, J.Mater. Res. 5, 2367(1990).
5. J. Narayan, A.R. Srivatsa, M. Peters S. Yokota and K.V. Ravi, Appl. Phys. Lett. 53, 1823(1988).
6. B.E. Williams, H.S. Kong and J.T. Glass, J. Mater. Res. 5, 801(1990).
7. K. Kobashi, K. Nishimura, K. Miyata, Y. Kawate, J.T. Glass and B.E. Williams, SPIE Diamond Optics, 969, 159(1988).
8. M.D Vaudin, B. Cunningham and D.G. Ast, Scripta Met. 17, 191(1983).
9. S. Iijima, Jpn. J. Appl. Phys. 26, 357(1987).
10. S. Iijima, Jpn. J. Appl. Phys. 26, 365(1987).
11. B.E. Williams, J.T. Glass, R.F. Davis, K. Kobashi and K.L. More, Proc. 1st Int. Symp. Diamond and Diamond Like Films, J.P. Dismukes, editor 202(1989).
12. S. Ranganathan, Acta Cryst. 21, 197(1966).

THE CRYSTALLIZATION OF AMORPHOUS Al:Ge THIN FILMS

Yossi Lereah[1] and Guy Deutscher[2]

[1]Faculty of Engineering
Department of Electrical Engineering - Physical Electronics
Tel-Aviv University, Tel-Aviv 69978, Israel

[2]Department of Condensed Matter Physics
School of Physics and Astronomy
Tel-Aviv University, Tel-Aviv 69978, Israel

INTRODUCTION

The subject of Pattern Formation is being studied intensively during the last years. The universality of the patterns which are created in different physical systems motivates studies which correlate the process of an object creation with its resulting morphology.

In the field of metallurgy, similar studies are well known, but due to technological reasons they main concentrate on the relation between the structure of the material and its relevant properties, like strength. However, the studies of phase transitions by direct observation are relatively rare and they are usually limited to optical microscopy, mainly because of the high velocity of the growing phase in its matrix.

We had found that the crystallization of amorphous Al:Ge thin film is slow enough to enable in-situ studies by transmission electron microscopy. We will present here the main established results and emphasize the potential for further studies.

EXPERIMENTAL PROCESS

Films of Al:Ge alloys, 250-500Å thick, were prepared by evaporation of the elements simultaneously from two sources. The alloys which contain about 50% of each element were found to be amorpuous. The thin films were crystallized by heating them to 250-300°C inside the microscope by a special furnace. Depending on the cyrstallization temperature, the growth velocity of the crystalline phase can be as slow as a few Angstrom/sec, so that the fine details of the process can be observed at high magnification. The process is video recorded by electron sensitive CCD camera and the growth of the crystalline phase can be stopped at any stage simply by stopping the heating. Thus the crystalline-amorphous interface can be studied by various methods, i.e. nanoprobe analysis, atomic scale microscopy and geometrical analysis.

THE GLOBAL STRUCTURE OF THE CRYSTALLINE PHASE

At the indicated range of temperature the amorphous films crystallize by nucleation and growth of colonies of about $50\mu m$ in diameter. The growth velocity is constant at a fixed temperature and increases exponentially with temperature. The colony (Fig.1) consists of a polycrystalline Ge core forming a Dense Branching Morphology embedded in single crystalline Al.

Soft Order in Physical Systems, Edited by Y. Rabin
and R. Bruinsma, Plenum Press, New York, 1994

Figure 1. An electron micrograph of a typical crystalline colony surrounded by the amorphous phase.

The creation of this unique morphology and the role played by diffusion of Ge atoms from the amorphous phase into the branched core was discussed in previous publications[1-3].

The following points have been studied:

a. Computer simulation of the process and comparison of the characteristic length scales of the simulated structure with the experimental one[5].

b. Crystallographic relationship between different Al and Ge crystals. The main results are that the Ge crystals are twin related and that there are preferred Al-Ge interfaces.[6].

We believe that the information coming from these simultaneous studies (when completed) will improve our understanding of the growth mechanism particularly concerning the factors which determine the length scale of the structure, such as branch's width, branching distance and so on.

THE AL CRYSTAL-AMORPHOUS PHASE INTERFACE

Despite the crystalline nature of the Al, its surface is extremely rough up to the scale of 100nm, as can be seen in Fig.2. High resolution microscopy of this interface indicates roughness down to atomic scale. Nanoprobe analysis (with probe size down to 4 nm) in the amorphous phase at increasing distances from the Al crystal gave no gradient of concentration indicating that this structure is not related to diffusion process. However, nanoprobe analysis at the Al rim combined with atomic resolution microscopy indicate that amorphous pockets of Ge exists in the Al rim[4].

Geometrical analysis of this interface gave a roughness exponent of 0.62[7]. Analysis of the interface for samples of various Al concentration which were crystallized at various temperatures is in progress in order to supply quantitative information on the interface and its growth mechanism.

The thickness of the film has a striking influence on the Al/amorphous interface, as can be seen by comparing Fig.2 (250Å thick film) with Fig.3 (500Å thick film). Qualitative analysis of the video records of this Al front propagation clearly shows that it involves two steps: fast growth of Al filaments and subsequently their slow broadening.

The growth dynamics of both types of interfaces, (Fig.2 and Fig.3) is presently tied by video records, for better characterization and understanding of the process.

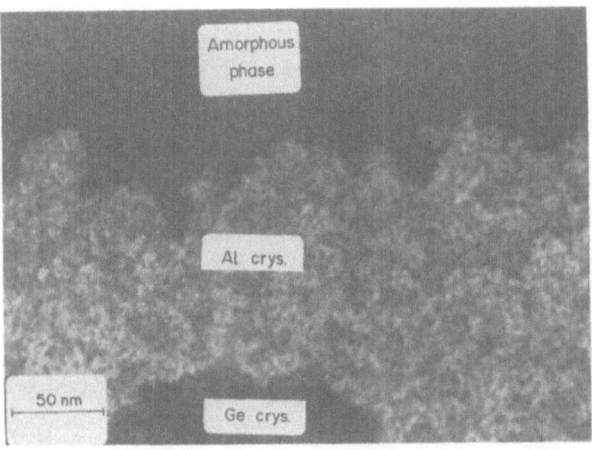

Figure 2. High magnification electron micrograph of the crystalline-amorphous interface typical for a 250Å thick films.

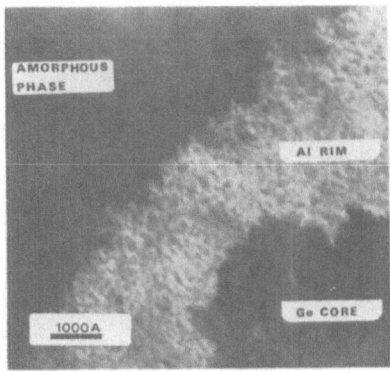

Figure 3. High magnification electron micrograph of the crystalline-amorphous phase typical for 500Å thick films.

REFERENCES

1. G. Deutscher and Y. Lereah, Phys. Rev. Lett., 60:1510 (1988).
2. S. Alexander, R. Bruinsma, R. Hilfer, G. Deutscher, and Y. Lereah, Phys. Rev. Lett., 60:1514 (1988).
3. Y. Lereah, E. Grunbaum, and G. Deutscher, Phys. Rev. A, 44:8316 (1991).
4. Y. Lereah, J.M. Penisson, and A. Bourret, Appl. Phys. Lett., 60:1682 (1992).
5. Y. Lereah, I. Zarudi, E. Grunbaum, G. Deutscher, S. Bulderyv, and H.E. Stanley, Phys. Rev. E, in press.
6. Y. Lereah and I. Zarudi, to be published.
7. Y. Lereah, S. Bulderyv, and H.E. Stanley, unpublished results.

GRAIN BOUNDARY MOTION DURING ANISOTROPIC GRAIN GROWTH

A. Brokman and A.J. Vilenkin

Division of Applied Physics
Graduate School of Applied Science and Technology
Hebrew University, Jerusalem 91904, Israel

The relation between the properties of polycrystalline materials (such as strength, effective diffusivity, electrical conductivity, etc.) and the grain size distribution has motivated many attempts to characterize and ultimately control crystalline dimension. This long-standing effort, starting more than forty years ago[1], has focused on many material systems, always reflecting the technological interest of the time. Such systems include: metals, ceramics, and it is not surprising that the efforts of last decade have been devoted primarily to the study of thin semiconducting metallic films for microelectronics applications[2]. These experiments provide a large data base which makes the characterization of the statistical and the related morphological properties of the grain growth process possible.

The explanation of these well characterized observations are based on the growth kinetics which relate the local thermodynamic driving force to the grain boundary velocity. This theoretical work was inspired by the developments in the general field of cellular growth, in particularly, the growth kinetics of soap froth. In these works it is assumed that the process of grain growth consists of two basic elements related to the motion of an individual grain boundary: a. growth is driven by local boundary curvature[1], and b. grain boundary triple junctions (vertices) remain in mechanical equilibrium. These assumptions have been successfully applied to grain growth of isotropic two dimensional systems[3]. The corresponding models, inspired by the soap froth analogy, predict the evolution of grain morphology with time. The most intriguing result of these studies is the von Neumann growth law[4], which became a subject of increasing interest during the last five years[5]. A two dimensional system provides a simple framework for the analysis of the basic physical rules governing the growth process, even in the complicated three dimensional systems. It is believed that the mean curvature driven growth model should explain the three dimensional morphological evolution. However, experimental observations in thin films and bulk specimens have not confirmed predictions of this theory. Most strikingly, many experiments on controlled bicrystals demonstrated that grain boundary steps, ledges, and facets appear during the growth process[6]. As the above mentioned fundamental of the physical processes governing grain growth must hold, the basic difference between the "real life" grain growth and various models could be the presence of anisotropy.

A long standing related problem is the morphological evolution during the process of recrystallization[7]. This process combines the grain nucleation and growth of a deformed material. In a unique experiment, Gottstein and co-workers succeeded in separating the growth process occurred during recrystallization. This experiment demonstrated that in this case the grown crystal exhibits strong anisotropic effects reflected by the development of boundary corners, facets, etc.[8]. Recently, similar morphological observations has been reported in the case of uncontrolled secondary recrystallization experiment in poly-crystalline material[9]. This observation has not been explained on a comprehensive kinetic basis. The present report, concerns with the role of anisotropy in the process of grain growth and recrystallization, by utilizing the progress achieved recently in the basic understanding of crystal form evolution during interface controlled growth. This progress, which was made possible due to the development of geometric models of crystal growth[10], have been employed for the study of the geometric constrained anisotropic growth[11-13]--a problem which is closely related to the present subject matter.

In frame of the geometric model, the normal velocity of a grain boundary element is the product of the its mobility $M(\mathbf{n})$ (which depends only on $\mathbf{n}$, the normal unit vectorand), and G, the driving force. For sake of simplicity, consider a two dimensional grain boundary with a shape function $y(x,t)$ which migrates during growth. In this process, the driving force for a displacement of a boundary element is given by the variation of the free energy $\int [F^B y + p\sigma(\partial_x y)]dx$ (σ being the boundary tension which for a given grain misorientation depends only on inclination, F^B the bulk driving force[7] when recrystallization is considered, and $p^2 = 1 + (\partial_x y)^2$) :

$$G = F^B + \partial_x z \cdot \partial_{zz}[p\sigma(z)], \qquad (1)$$

where $z = \partial_x y$. Therefore, the kinetic equation for the boundary motion is expressed as:

$$\partial_t y = p\, M(z)\, [F^B + \kappa(z)\partial_x z]. \qquad (2)$$

Here, p resolves the normal velocity in the y direction, and   from eq. (1) $\kappa = \partial_{zz}[p\,\sigma(z)]$. In the case of grain growth ($F^B = 0$) in isotropic system ($\sigma(z) = const$, $M(z) = const$), eq.(1) reduces to the familiar Mullins equation[4] of boundary motion. Eq. (2), which generalizes this case to the problem of anisotropic growth with the possible application to recrystallization is the backbone of our present discussion.

In Eq. (2) we employ boundary conditions which depend on the vertex state. Conventionally, mechanical equilibrium at the vertex is assumed, which is equivalent to the postulation that the vertex mobility is infinite. In contrast to isotropic case, the equilibrium boundary inclination at the vertex is fixed by torque balance of the three boundaries. For this purpose, the Herring equation[14] can be used to extract a discrete set of slopes at the vertex . These slopes may be used as boundary conditions for our kinetic equation. The knowledge of these slopes, together with the vertex trajectory in phase space provide the complete set of boundary conditions required for the solution of eq. (2). However, the vertex trajectory depends on the boundary state, and therefore, a full account of the many-boundary and their vertices motion becomes a self-consistent bothersome problem. The analysis of local boundary slope is possible if we consider the fixed vertex inclination (Neumann condition). Alternatively, we may impose the conditions of a fixed average inclination $< z >$ over an extended (mathematically infinite) boundary section. Such analysis enables the investigation of a model system consisting of a single boundary, the understanding of which may provide a physical basis for the numerical solution of a conjugated many grains problem. Furthermore, most of the morphological observations in the above mentioned experiments were inspected in bicrystal specimens which can be analyzed by one of the two fixed inclination conditions. Therefore, in order to elucidate the possible (in)stability during growth, we firstly consider the fixed vertex inclination of a bicrystal system. In this system, a bicrystal is formed by "welding" together two crystals with a fixed misorientation to form a two-block bicrystal as seen in Fig. 1a. One of the crystals is deformed in the case of the recrystallization, or the two crystals are free of deformation in case of grain growth process. The grain boundary is inclined to the surface at a fixed slope at which it balances the two surfaces torque. (The two surfaces have different crystallographic axis due to difference in the grain orientation that remains unchanged during growth. The associated difference in surface tensions, scaled by surface-to-surface distance, may be added to the bulk driving force F^B in eq. (1)). The vertex equilibrium may incorporate surface grooving[15]. In the present discussion we neglect the groove growth that may be of a essential importance in case of thin film recrystallization.

Formally, eq. (2) with $F^B \neq 0$ and fixed two-vertex inclination as seen in Fig.1a, is similar to the problem of a crystal growth between parallel rigid walls. This problem is considered in Ref. 11-12, and it is shown that the steady state solution $y(x,t) = Wt + g(x)$ is characterized by: a. The steady state velocity of the grain boundary is: $W \cong [pMF^B]_{z=z*}$ where $z*$ minimizes (maximizes) pMF^B, the surface component of the planar boundary velocity, in the vertex-to-vertex z interval when the surface of lower (higher) tension is consumed during growth; b. the boundary is <u>faceted</u> with a slope z^*, that terminates at a boundary layer of thickness comparable to the critical nuclei radius, in which the curvature is localized to match the vertex inclination (Fig 1b); and, c. the kinetic faceting and the steady state velocity found above solely depend on the grain boundary (an)isotropy, as long as $\kappa > 0$.

An instability is developed when $\kappa < 0$, which is the case when the grain boundary's Wulff plot consists of preferred inclination[12]. This instability yields the decomposition of the boundary slope--i.e. the formation of a corner which modulates the surface boundary layer. In case of grain growth ($F^B = 0$), the instability ($\kappa < 0$) reflects a motion of the boundary element <u>away from</u> its local center of curvature[16], ultimately generating a stable corner as drawn in Fig. 1c. The detailed steady state shape function of the migrating boundary requires the study of the bulk resistance to deformation due to strongly curved boundary[17]. This was done[13] for the case of a crystal growth (where F^B represents chemical potential difference between the growing stable phase and the consumed metastable phase), by using an ad-hoc Landau-Ginzburg Hamiltonian for the two-phase system. It is shown there that the character of the instability is spinodal, and the case $\kappa = 0$ represents the Gibbs instability limit. The boundary slope decomposes at the corner to two slopes which are found by the Maxwell common tangent construction to the $p\sigma(z)$ curve. Independently, Weirstrass-Erdmann analysis of the stationary solution of eq. (2) consists of the same common tangent corner[18,16]. Therefore, we conclude that as the corner vertex free to rotate, the adjacent boundary torque is balanced, hence the steady state corner maintains a local equilibrium.

The fixed average inclination condition may regard as the case considered above with infinite surface-to-surface distance. In this case the steady state solution of eq. (2), $y(x,t) = W_y t + g(x - W_x t)$ simply represents a straight boundary with a slope $<z>$ and velocity $W_y = [pMF^B]_{z=<z>}$ when $\kappa(<z>) > 0$. However, when $<z>$ appears between two inflection points in $p\sigma(z)$ (i.e., $\kappa(<z>) < 0$), a slope decomposition occurs. This leads to infinite number of possible stable (stationary[18] if $F^B = 0$, steady state otherwise) configurations as in Fig. 1d, with different facet lengths distributions. Transition between different facet configurations may occur by means of coarsening[6], which ultimately yields a single corner morphology. This process requires the consideration of the corner interaction contribution to the boundary energy, and is beyond the present scope. Finally, it should be stressed, that while the structure shown in Fig. 1d is frequently observed, we are not aware of experimental evidence for the structure shown in Fig. 1c. In fact, most of these works investigated bicrystal with a large vertex-to-vertex distance, hence may be considered as fixed average slope experiments. The examination of the surface induced boundary corner, thus remains as a future experimental challenge. This work was supported by the US-Israel Binational Science Foundation.

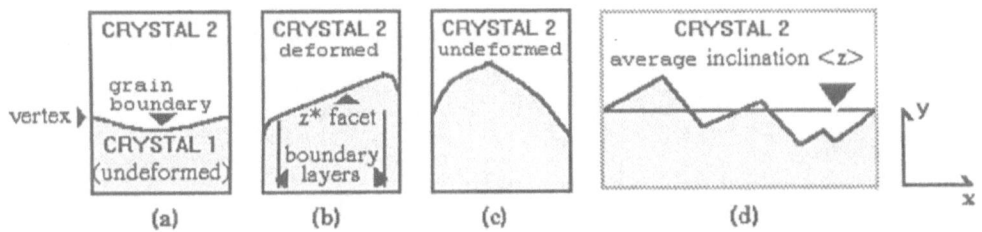

Figure 1. (a) bicrystal setup (fixed vertex configiration); (b) recrystallization experiment: "Crystal 1" is undeformed, "Crystal 2" is deformed; (c) grain growth experiment: both crystals are undeformed: (d). A fixed average inclination experiment representing the limit case of infinite vertex-to-vertex spacing (see text for the different steady state morphologies)

REFERENCES

1. C. S. Smith, p.65, in "Metal Interfaces" (ASM, Cleveland, OH,1952).
2. C.V. Thompson, Mat. Sci. Forum **94**, 245 (1992).
3. K. Lucke, G. Abbruzzese and I. Heckelmann, Mat. Sci. Forum **94**, 1 (1992).
4. W.W. Mullins, J. Appl. Phys., **27**, 900 (1956).
5. J.A. Glaizer and Denis Weaire, J. Phys. (Cond. Mat.) **4** .,1867 (1992).
6. e.g., D.A. Smith, C.M.F. Rae, C.R.M. Grovenor, p.337 in "Grain Boundary Structure and Kinetics" (ASM, Ohio, 1979).
8. G. Gottstein, H.C. Murmann, G. Renner. C. Simpson and K. Lucke, ICOTOM **5**, 521 (1978).
9. J. Gastaldi, C. Jourdan and G. Grange, Mat. Sci. Forum **94**, 17 (1992).
10. J.W. Cahn, J.E. Taylor and C.A. Handwerker, p. 88 in the Eightieth Birthday Tribute to Sir Charles Frank, ed. by R.G. Chambers, J.E. Enderby, A. Keller, A.R. Lang, and J.W. Steed, (Adam Hilger Press, Bristol, 1991).
11. A.J. Vilenkin and A. Brokman, J. Cryst. Growth, **123**, 261 (1992).
12. A.J. Vilenkin and A. Brokman, J. Cryst. Growth, **129**, 67 (1993).
13. A.J. Vilenkin and A. Brokman, J. Cryst. Growth, **131**, 239 (1993).
14. C. Herring, p. 143 in "The Physics of Powder Metallurgy", ed. W.E. Kingston (McGrow-Hill, N.Y., 1953).
15. W.W. Mullins, Acta. metall. **6**, 414 (1958).
16. A. Brokman and A.J. Vilenkin, p. 195 in "Modeling of Coarsening and Grain Growth", ed. S.P. Marsh and C.S. Pande (TMS Pub., Pennsylvania 1993)
17. V.I. Marchenko, Sov. Phys. JETP, **54**, 605 (1981).
18 J.W. Cahn, J. Physique **43**, C6-199 (1982)

SOFT ATOMIC POTENTIALS AND LOW-FREQUENCY RAMAN SCATTERING IN GLASSES

D.A. Parshin*

Laboratoire de Science des Materiaux Vitreux, URA 1119, Case 069
Université Montpellier II, 34095, Montpellier Cedex 5, France

INTRODUCTION

During the last 20 years it was established that different glasses exhibit universal properties, which are usually regarded as anomalous compared to those of the crystalline counterparts [1]. The universal low-temperature properties of glasses (below a few kelvin) have been understood well in the framework of the model of Anderson, Halperin, Varma and Phillips (AHVP model) [2,3]. This model postulates the existence of two-level systems (TLS's) in glasses with an almost a constant density of states $\overline{P}$. There are several review articles where experimental data and their interpretation in the framework of the AHVP model are given [4-8].

However, above a few kelvin the universal properties of glasses deviate from the predictions of the AHVP model. The thermal conductivity shows a plateau around $10\,\mathrm{K}$, which cannot be understood in terms of a constant density of tunneling states [9]. The sound velocity decreases linearly with temperature above a few kelvin [10]. Furthermore, there are an additional increase in the specific heat, and in the low-frequency Raman scattering [11] indicating the existence of still another kind of low-frequency modes. Recent neutron measurements [12] have shown these to be soft harmonic vibrations with a crossover to anharmonicity at the low-frequency end (at frequencies corresponding to several kelvin).

All these experimental facts indicate that there is a common basis for the low-temperature (below a few kelvin) and the higher-temperature (above a few kelvin) universal properties of glasses. And that the AHVP model describes only one part of it. In addition to the TLS's there are other low-energy excitations in glasses, which are also responsible for their universal properties at higher temperatures. In experiments of Grace and Anderson [13] and Brand and Löhneysen [14] it has been demonstrated that TLS's and these additional excitations have the same basic vibrational structure.

That is what just follows from the predictions of the soft potential model (SPM) which was proposed in [15] and developed further in [16-21]. The SPM explains all the universal low-temperature properties of glasses as well as the AHVP model [19]. But in this model besides the TLS's, soft harmonic oscillators (HO) exist too and both these types of low-energy excitations have a common basis, namely soft atomic potentials. The TLS's exist in the soft double-well potentials and the HO in the soft single-well ones. The soft harmonic excitations are just the low-energy excitations, which are responsible for some universal properties of glasses at higher temperatures [20,21]. Here we consider the low-frequency Raman scattering in glasses in the framework of the SPM.

Soft Order in Physical Systems, Edited by Y. Rabin
and R. Bruinsma, Plenum Press, New York, 1994

SOFT POTENTIAL MODEL (SPM)

According to the SPM [15], the quasilocal low-frequency modes in glasses are described by the soft anharmonic oscillator potential

$$V(x) = \mathcal{E}_0 \left[\eta \left(\frac{x}{a} \right)^2 + \xi \left(\frac{x}{a} \right)^3 + \left(\frac{x}{a} \right)^4 \right] . \tag{1}$$

Here x is the generalized coordinate of the soft mode having units of length, a is the characteristic length of the order of the interatomic spacing ($a \simeq 1$ Å), $\mathcal{E}_0$ is the binding energy of the order of $\overline{M}v^2 \simeq 10$ eV, $\overline{M}$ being the average mass of atoms constituting the glass, v is the sound velocity. The values of the dimensionless parameters η and ξ are random due to fluctuations of the structural parameters of a glass. The soft potentials correspond to $|\eta|, |\xi| \ll 1$. The distribution function of these parameters for $|\eta|, |\xi| \ll 1$ is given by [17]

$$P(\eta, \xi) = \frac{|\eta|}{2} \mathcal{P}_0 , \tag{2}$$

where $\mathcal{P}_0$ is a constant and the factor $|\eta|$ describes so-called "sea-gull" singularity in the distribution of the parameter η. The interaction of the soft atomic potentials (1) with a deformation ϵ is described by the bilinear term [19,20]

$$V_{\text{int}}(x) = \mathcal{E}_0 \acute{H} \left(\frac{x}{a} \right) \epsilon . \tag{3}$$

The dimensionless coefficient $\acute{H} \simeq 1$. This term describes both the interactions of TLS's and of HO with the strain field with the same coupling constant. Dimensionless parameter η_L is an important small parameter of the model, and W is a characteristic energy in the potential (1) for $\eta = \xi = 0$

$$\eta_L = \left(\hbar^2 / 2M a^2 \mathcal{E}_0 \right)^{1/3} \approx 10^{-2} , \qquad W = \mathcal{E}_0 \eta_L{}^2 \approx k \cdot 10 \, K . \tag{4}$$

Here M is an effective mass of the tunneling entity.

LOW-FREQUENCY RAMAN SCATTERING

Quasilocal soft modes modulate the local dielectric susceptibility of the glass $\delta\epsilon_{ik} \propto \alpha_{ik} x$ and therefore, causes the inelastic light scattering. As a consequence of the fluctuation-dissipation theorem the frequency and temperature dependence of the stokes Raman scattering intensity are determined by

$$I(\omega, T) \propto \frac{[1 + n(\omega)]}{\omega} l^{-1}(\omega, T) . \tag{5}$$

Here $n(\omega)$ is the Bose distribution function $n(\omega) = [\exp(\hbar\omega/kT) - 1]^{-1}$ for temperature T and $l(\omega, T)$ is the mean free path of phonons with frequency ω at temperature T.

According to the SPM the inverse mean free path l^{-1} for $\hbar\omega, kT \gg W$ are determined by sum of three processes

$$l^{-1} = l_{\text{rel,TLS}}^{-1} + l_{\text{res,HO}}^{-1} + l_{\text{rel,HO}}^{-1} . \tag{6}$$

The first one is the relaxation absorption due to TLS's [20]

$$l_{\text{rel,TLS}}^{-1} = \frac{\pi\omega C}{v} \left(\frac{kT}{W} \right)^{3/4} \ln^{-1/4} \frac{1}{\omega\tau_0} , \qquad C = \frac{\mathcal{P}_0 \acute{H}^2 W}{\rho v^2 \sqrt{\eta_L}} , \tag{7}$$

where ρ is the mass density of the glass. The relaxation processes are due to thermal activation hopping over the barrier with relaxation time $\tau = \tau_0 \exp(V/kT)$. The dimensionless parameter C is equivalent to the parameter $\overline{P}\gamma^2/\rho v^2$ in the AHVP model, where γ is deformation potential of the TLS's. The second contribution to (6) is resonant absorption due to HO [20]

$$l_{\text{res,HO}}^{-1} = \frac{\pi}{6\sqrt{2}} \frac{C\omega}{v} \left(\frac{\hbar\omega}{W}\right)^3 \propto \omega^4 \tag{8}$$

and the third one is the relaxation absorption due to HO. In the frequency region $\hbar\omega \ll W(2kT/E_c)^2$ [21]

$$l_{\text{rel,HO}}^{-1} = \frac{16}{9} \frac{C\omega}{v} \frac{kT}{E_0} F_1\left(\frac{E_0}{E_c}\sqrt{\frac{W}{\hbar\omega}}\right) \ ; \qquad E_c = \frac{\sqrt{2\pi\rho\hbar^3 v^5}}{\left(|\acute{H}|W/\eta_L^{3/2}\right)} \ , \tag{9}$$

where

$$F_1(x) = x\left[\arccos\frac{x^2 - 1}{\sqrt{1 + x^4}} - \ln\frac{x^2 + \sqrt{2}\,x + 1}{\sqrt{1 + x^4}}\right] \ . \tag{10}$$

Here $E_0 \approx 3W$ is a crossover energy between TLS and HO description of the spectrum in the potential (1). At still higher frequencies $\hbar\omega \gg W(2kT/E_c)^2$ the absorption is proportional to the temperature squared and does not depend on the frequency.

$$l_{\text{rel,HO}}^{-1} = \frac{32\pi^2}{27} \frac{C\omega}{v} \underbrace{\frac{W}{\hbar\omega}\left(\frac{kT}{E_c}\right)^2}_{\ll 1} \propto T^2 \ . \tag{11}$$

As a result for reduced Raman scattering intensity $I\omega^{-1}[n(\omega)+1]^{-1} \propto l^{-1}(\omega,T)/\omega^2$ we have a minimum in the frequency dependence. Position of this minimum is determined by value of W and slightly depends on temperature. The value of characteristic energy W we can estimate from position of the minimum $T_{\min}$ in the temperature dependence of the specific heat $C(T)/T^3$: $W \approx (2-2.5)\,kT_{\min}$ [17,18]. For example, in $a - SiO_2$, where one can neglect relaxation contribution of HO (9), (11) at these frequencies we obtain $\hbar\omega_{\min} \approx 1.6W(kT/W)^{1/4}$. For $T = 80$ K and $W/k = 4$ K the minimum situated at $\omega/2\pi \approx 10$ cm^{-1} what coincides with experiment [11]. This approach gives also the same depolarization ratio D

$$D = \frac{I_{\text{HV}}}{I_{\text{VV}}} = \frac{1}{2}\frac{3\operatorname{Tr}\alpha^2 - (\operatorname{Tr}\alpha)^2}{2\operatorname{Tr}\alpha^2 + (\operatorname{Tr}\alpha)^2} \leq \frac{3}{4} \tag{12}$$

in the whole frequency region because tensor α_{ik} is assumed to be nearly the same for all soft modes both TLS's and HO (compare with (3)).

Strongly rising resonant scattering of phonons due to HO (8) leads to that at some energy E_d the mean free path of phonons becomes equal to their wave length [17,22]. The value of this energy is determined by expression [22]

$$E_d \approx (0.6 - 0.75)WC^{-1/3} \ . \tag{13}$$

For example, for $a - SiO_2$ $E_d/k \approx 45$ K. At higher energies excitation from one oscillator can jumps to other oscillators directly on the distance of the wave length (compare with Einstein model [23]). The picture of independent quasilocal harmonic excitation in this case is lossed. Phonons and HO with energies $E > E_d$ cannot be considered independently any longer. They become intermixed with each other. Above this energy the total density of states should be reconstructed. One can believe that just this phenomenon responsible for the bump in the specific heat $C(T)/T^3$ at temperature of the order $E_d/5k$, for the rise of the thermal conductivity above the plateau and for the "boson" peak at frequency $\hbar\omega \simeq E_d$ in the Raman scattering in glasses.

ACKNOWLEDGMENTS

Useful discussions with U.Buchenau, V.L.Gurevich, G.Kasper, V.I.Mel'nikov, J.Pelous, H.R.Schober and R.Vacher are gratefully acknowledged. I thank the C.N.R.S. (France) for financial support and the Laboratoire de Science des Materiaux Vitreux, Universite Montpellier II for hospitality during my stay in France.

* On leave from: St. Petersburg State Technical University, St. Petersburg, Politechnicheskaya 29, 195251, Russia.

REFERENCES

1. Zeller, R.C. and Pohl, R.O. *Phys.Rev. B*, **4**, 2029 (1971).
2. Anderson, P.W., Halperin, B.I. and Varma, C.M., *Phil Mag*, **25**, 1 (1972).
3. Phillips, W.A., *J. Low Temp. Phys.*, **7**, 351 (1972).
4. Hunklinger, S. and Arnold, W., *Physical Acoustics* (edited by W.P.Mason and R.N.Thurston) (New York: Academic Press, 1976), Vol.XII, p.155.
5. Hunklinger, S. and Raychaudhuri, A.K., *Progress in Low Temperature Physics* (edited by D.F.Brewer) (Amsterdam: Elsevier, 1986) Vol.IX, p.267.
6. Phillips, W.A., *Rep. Prog. Phys.*, **50**, 1657 (1987).
7. *Amorphous Solids. Low-Temperature Properties* (edited by W.A.Phillips) (Springer-Verlag Berlin Heidelberg New York, 1981).
8. Black, J.L., *Glassy Metals I* (edited by H.J. Günterodt and H. Beck) (Berlin: Springer, 1981), p.245.
9. Freeman, J.J. and Anderson, A.C., *Phys.Rev.B*, **34**, 5684 (1986).
10. Bellessa, G., *Phys.Rev.Lett.*, **40**, 1456 (1978).
11. Jäckle, J., in Ref.[7], p.135.
12. Buchenau, U., Zhou, H.M., Nücker, N., Gilroy, K.S. and Phillips, W.A., *Phys.Rev.Lett.*, **60**, 1318 (1988).
13. Grace, J.M. and Anderson, A.C., *Phys.Rev.B*, **40**, 1901 (1989).
14. Brand, O. and H.v.Löhneysen, *Europhys.Lett.*, **16**, 455 (1991).
15. Karpov, V.G., Klinger, M.I. and Ignatiev, F.N., *Sov.Phys.JETP*, **57**, 439 (1983).
16. Karpov, V.G. and Parshin, D.A., *Sov.Phys.JETP Lett.*, **38**, 648 (1983).
17. Il'in, M.A., Karpow, V.G. and Parshin, D.A., *Sov.Phys.JETP*, **65**, 165 (1987).
18. Buchenau, U., Galperin, Yu.M., Gurevich, V.L. and Schober, H.R., Phys.Rev.B, **43**, 5039 (1991).
19. Parshin, D.A. *Interactions of soft atomic potentials and universality of low-temperature properties of glasses*, Submitted to Phys.Rev.B.
20. Buchenau, U., Galperin, Yu.M., Gurevich, V.L., Parshin, D.A., Ramos, M.A. and Schober, H.R., *Phys.Rev.B*, **46**, 2798 (1992).
21. Parshin, D.A., *Anharmonic Oscillators in Glasses: Relaxation Ultrasound Absorption and Sound Velocity*, Submitted to Phys.Rev.B.
22. Liu, X., H.v.Löhneysen and Parshin, D.A., to be published.
23. Cahill, D.G., Watson, S.K. and Pohl, R.O., *Phys.Rev.B.*, **46**, 6131 (1992).

DYNAMICS OF INTERFACE DEPINNING

IN A DISORDERED MEDIUM

Semjon Stepanow[1], Thomas Nattermann[1], Lei-Han Tang[1]
and Heiko Leschhorn[2]

[1]Institut für Theoretische Physik, Universität zu Köln,
Zülpicher Str. 77, D-5000 Köln 41, Germany
[2]Theoretische Physik III, Ruhr-Universität Bochum,
Postfach 102148, D-4630 Bochum, Germany

INTRODUCTION

The motion of an interface in a disordered medium driven by a constant force is one of the paradigms of condensed matter physics. Well known examples are domain walls in random magnets[1] and interfaces between two immiscible fluids pushed through a porous medium[2]. Closely related problems include impurity pinning in type-II superconductors[3] and in charge-density waves (CDWs)[4]. Despite the importance of these problems, significant progress has been made only recently by Nattermann et al. for interface motion[5] (see also[6]) and by Narayan and Fisher for CDWs[7].

The d-dimensional interface profile $z(\mathbf{x}, t)$ obeys the following equation[8-12]

$$\mu^{-1} \, \partial z(\mathbf{x}, t)/\partial t = \gamma \nabla^2 z(\mathbf{x}, t) + F + g(\mathbf{x}, z), \tag{1}$$

where μ is the mobility, γ is the stiffness constant, and F is the driving force. The quenched random force $g(\mathbf{x}, z)$ is Gaussian distributed with $< g(\mathbf{x}, z) >= 0$ and

$$< g(\mathbf{x}, z)g(\mathbf{x}', z') >= \delta^d(\mathbf{x} - \mathbf{x}')\Delta(z - z'). \tag{2}$$

We will be concerned with the random-field case where the correlator $\Delta(z) = \Delta(-z)$ is a monotonically decreasing function of z for $z > 0$ and decays rapidly to zero over a finite distance a. Unless otherwise specified, the width of the correlator (2) along the interface is taken to be much smaller than any other characteristic length of the problem.

Soft Order in Physical Systems, Edited by Y. Rabin
and R. Bruinsma, Plenum Press, New York, 1994

As pointed out by Bruinsma and Aeppli[9], an important length of this model is $L_c = [\gamma^2 a^2 / \Delta(0)]^{1/\epsilon}$, where $\epsilon = 4 - d$. For $d < d_c = 4$, the interface is kept smooth (i.e., fluctuations in z are limited to a or smaller) on length scales $L < L_c$, but is able to explore the inhomogeneous force field on larger length scales. It follows that the maximum pinning force on a piece of interface of linear dimension $L > L_c$ is of the order of $(L/L_c)^d [\Delta(0) L_c^d]^{1/2}$, which leads to the estimate $F_c \simeq [\Delta(0)/L_c^d]^{1/2} \sim \Delta(0)^{2/\epsilon}$ for the critical driving force of a depinning transition[8,9]. For $F > F_c$ we expect a steady-state moving solution to (1) while for $F < F_c$ the interface at long times is pinned. In the vicinity of the threshold the velocity v behaves as $v \sim (F - F_c)^\theta$, where θ is a critical exponent. The correlation length ξ can be introduced as the characteristic length above which the term vt in the random force $g(\mathbf{x}, vt + h(\mathbf{x}, t))$ dominates the term $h(\mathbf{x}, t)$. Using the relations $t \sim L^z$ and $h \sim L^\zeta$ we get $\xi \sim v^{-1/(z-\zeta)}$. Here z and ζ are the dynamic exponent and the roughness exponent, respectively. Combining the power-law dependence of the correlation length on v and of the velocity on $F - F_c$ we get for the correlation length $\xi \sim f^{-\nu}$ $(f = F - F_c)$, with

$$\nu = \theta/(z - \zeta). \tag{3}$$

RENORMALIZATION-GROUP ANALYSIS

We now will outline the formalism which we used to derive the renormalization group flow equations. First we reformulate (1) by using path integrals[13]. To do this we rewrite (1) as a Fokker-Planck equation for the probability density $P(z(\mathbf{x}), t)$ for the interface profile $z(\mathbf{x})$ at time t. The conditional probability $P(z(\mathbf{x}), t; z^0(\mathbf{x}), t^0)$ to have the profile $z(\mathbf{x})$ at time t having the profile $z^0(\mathbf{x})$ at time t^0 averaged over the disorder can be represented by a path integral as

$$P(z(\mathbf{x}), t; z^0(\mathbf{x}), t^0) = \int \int Dp Dz \; exp(-S), \tag{4}$$

where the 'action' $S = S_0 + S_i$ is given by

$$S_0 = -i \int_{t_0}^{t} dt' \int d\mathbf{x} \; p(\mathbf{x}, t')(\mu^{-1} \partial z(\mathbf{x}, t')/\partial t' - F - \gamma \nabla^2 z(\mathbf{x}, t')), \tag{5}$$

$$S_i = \frac{1}{2} \int_{t_0}^{t} dt' \int_{t_0}^{t} dt' \int d\mathbf{x} \; p(\mathbf{x}, t')\Delta(z(\mathbf{x}, t') - z(\mathbf{x}, t'')) \; p(\mathbf{x}, t''), \tag{6}$$

where $p(\mathbf{x}, t)$ is the momentum conjugated to the interface height $z(\mathbf{x}, t)$.

The procedure of the shell integration can be carried out within the path integral (4). First we separate $z(\mathbf{x}, t) = z^<(\mathbf{x}, t) + z^>(\mathbf{x}, t)$ and $p(\mathbf{x}, t) = p^<(\mathbf{x}, t) + p^>(\mathbf{x}, t)$, where $z^>(\mathbf{x}, t)$ and $p^>(\mathbf{x}, t)$ contain Fourier components only in the momentum shell $\Lambda < k < \Lambda_0$. Then we expand (4) up to the second order in powers of S_i , carry out the average over $z^>(\mathbf{x}, t)$ and $p^>(\mathbf{x}, t)$ and rewrite the result of the average as $exp(-S(z^<, p^<))$ with renormalized quantities μ, $\Delta(z)$ and etc. The flow equations for quantities under renormalization are obtained in the one-loop approximation as follows

$$\frac{d \; ln\mu}{d \; lnL} = g, \tag{7}$$

$$\frac{d\,\Delta(z)}{d\,\ln L} = -L^\epsilon \frac{d^2}{dz^2}\left[\frac{1}{2}\Delta^2(z) - \Delta(z)\Delta(0)\right], \tag{8}$$

where $L = 1/\Lambda$. Interestingly, Eq.(8) appears also in the treatment of an equilibrium interface in a random system by Fisher[14]. The stiffness constant γ does not renormalize to this order. The factor $(2\pi)^{-d}S_d\gamma^{-2}$ (S_d is the surface of the unit sphere) is absorbed into $\Delta(z)$. Here $g = \Delta''(0)L^\epsilon$ is the dimensionless coupling constant. Eq.(8) can be integrated to give $g(L) = g_0/(1 + (3/\epsilon)g_0(L^\epsilon - L_0^\epsilon))$, where $g_0 = g(L_0)$. A negative g_0, which appears to be a natural choice if an analytic $\Delta(z)$ is assumed at L_0, leads to a diverging $g(L)$ and hence $\Delta''(0)$ at a finite length $L \simeq (\epsilon/3|g_0|)^{1/\epsilon} \simeq L_c$. But the divergence of $\Delta''(0)$ can be associated with the nonanalytic behaviour of $\Delta(z)$ at the origin. Nevertheless, Eq. (8) is still well defined away from $z = 0$ and can thus be followed. To look for a fixed point solution, we make the scaling ansatz $\Delta(L, z) = A^{2/3}L^{-a}\Delta_L^*(zA^{-1/3}L^{-\zeta})$, where $\lim_{L\to\infty}\Delta_L^*(y) = \Delta^*(y)$, $a = \epsilon - 2\zeta$, and A is chosen such that $\Delta^*(0) = 1$. As for Δ, Δ^* is an even function of its argument. Inserting the scaling ansatz for $\Delta(L, z)$ into (8) and taking the limit $L \to \infty$ yields,

$$(\epsilon - 2\zeta)\Delta^*(y) + \zeta y\Delta^{*\prime}(y) - [\Delta^{*\prime}(y)]^2 - \Delta^{*\prime\prime}(y)[\Delta^*(y) - 1] = 0. \tag{9}$$

In order to find the fixed-point solution of (9) possessing the singularity at the origin we use the ansatz

$$\Delta^*(y) = 1 + a_1|y| + \frac{1}{2}a_2 y^2 + \dots \tag{10}$$

and get from (9) and (10) $a_1^2 = \epsilon - 2\zeta$ and $a_2 = (\epsilon - \zeta)/3$. Here $\Delta^{*\prime\prime}(0+) = a_2$ is finite. Using an expansion of the type (10) for $\Delta(z)$, we find that Eq.(7) (but not (8)) remain valid to the first order in ϵ, with the understanding that $g = \Delta''(0+)$. The singular term $|z|$, however, yields a reduction of the driving force, $F \to f = F - F_c$. Here $F_c \simeq -(16\pi^2\gamma)^{-1}\Lambda_0^{d-2}\Delta'(L_0, 0+)$ depends on the lower cutoff $\Lambda_0 \simeq \pi/L_0$ of the momentum space integration. Using the scaling form for Δ and identifying L_0 with L_c yields an F_c in agreement with the estimate given earlier[8,9].

Integrating (7) at the fixed-point $g^* = a_2$ we get the renormalized mobility as

$$\mu(L) = \mu_0(L/L_0)^{(\epsilon-\zeta)/3}, \tag{11}$$

where $\mu_0 = \mu(L_0)$. The dynamical exponent z is obtained by combining (11) with the relation $\gamma\mu(L)t \sim L^2$ as $z = 2 - (\epsilon - \zeta)/3$. The velocity exponent θ is derived from $v = \mu(L)f$ by using (11) with $L \simeq \xi$ as $\theta = 1 - \frac{1}{3}\frac{\epsilon-\zeta}{2-\zeta}$. The correlation length exponent ν is obtained as $\nu = 1/(2 - \zeta)$. Our final task is to determine the exponent ζ from (8). For this purpose it is useful to consider the integral $I_\Delta = \int_{-\infty}^{\infty}\Delta(L, z)dz$, which is an invariant of the flow equation (8). Provided that $I_\Delta \neq 0$ we get $\zeta = \epsilon/3$. Using $\zeta = \epsilon/3$ we have the following results for the exponents to the first order in ϵ,

$$\zeta \simeq \frac{1}{3}\epsilon, \qquad z \simeq 2 - \frac{2}{9}\epsilon, \qquad \theta \simeq 1 - \frac{1}{9}\epsilon, \qquad \nu \simeq \frac{1}{2} + \frac{1}{12}\epsilon. \tag{12}$$

At $d = 2$ the correlation length exponent ν is $\nu = 1/(2 - \zeta) = 3/4$. This value is in good agreement with the result of numerical simulations of Ji and Robbins[15] for an interface in the random-field Ising model. Our results are also supported by recent numerical simulations[16] of a lattice model which is expected to be in the same universality class as Eq.(1). The deviation between these numerical results and those given in (12) decreases when the interface dimension increases from $d = 1$ to $d = 2$ and reach a few percent for $d = 3$.

SUMMARY

We have shown that a simple extension of the perturbation theory carried out to the lowest order runs into difficulty on the length scale L_c where pinning effects become significant. When the renormalization procedure is extended to the whole function $\Delta(z)$ (random force correlator in the moving direction), the divergent behavior of the coupling constant can be attributed to the nonanalyticity of $\Delta(z)$ at the origin. By isolating the singular term the divergence can be formally removed and a consistent renormalization scheme is found at the transition $F = F_c$. The interface roughness exponent $\zeta = \epsilon/3 + O(\epsilon^2)$ so obtained coincides with that of the equilibrium random-field problem[1].

REFERENCES

1. For recent reviews on the closely related equilibrium problem see, e.g., T. Nattermann and P. Rujan, Int. J. Mod. Phys. B **3**, 1597 (1989); G. Forgacs, R. Lipowsky, and Th. M. Nieuwenhuizen, in *Phase transitions and critical phenomena*, Vol. 14, edited by C. Domb and J. L. Lebowitz (Academic Press, London, 1991), p.135.
2. M. A. Rubio, C. A. Edwards, A. Dougherty, and J. P. Gollub, Phys. Rev. Lett. **63**, 1685 (1989).
3. A. I. Larkin and Yu. N. Ovchinikov, J. Low Temp. Phys. **34**, 409 (1979).
4. K. B. Efetov and A. I. Larkin, Sov. Phys. JETP **45**, 1236 (1977).
5. T. Nattermann, S. Stepanow, L.-H. Tang and H. Leschhorn, J. Phys. II (France) **2**, 1483 (1992).
6. O. Narayan and D. S. Fisher, preprint.
7. O. Narayan and D. S. Fisher, Phys. Rev. Lett. **68**, 3615 (1992); Phys.Rev. B, in press.
8. M. V. Feigel'man, Sov. Phys. JETP **58**, 1076 (1983).
9. R. Bruinsma and G. Aeppli, Phys. Rev. Lett. **52**, 1547 (1984).
10. J. Koplik and H. Levine, Phys. Rev. B **32**, 280 (1985); D. A. Kessler, H. Levine, and Y. Tu, Phys. Rev. A **43**, 4551 (1991).
11. G. Parisi, Europhys. Lett. **17**, 673 (1992).
12. S. F. Edwards and D. R. Wilkinson, Proc. R. Soc. London, Ser. A **381**, 17 (1982).
13. See, e.g., J. Zinn-Justin, *Quantum Field Theory and Critical Phenomena* (Oxford Univ. Press, Oxford, 1989).
14. D. S. Fisher, Phys. Rev. Lett. **56**, 1964 (1986).
15. H. Ji and M. Robbins, Phys. Rev. B **46**, 5258 (1992).
16. H. Leschhorn, Physica A, in press; unpublished.

HULL OF PERCOLATION CLUSTERS

IN THREE DIMENSIONS

Jean-Marc Debierre

Laboratoire de Physique du Solide
Université de Nancy I
B P 239
F-54506 Vandœuvre-lès-Nancy
France

INTRODUCTION

Percolation has been introduced several decades ago and the related scientific activity has grown steadily since.[1] In regular percolation, black (resp. white) particles are randomly distributed on the sites of a periodic lattice with probability p (resp. $1 - p$). Two black particles sitting on first-neighbor sites belong to the same percolation cluster. As p is increased from zero, an infinite cluster spans the (infinite) lattice for the first time when p reaches a critical value p_c. The clusters are self-similar fractals at $p = p_c$, with a fractal dimension $d_f = \frac{91}{48}$.[2,3] In order to investigate the geometry of percolation clusters in more details, it proved nesserary to define several subsets for the particles inside a cluster.[1] The hull, i.e., the ensemble of the cluster particles in contact with the surrounding medium of white particles is one of these subsets.

In two dimensions (2D), the exact value[4] for the hull fractal dimension, $d'_f = \frac{7}{4}$, differs from the bulk fractal dimension. Although percolation in 3D is of greater practical interest, no exact result is known in this case. The aim of the present paper is to review briefly the numerical results obtained so far for the hulls of 3D percolation clusters and to discuss the possibility that the hull and bulk critical behaviors are the same in this case.

NUMERICAL RESULTS FOR THE CRITICAL EXPONENTS

Diffusion fronts in gradient percolation

Hulls of 3D percolation clusters were first studied by Gouyet *et al*[5,6] as diffusion fronts simulated by a gradient percolation technique. This approach consists in performing numerical simulations of 3D percolation on a $L' \times L \times L$ cubic lattice of length L' in the x direction. The simple case of a constant gradient along the x axis is considered. The concentration p of the black particles is thus constant in a given $x = cst$ plane and it varies linearly with abscissa x as $p(x) = \frac{(L'-x)}{L'}$. Attention is focussed on the 'infinite' black cluster defined as the set of black particles connected to the $p = 1$ ($x = 0$) plane through first-neighbor connections. For the complementary infinite white cluster, connections to first, second and third neighbors are used.

To characterize the geometry of the hull, a pair correlation function $C(r, x)$ was computed in the planes $x = cst$ (i.e., $p = cst$). This function is averaged over pairs of black particles which belong to the intersection of the hull with the plane at abscissa x and are a distance r apart. In the plane $x = x_c$ where $p(x_c) = p_c$, the correlation function was found to behave as $C(r, x_c) \sim r^{d'_f - 3} \sim r^{-05}$ for small r, indicating that the hull fractal dimension is $d'_f \simeq 2.5$. Gouyet *et al* finally argued that the result $d'_f \simeq d_f$ in 3D is not so surprising because of the similitude they observed numerically between the hull and the bulk of the infinite black cluster.

Growing the 3D hull as a kinetic surface

In 2D, an ingenious way to construct directly a percolation hull consists in growing a kinetic walk around the cluster.[7] We extended this idea to 3D by using a growing kinetic surface to construct directly the hulls of percolation clusters on the body-centered-cubic (bcc) lattice.[8,9] On the bcc lattice, the complementary problem of site percolation for the black particles is site percolation with first- and second-neighbor connections for the white particles. The critical concentration for the black sites[8,9] is $p_c \simeq 0.246$ whereas it is $p'_c \simeq 0.175$ for the white sites.[10] For a given black cluster, two equivalent hulls can thus be defined: a white particle belongs to the white hull H_w if it is connected to the surrounding white medium by first- and second-neighbor connections and if one of its first-neighbor sites is occupied by a black particle in the cluster. Conversely, a black particle in the cluster belongs to the black hull H_b if one of its first-neighbor sites is occupied by a white particle belonging to H_w. Since we used a kinetic surface to construct the hull, only black and white particles belonging respectively to H_b and H_w were actually generated (figure1).

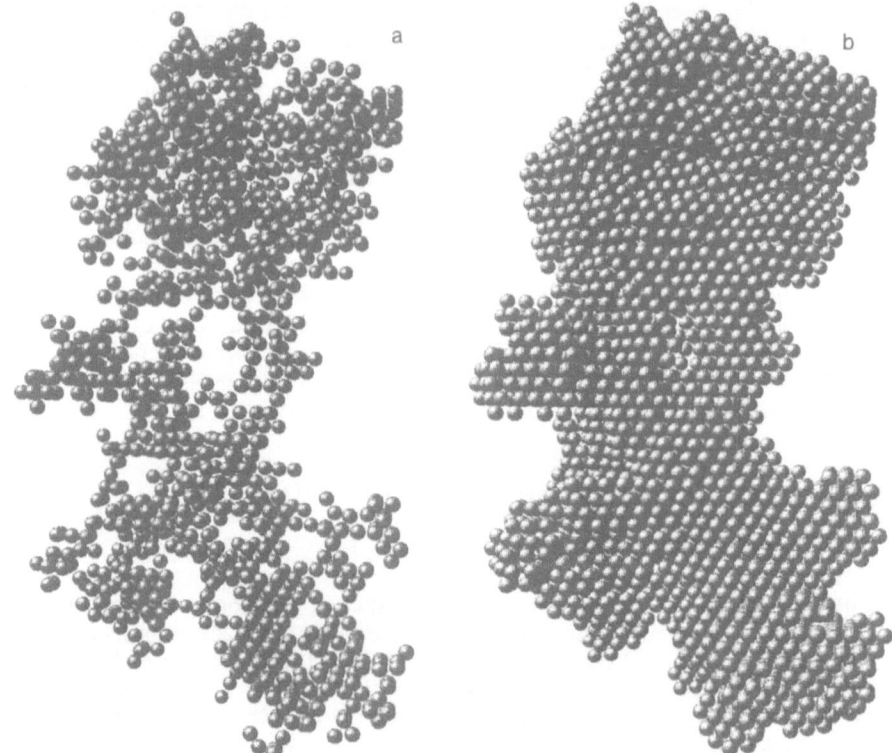

Figure 1. Two hulls, for a cluster of 1459 black particles: the black hull (a) contains 1458 particle and the white hull (b) contains 4572 particles.

The mass distribution of black hulls containing up to 5000 particles was computed and we obtained the estimate $\tau' = 2.19 \pm 0.01$ for the distribution exponent. On the other hand, a finite-size study of the data for the radius of gyration gave $d'_f = 2.548 \pm 0.014$ for the hull fractal dimension. Finally, from our data for $p < p_c$, the mean number χ of black particles in H_b was calculated and we found numerically $\chi \sim (p - p_c)^{-\gamma'}$, with $\gamma' = 1.77 \pm 0.02$.

The equality of these three critical exponents (within numerical accuracy) with the corresponding exponents for percolation clusters[11] led us to conjecture that the hull and bulk critical behaviors are indeed identical for 3D percolation. We proposed an argument in favor of this conjecture which is discussed in the next section.

DISCUSSION

We consider a 3D percolation cluster containing s black particles. Its external surface S_e is defined as the set of white particles which sit on first-neighbor sites of the black particles in the cluster. Note that, contrary to the case of the white hull H_w defined in the previous section, S_e may contain particles which are not connected to the exterior of the cluster. The average number t_s of white particles in S_e verifies

$$\frac{t_s}{s} = \frac{1-p}{p} + A s^{-\psi}, \qquad (E)$$

so that $t_s \sim s$, for large s.[1]

At $p = p_c$, there is an infinite black cluster, B_∞, coexisting with an infinite white cluster, W_∞, since the concentration $1 - p_c$ of the white particles is much greater than its critical value p'_c (see previous section). As a consequence, a finite fraction of the white particles on the lattice belongs to W_∞. Considering the infinite black cluster B_∞, it is reasonable (although not rigorously proven) to think that a finite fraction of its external surface also belongs to W_∞. This fraction is precisely the white hull for B_∞, so that the hull and bulk properties should be the same, according to equation (E). This argument can actually be extended to any p value in the interval $p_c \leq p < 1 - p'_c$. It can also be extended to the black hull H_b which is statistically equivalent to H_w.

An independent test of this conjecture has been proposed recently by Cao and Wong.[12] These authors performed numerical simulations for percolation on the simple cubic lattice. Several subsets of H_w and H_b were computed at $p = p_c$. In every case, the ratio of the number of particles in these subsets to the total number of particles in the cluster was found to tend asymptotically to $\frac{1-p}{p}$, in the same way as the ratio $\frac{t_s}{s}$ in equation (E). Thus, these hull subsets contain a finite fraction of the bulk and this is a fortiori true for the hulls themselves.

As a conclusion, we have shown that there is now a number of numerical evidences indicating that the hull and bulk properties for 3D percolation clusters are identical.

AKNOWLEDGMENTS

I would like to thank R.M. Bradley, J.F. Gouyet and L. Turban for valuable discussions. I would also like to thank M. Dubuit and O. Martischang (CNI/MAT-Nancy) for assistance with the graphics.

REFERENCES

1. D. Stauffer and A. Aharony. "Introduction to Percolation Theory", Taylor and Francis, London (1992).
2. M.P. den Nijs, A relation between the temperature exponents of the eight-vertex and q-state Potts model, J. Phys. A 12:1857 (1979).

3. B. Nienhuis, E.K. Riedel, and M. Schick, Magnetic exponents of the two-dimensional q-state Potts model, *J. Phys. A* 13:L189 (1980).

4. H. Saleur and B. Duplantier, Exact determination of the percolation hull exponent in two dimensions, *Phys. Rev. Lett.* 58:2325 (1987).

5. M. Rosso, J.F. Gouyet, and B. Sapoval, Gradient percolation in three dimensions and relation to diffusion fronts, *Phys. Rev. Lett.* 57:3195 (1986).

6. J.F. Gouyet, M. Rosso, and B. Sapoval, Fractal structure of the diffusion and invasion fronts in three-dimensional lattices through the gradient percolation approach, *Phys. Rev. B* 37:1832 (1988).

7. R.M. Ziff, Hull generating walks, *Physica D* 38:377 (1989).

8. P.N. Strenski, R.M. Bradley, and J.M. Debierre, Scaling behavior of percolation surfaces in three dimensions, *Phys. Rev. Lett.* 66:1330 (1991).

9. R.M. Bradley, P.N. Strenski, and J.M. Debierre, Surfaces of percolation clusters in three dimensions, *Phys. Rev. B* 44:76 (1991).

10. J.W. Essam, Percolation and cluster size, *in:* "Phase Transitions and Critical Phenomena, "C. Domb and M.S. Green eds., Academic Press, London, vol. 2 (1972).

11. R.M. Ziff and G. Stell, unpublished.

12. Q.Z. Cao and P.Z. Wong, External surface of site percolation clusters in three dimensions, *J. Phys. A* 25:L69 (1992).

DYNAMICS OF DIFFUSION AND INVASION FRONTS: ON THE DISCONNECTION-RECONNECTION EXPONENTS OF PERCOLATION CLUSTERS

J.-F. Gouyet

Laboratoire de Physique de la Matière Condensée
Ecole Polytechnique, F-91128, France

We present the main lines of the connection-disconnection behavior of finite clusters during diffusion, invasion or fragmentation processes, correcting a previous incorrect derivation of the anti-red bonds distribution. Using Bethe lattice as a probe, we confirm Coniglio and Wright et al. anti-red bonds exponent. The Roux-Guyon result giving the invasion probability of finite clusters during a slow drainage process is shown to be valid only in d=2 dimensions. The disconnection-reconnection exponent $\mathcal{Y}_s$ valid in d = 1, 2 and 6 dimensions, is expected to be correct in any dimensions. A rate equation of the cluster distribution is proposed which incorporates dependence of the bond creation and removal rates on cluster size.

PACS numbers: 05.40.+j, 64.60.Ak, 47.55.Mh

INTRODUCTION

One very specific characteristic of the interfaces created by diffusion is their instability [1]. They fluctuate at a frequency much higher than the inverse of the average jump time of the particles. A very similar behavior is found during very slow drainage of a non wetting fluid displacing a wetting one [2].

In both cases, a convenient model is percolation, and theoretical models have been elaborated to describe the dynamics of these fronts. The dynamics of diffusion fronts was first studied [3] in the case of length (or surface) fluctuations while a model for the temporal development of invasion percolation was published at the same time [4]. Dynamics of invasion percolation, directly derived from the theory of the dynamics of diffusion fronts was also studied when a gravity field is present [5]. Two different expressions were found in [4] and [5] for the number of events of a given size, per unit time. Very surprisingly these expressions give the same result in d = 2 but not in d = 3.

Independently, a two dimensional approach of the interface motion through a disorder media, led Martys, Robbins and Cieplak [6] to an expression of the number of events similar to the one obtained in [5] (their study includes invasion percolation and compact growth).

A two dimensional experiment on the dynamics of slow drainage in porous media [7] confirms the numerical value given by all these studies [4-6], with the particularity that for an uncompressible displaced fluid in d=2, the fractal dimension of the external perimeter which is observable is the Grossman-Aharony perimeter for which $D_f = D_e(d=2) = 4/3$ (see [2] ref.[14]).

Recently, Edwards, Gyure and Ferer considered the scaling of fragmentation of percolation clusters [8]. Again the exponent they obtain for the fragmentation enter in the frame of the preceding expressions, but now the fractal dimension of the front is replaced by the fractal dimension of the (external and internal) perimeters because fragmentation does not distinguish these perimeters. They checked their results in d = 1, d = 2 and also for the Bethe lattice (corresponding to mean field approximation and $d > d_c = 6$).

Now, what can be the expected situation in d = 3 ?

AN IMPROPER RESULT AND A PLAUSIBLE CONJECTURE

In a recent paper [9], the present author has considered a model for the dynamics of percolation clusters with the aim of harmonizing the two independent results obtained in references [4] and [5] for the connection-disconnection probability $N_{ev}(s)$ of a finite cluster of size s. These two results are in mutual agreement in d=2 but not in d > 2 and the only way to obtain coherent results is to consider that the scaling of the anti-red bonds (or reconnecting bonds) is different from that of the red bonds (or disconnecting bonds). The general expression, supposed valid in any case is,

$$N_{ev}(s) \propto s^{-\mathcal{Y}_s} \tag{1}$$

where $\quad \mathcal{Y}_s = 1 + \dfrac{D_f - D_{anti\text{-}red}}{D}$

D is the fractal dimension of the finite percolation clusters, $D_{anti\text{-}red}$ is the fractal dimension of the anti-red bonds and D_f the fractal dimension of the interface seat of the connection-fragmentation process.

$D_f = D_h$ if the connection process is on the external parameter,
$D_f = D_e$ if the connection process is on the "Grossman-Aharony" perimeter and
$D_f = D$ if the connection process is on any internal or external perimeters.
Identification of this expression with the Roux-Guyon result [4] would lead to the relation :

$$D_h = \frac{1}{2} (d + D_{red} + D_{anti\text{-}red}) \tag{2}$$

that is to say in d > 3, $D_{anti\text{-}red} = 2D - d - 1/v$. We will show that this result (2) is incorrect. As a matter of fact, Coniglio [10] and Wright et al.[11] have recently shown that the fractal dimensions of the disconnecting and reconnecting bonds should be equal :

$$D_{red} = D_{anti\text{-}red} = 1/v \tag{3}$$

Numerically, in d=3, the results of equation (2) is relatively close to $1/v_{3d}$. The above equation would give $D_{anti\text{-}red}[equ.(2)] \approx 0.90$ while $D_{red} \approx 1.14$ and a numerical simulation of this behaviour appeared delicate.

Curiously, at the upper critical dimension $d_c = 6$, equ.(2) gives zero for $D_{anti\text{-}red}$. Unfortunately the reason given in proof in reference [9] to support this result, is also not correct. At the upper critical dimension, as the loops are irrelevant in the cluster structure, the problem can be replaced by a Bethe lattice problem. In this case, there effectively exists only one bond to reconnect a cluster of size s, but this reconnecting bond (anti-red bond) is highly degenerated. Hence to confirm or to infirm the above assumption it is important to examine more in details the case of the Bethe lattice. This is the purpose of the following part.

The disconnection probability, per fragmenting bond, $b_{s's}(p_c)$ of a cluster of size s' from a cluster of size s in a Bethe lattice with coordination z, has been recently given by Edwards et al. [8],

$$b_{s's}(p_c) = s'^{-3/2} \widetilde{G}(s'/s) \tag{4}$$

where $\quad \widetilde{G}(x) = \left(\dfrac{1}{2\pi} \dfrac{z-1}{z-2} \right)^{1/2} \left(1 - \left(\dfrac{z-2}{z-1} \right)^{z-1} \right)^{-1} (1-x)^{-3/2}$

The probability to create an s'-cluster from an s-cluster in a lattice with a concentration p of occupied bonds, is then,

$$d_{s's}(p) = n_s(p) a_s(p) b_{s's}(p) = 2 Q(s'+1) Q(s-s') p^s (1-p)^{t(s)} \tag{5}$$

where t(s) is the number of bonds along the perimeter of a cluster of size s, t(s) = z + s(z−2), Q(s) is the number of configurations of an s-cluster linked to a given bond (counted in s) (Kerstein [12]). $a_s(p)$, the number of fragmenting bonds, is proportional to s. It is independent on p for the Bethe lattice. The coefficient 2 comes from the symmetry between the two fragments.

On the other hand, the reconnection probability $r_{s's}(p)$ (two clusters s' and (s−s'−1) leading to an s-cluster) can also be easily calculated. The behavior is analogous to the disconnection case and

$$r_{s's}(p) = \frac{1-p}{p} d_{s's}(p) \tag{6}$$

This expression clearly shows that $r_{s's}(p_c)$ has the same behavior in $s'^{-3/2}$ as $b_{s's}(p_c)$.
Both results (4) and (6) are in agreement with the expression of the disconnection-reconnection exponent,

$$\mathcal{Y}_s = 1 + \frac{D_f - D_{red}}{D} \tag{1}$$

when $D_f = D_h = D = 4$, and $D_{anti\text{-}red} = D_{red} = 2$ (Bethe lattice or mean field).

For the same system, the Roux and Guyon result [4] would lead to

$$r_{s's} \propto s'^{-\tau'}$$

with $\tau' = \tau + \sigma - (D_h/D) = (5/2) + (1/2) - 1 = 2$, which is not correct. There probably exists two compensating contributions making this expression for τ' valid in $d = 2$, because nothing in the calculation indicates a limitation to a particular d, and it would be of great interest to understand the reasons of this behavior.

To summarize, the exponents of the disconnecting and reconnecting bonds are well both equal to $1/\nu$ as shown by Coniglio [10] and Wright et al.[11]:

$$D_{red} = D_{anti\text{-}red} = 1/\nu$$

The expression obtained by Roux and Guyon in their study of temporal development of invasion percolation [4], to determine the bursts distribution is only valid in d=2, and the expression (1), correct in $d = 1$ and in $d = 2$ and in $d \geq 6$ is expected to be general (its proof does not involve any restriction on the dimensionality). A precise numerical confirmation remains necessary in $d = 3$. In $d = 2$, numerical calculations have been performed in the bond percolation case [4,6,8], and in the site percolation case [3,5], leading to results compatible with equation (1). This supposes that for site percolation, when removal a site, the contribution due to many daughter clusters remains marginal [8].

The value for D_f depends on the type of disconnection-reconnection considered as shown in the beginning of this article. In addition, the case $D_f = D$ agrees perfectly with the Edwards et al. scaling for fragmentation of percolation clusters [12].

RATE EQUATION OF THE CLUSTER DISTRIBUTION

Using the connection-disconnection probabilities, it is in principle possible to formulate a Smoluchowski's rate equation for the distribution of clusters during a diffusion process:

$$\frac{\partial n_k}{\partial t} = \frac{1}{2} \sum_{i+j+1=k} (K_{ij}\, n_i\, n_j - F_{ij}\, n_{i+j+1}) - \sum_{j=0}^{\infty} (K_{kj}\, n_k\, n_j - F_{kj}\, n_{k+j+1}) - K_k\, n_k\, n_\infty + F_k\, n_\infty \tag{7}$$

n_i is the probability to be on a cluster of size i, n_∞ the probability to be on the infinite cluster.

The corresponding expressions for K_{ij} ($i \oplus j \Rightarrow i+j+1$) and F_{ij} ($i \oplus j \Leftarrow i+j+1$) are obtained by symmetrization of the preceding results. In the case the bonds are added or removed at random with the same probability, we find at the critical concentration p_c,

$$F_{ij} = (i+j+1) \left(\frac{i\,j}{i+j+1} \right)^{-\mathcal{Y}_s} S\left(\frac{i}{i+j+1}, \frac{j}{i+j+1} \right) \tag{8}$$

where S is a symmetrical scaling function, and using the fact that the steady state corresponds to the percolation cluster distribution,

$$K_{ij} = F_{ij} \frac{n^{eq}_{i+j+1}}{n^{eq}_i\, n^{eq}_j} = i\,j \left(\frac{i\,j}{i+j+1} \right)^{\frac{d}{D} - \mathcal{Y}_s} S\left(\frac{i}{i+j+1}, \frac{j}{i+j+1} \right) \frac{1}{f(0)} \tag{9}$$

where f is the scaling function of the cluster distribution, $n^{eq}_s (p) \approx s^{-1-d/D} f[\, (p-p_c)s^{1/(\nu D)}\,]$.

K_k and F_k have a similar behavior and represent for instance the connection-disconnection probabilities from a front of diffusion. Equation (7) is valid as long as the initial distribution is already a random distribution of bonds (or as expected, of sites).

An interesting behavior concerns the case when the rate of removing bonds is different from the rate of adding bonds. Then, p becomes time dependent, but the cluster distribution follows adiabatically the percolation distribution if removing and adding bonds is made at random.

The above equation must first be generalized to arbitrary concentration. A scaling factor, $u[(p-p_c)i^\sigma]\, u[(p-p_c)j^\sigma] / u[(p-p_c)(i+j+1)^\sigma]$ is then expected in F ($\sigma = 1/\nu D$), while in equation (9), $1/f(0)$ must be replaced by the p dependent scaling functions f, which insure the stationary equilibrium constraint.

If τ is the relative rate of removing particles, and R(t) the general time dependent rate of removing or adding particles, the Smoluchowski coefficients now become,

$$\tilde{K}_{ij} = (1-\tau)\, R(t)\, K_{ij}\,;\quad \tilde{F}_{ij} = \tau\, R(t)\, F_{ij} \qquad (10)$$

Such an equation has been studied by Kerstein [12] in the case $\tau = 1$, for the Bethe lattice.

CONCLUSION

We have presented here the main lines of the connection-disconnection behavior of finite clusters during diffusion, invasion or fragmentation processes, correcting some previous incorrect derivations of the anti-red bonds distribution. In particular, the derivation given in [3], equ.(18), for the fluctuations of front perimeter, in which both connection and disconnection probabilities where taken equal, appears correct in any dimensions. Many questions remain open, like the numerical confirmation of the 3d behavior or the generalization of this dynamics to diffusion fronts in presence of attractive interactions [13]. Interesting problems are also related with the noise generation in this self-organized critical phenomenon [3].

ACKNOWLEDGMENTS

I would like to acknowledge the organizers for their invitation to participate to this conference on "Soft Order", in the honor of Shlomo Alexander. I also acknowledge Antonio Coniglio for a fruitful discussion. Laboratoire de Physique de la Matière Condensée is Unité de Recherches Associée au Centre National de la Recherche Scientifique.
I dedicate this paper to the memory of S. Reich's wife.

APPENDIX

Except for the explicit expression of $D_{anti\text{-}red}$ in $d > 2$, the main lines of reference [9] are correct. The equality of the disconnecting and reconnecting bonds leads to an equivalent number of clusters containing connected and disconnected clusters of size s': $N_{cp} \propto N_{dp} \propto N$; N being the number of boxes of side R_s, mean radius of s-clusters covering the perimeter. Applied to the Bethe lattice this gives,

$$N \propto \left(\frac{R_s}{R_{s'}}\right)^{D_f} \propto \left(\frac{s^{1/D}}{s'^{1/D}}\right)^{D_f} = \frac{s}{s'} \quad \text{boxes along the perimeter of an s-cluster.}$$

$N_{dp}(s') \approx N$ boxes contains disconnected clusters with a size $\tilde{s}'$, $(1-\delta)s' < \tilde{s}' < (1+\delta)s'$, so that the number of events (connection or disconnection) of size s' per s-cluster is given by,

$$N_{ev}(s') \propto (R_s)^{1/\nu}\, (1/s')\, N_{dp}(s') \propto s\, s'^{-2+1/2}$$

in agreement with equation (1), and with $a_s\, b_{s'}$s of ref.[8].

REFERENCES

1. B. Sapoval, M. Rosso, J.F. Gouyet and J.F. Colonna, Fractal structure of a diffusion front, Video 12', (1985), Imagiciel, F-91128 Palaiseau Cedex; B. Sapoval, M. Rosso, J.F. Gouyet and J.F. Colonna, *Solid State Ionics* 18&19:21 (1986).
2. L. Furuberg, J. Feder, A. Aharony, and T. Jossang, *Phys. Rev. Lett.* 61:2117 (1988).
3. J.F. Gouyet and Y. Boughaleb, *Phys. Rev. B* 40:4760 (1989) and a simpler version in, B. Sapoval, M. Rosso, J.F. Gouyet and Y. Boughaleb, *in* Proceedings of the special seminar on fractals, L. Pietronero ed., Plenum, New York (1989).
4. S. Roux and E. Guyon, *J. Phys. A* 22:3693 (1989).
5. J.F. Gouyet, *Physica A* 168:581 (1990).
6. N. Martys, M.O. Robbins, and M. Cieplak, *Phys. Rev. B.*44:12294 (1991).
7. K.J. Måløy, L. Furuberg, J. Feder, and T. Jøssang, *Phys. Rev. Lett.* 68:2161 (1992).
8. B.F. Edwards, M.F. Gyure, and M. Ferer, *Phys. Rev. A.* 46:6252 (1992).
9. J.F. Gouyet, *Physica A* 191:301 (1992).
10. A. Coniglio, in *Physics of Finely Divided Matter*, N. Boccara and M. Daoud (Springer-Verlag, Berlin, 1985); *Phys. Rev. Lett.* 62:3054 (1989).
11. D. Wright, D. Bergman and, Y. Kantor, *Phys. Rev. B* 33:396 (1986), appendix A.
12. A.R. Kerstein, *J. Phys. A* 22:3371 (1989).
13. M. Kolb, T. Gobron, J.-F. Gouyet, and B. Sapoval, *Europhys. Lett.* 11:601 (1990).

DIFFUSION REACTION A + B → C, WITH A & B INITIALLY-SEPARATED

S. Havlin,[1,2] M. Araujo, [1] H. Larralde, [1]
H.E. Stanley,[1] & P. Trunfio [1]

[1] Centre for Polymer Studies and Department of Physics
Boston University, Boston, MA 02215, USA
[2] Department of Physics Bar-Ilan University Ramat-Gan Israel

Introduction

The dynamics of diffusion controlled reactions of the type $A + B \to C$ has been studied extensively since the pioneering work of Smoluchowski [1,2]. Most studies have focused on homogeneous systems, i.e., when both reactants are initially uniformly mixed in a d-dimensional space, and interesting theoretical results have been obtained. When the concentrations of the A and B reactants are initially equal, i.e., $c_A(0) = c_B(0) = c(0)$, the concentration of both species is found to decay with time as, $c(t) \sim t^{-d/4}$ for Euclidean $d \leq 4$-dimensional systems [3-10] and as $c(t) \sim t^{-d_s/4}$ for fractals [5,6] with fracton dimension $d_s \leq 2$. Also, self-segregated regions of A and B in low dimensions ($d \leq 3$) [4] and in fractals [9] have been found. Quantities such as the distributions of domain sizes of segregated regions and interparticle distances between species of the same type and different types have been calculated [11–13]. These systems were also studied theoretically and numerically under steady state conditions and interesting predictions have been obtained [14–17]. However, the above numerical and theoretical predictions have not been observed in experiments, in part because of difficulties to implement the initially uniformly-mixed distributions of reactants.

In recent years it was realized that diffusion reaction systems in which the reactants are initially separated [18], can be studied experimentally [19,20] and that the dynamics of such a system have many surprising features [20–26]. These systems are characterized by the presence of a dynamical interface or a front that separates the reactants. Such a reaction front appears in many biological, chemical and physical processes [27].

Gàlfi and Ràcz [18] were the first to study diffusion controlled reactions with initially separated reactants. They studied the kinetics of the reaction diffusion process by a set of mean-field (MF) type equations,

$$\frac{\partial c_A}{\partial t} = D_A \frac{\partial^2 c_A}{\partial x^2} - k c_A c_B. \tag{1a}$$

$$\frac{\partial c_B}{\partial t} = D_B \frac{\partial^2 c_B}{\partial x^2} - k c_A c_B, \tag{1b}$$

where $c_A \equiv c_A(x,t)$ and $c_B \equiv c_B(x,t)$ are the concentrations of A and B particles at position x at time t respectively, D_i are the diffusion constants and k is the reaction constant. The rate of production of the C-particles at site x and time t, which we call the reaction-front profile, is given by $R(x,t) \equiv k c_A c_B$. The initial conditions are that the A species are uniformly distributed on the right-hand side of $x = 0$ and the B species are uniformly distributed on the left-hand side.

Soft Order in Physical Systems, Edited by Y. Rabin
and R. Bruinsma, Plenum Press, New York, 1994

Using scaling arguments Gàlfi and Ràcz [18] find that the width w of the reaction front $R(x,t)$ scales with time as, $w \sim t^\alpha$ with $\alpha = 1/6$ and the reaction rate at the center of the front, which is called the reaction height, scales as $h \sim t^{-\beta}$ with $\beta = 2/3$.

Experiments [19] and simulations [19,21–24] for $d \geq 2$ systems in which both reactants diffuse, support the above predicted values for α and β. Indeed, Cornell et al [23] argue that the upper critical dimension is $d = 2$ and the MF approach should therefore be valid for $d \geq 2$. However, numerical simulations of 1D systems show that the width exponent appears to be $\alpha \simeq 0.3$ and the height exponent $\beta \simeq 0.8$ [23,24]. The origin of the difference between the exponents of 1D systems and those of higher dimensional systems is due to fluctuations in the location of the front which are important in low dimensions and are neglected in the MF approach.

Taitelbaum et al [20,22] studied analytically Eqs. (1) and presented experiments for the limit of small reaction constant or short time. The main results are that several measurable quantities undergo interesting crossovers. For example, the global reaction rate changes from $t^{1/2}$ in the short time limit to $t^{-1/2}$ at the assymtotic time regime. The center of the front can change its direction of motion as found in experiments [25]. Ben-Naim and Redner [22] studied the solution of (1) under steady-state conditions.

The Form of the Reaction-Front Profile, $R(x,t)$, in the Mean-Field Approach

In a recent work [26] we consider the symmetric case in which both diffusion constants and concentrations are equal, i.e., $D_A = D_B \equiv D$ and $c_A(x,0) = c_B(x,0) = c_0$.

We find an expression for the asymptotic reaction-front profile $R(x,t)$ defined in (1) as,

$$R(x,t) \simeq \frac{kax}{t^{1/2}}(\delta c) \sim t^{-2/3}\left(\frac{x}{t^{1/6}}\right)^{3/4}\exp\left[-\frac{2}{3}\left(\frac{\lambda x}{t^{1/6}}\right)^{3/2}\right], t^{1/6} \ll x \ll (4Dt)^{1/2}, \qquad (2)$$

where $\lambda = (ka/D)^{1/3}$, $a \equiv c_0/(\pi D)^{1/2}$.

It is seen that the width of the reaction front grows as $t^{1/6}$, whereas the height can be identified with the prefactor $t^{-2/3}$ in Eq. (2), consistent with the exponents found by Gàlfi and Ràcz [18]. Equation (2) provides a more quantitative solution of Eqs. (1) than the previous scaling arguments [18], as well as information on the dependence of the form of the reaction front on the parameters c_0, k, and D, for the symmetric case.

For the case in which one reactant is static no analytical solution (of Eq. 1) exists for the form of the reaction front. However, numerical solutions [28] of Eq. (1) with $D_B = 0$, suggest that $R(x,t) \sim t^{-\beta}g(x/t^\alpha)\exp(-|x|/t^\alpha)$. The excellent scaling suggests that the width does not increase with time, i.e., $w \sim t^\alpha$ with $\alpha = 0$ and $h \sim t^{-\beta}$ with $\beta = 1/2$.

The Front, $R(x,t)$, in $d = 1$

The spatial distribution of the C particles in $d = 1$ systems, $R(x,t)$, when both reactants are diffusing with the same diffusion constant, $D_A = D_B \neq 0$, has been calculated numerically [24]. The data suggests that

$$c_C(x,t) \equiv \int_0^t R(x,t')dt' \sim t^{1-\beta}\exp(-a|x|/t^\alpha), \qquad (3)$$

with $\alpha = 0.30 \pm 0.02$ and $\beta = 0.80 \pm 0.02$. These values are in agreement with numerical simulations obtained using a cellular automata algorithm [23].

In a recent study [29] it is found that the spatial moments of the reaction front, are characterized by a hierachy of exponents, $< x^q >^{1/q} \sim t^{\alpha(q)}$, bounded by the exponents $\sigma = 1/4$ and $\delta = 3/8$ characterizing the asymptotic time dependence of the distance $l_{AB}(t)$ between nearest neighbor A and B particles and the midpoint $m(t)$ between them, respectively. Thus Eq. (3) is only an approximate result and $\alpha \simeq 0.3$ represents the scaling of the second moment i.e., $\alpha(2) \simeq 0.3$.

For the case $D_A \neq 0, D_B = 0$, analytical and numerical studies [30] yield for the reaction front

$$R(x,t) = \frac{1}{4t^{3/4}}\left(\frac{2\gamma^2}{\mu\pi}\right)^{1/2}\exp\left[-\frac{(x - \gamma t^{1/2})^2}{2\mu t^{1/2}}\right]\left[1 + \frac{x - \gamma t^{1/2}}{2\gamma t^{1/2}}\right], \qquad (4)$$

where γ and μ are constants. From Eq. (4) follows that $\alpha = 1/4$ and $\beta = 3/4$. It is interesting to note that the time integral of $R(x,t)$, which is the total production of the C particles at x up to time t is given by,

$$c_C(x,t) = \int_o^t R(x,\tau)d\tau = \frac{1}{2}\mathrm{erfc}\left(\frac{x - \gamma t^{1/2}}{\sqrt{2\mu t^{1/2}}}\right). \tag{5}$$

Note that for the short time regime α and β have different values from those given above and, as was shown by Taitelbaum et al [20], this short time regime can be observed in experiments.

The A + B (static) → C (inert): Localized Source of A

Another system in which the reactants are initially separated and which is amenable to experiments is the reaction A + B (static) → C (inert) with a localized source of A species. There exist many systems in nature in which a reactant A is "injected" into a d-dimensional substrate B whereupon it reacts to form an inert product C. Recently such an experiment has been performed [31] by injecting iodine at a point of a large silver plate and measuring quantities of the reaction $I_{2(gas)} + 2Ag_{(solid)} \rightarrow 2AgI_{(solid)}$.

First we consider N particles of type A that are initially at the origin of a lattice. The B particles are static and distributed uniformly on the lattice sites. Using an approximate quasistatic [32] analytical approach for trapping in a moving boundary we derived expressions for $C(t)$, the time-dependent growth size of the C-region and for $S(t)$ the number of surviving A particles at time t. For extremely short time $t < t_x \sim \ln N$ we find $C(t) \sim t^d$. For $t > t_x$ we find [33]

$$C(t) \sim Nf\left(\frac{t}{N^{2/d}}\right) \qquad \text{and} \qquad S(t) = N - C(t). \tag{6a}$$

The scaling function $f(u)$ is the solution to the differential equation

$$\frac{df}{d\tau} \sim k_d f^{-2/d}[1 - f], \tag{6b}$$

and k_d is a constant, depending only on dimension.

Now consider the case in which λ particles of type A are injected per unit time at the origin of the lattice. For this case we find [31]

$$C(t) \sim \begin{cases} \sqrt{8Dt \ln(\lambda^2 t/2D)} & d = 1 \\ \pi\alpha t & d = 2 \\ \lambda t & d = 3, \end{cases} \tag{7}$$

and

$$S(t) \sim \begin{cases} \lambda t & d = 1 \\ (\lambda - \pi\alpha)t & d = 2 \\ C_3(\lambda)t^{2/3} & d = 3. \end{cases} \tag{8}$$

In (7) and (8), α is the solution of $\alpha\pi = \lambda\exp(-\alpha/4D)$ and $C_3(\lambda) = (\lambda/4D)(3\lambda/4\pi)^{2/3}$. Moreover, we find that for both one and three-dimensional systems $C(t)$ satisfies the scaling relation

$$C(t) \sim \lambda^{d/(d-2)}g\left(\frac{t}{\lambda^{2/(d-2)}}\right). \tag{9}$$

Equations (7)–(9) have been supported by numerical simulations [31].

Equations (6) can be generalized for fractals;

$$C(t) \sim Nf\left(\frac{t}{N^{2/d_s}}\right), \tag{10a}$$

where $f(u)$ is the solution of the differential equation,

$$\frac{df}{du} \sim k_{d_s} f^{-2/d_s}[1 - f]. \tag{10b}$$

Here d_s is the fracton dimension defined by $d_s = 2d_f/d_w$, in which d_f is the fractal dimension and d_w the diffusion exponent [34]. For the case of constant injection rate on a fractal we do not have an analytical derivation. However, we recently calculated [35] the number of distinct sites visited on a fractal by N random walkers starting from the origin, $S_N(t) \sim (\ln N)^{d_f/\delta} t^{d_s/2}$ with $\delta = d_w/(d_w - 1)$. This result can be shown to be valid also for the number of distinct sites visited by random walkers injected at the origin with a constant rate λ when replacing $N = \lambda t$. Thus we obtain that $(\ln \lambda t)^{d_f/\delta} t^{d_s/2}$ is an upper bound for $C(t)$.

Acknowledgements

We wish to thank H. Taitelbaum and G. H. Weiss for useful discussions.

References

[1] M. V. Smoluchowski, Z. Phys. Chem. 92, 129 (1917).
[2] See, e.g., S. A. Rice, *Diffusion-Limited Reactions* (Elsevier, Amsterdam, 1985).
[3] A. A Ovchinnikov and Y. B. Zeldovich, Chem. Phys. 28, 215 (1978).
[4] D. Toussiant and F. Wilzek, J. Chem. Phys. 78 , 2642 (1983).
[5] P. Meakin and H. E. Stanley, J. Phys. A 17, L173 (1984).
[6] K. Kang and S. Redner, Phys. Rev. Lett. 52, 955 (1984); Phys. Rev. 32, 435 (1985).
[7] K. Lee and E. J. Weinberg, Nucl. Phys. B 246, 354 (1984).
[8] G. Zumofen, A. Blumen and J. Klafter, J. Chem. Phys. 82, 3198 (1985).
[9] R. Kopelman, Science, 241, 1620 (1988).
[10] M. Bramson and J. L. Lebowitz, Phys. Rev. Lett. 61, 2397 (1988); J. Stat. Phys. 65, 941 (1991).
[11] P. Argyrakis and R. Kopelman, Phys. Rev. A 41, 2121 (1990).
[12] G. H. Weiss, R. Kopelman and S. Havlin, Phys. Rev. A 33, 466 (1989).
[13] F. Leyvraz and S. Redner, Phys. Rev. Lett. 66, 2168 (1991); S. Redner and F. Leyvraz, J. Stat. Phys. 65, 1043 (1991).
[14] L. W. Anacker and R. Kopelman, Phys. Rev. Lett. 58, 289 (1987); J. Chem. Phys. 91, 5555 (1987).
[15] K. Lindenberg, B. J. West and R. Kopelman, Phys. Rev. Lett. 60, 1777 (1988).
[16] D. ben-Avraham and C. R. Doering, Phys. Rev. A37, 5007 (1988).
[17] E. Clément, L. M. Sander and R. Kopelman, Phys. Rev. A 39, 6455, (1989).
[18] L. Gálfi and Z. Rácz, Phys. Rev. A 38, 3151 (1988).
[19] Y.E. Koo, L. Li, and R. Kopelman, Mol. Cryst. Liq. Cryst. 183, 187 (1990); Y.E. Koo and R. Kopelman, J. Stat. Phys. 65, 893 (1991).
[20] H. Taitelbaum, Y.E Koo, S. Havlin, R. Kopelman and G.H. Weiss, Phys. Rev. A 46 (15 August 1992).
[21] Z. Jiang and C. Ebner, Phys. Rev. A 42, 7483 (1990).
[22] H. Taitelbaum, S. Havlin, J. Kiefer, B. L. Trus, and G.H. Weiss, J. Stat. Phys. 65, 873 (1991).
[23] S. Cornell, M. Droz, and B. Choppard, Phys. Rev. A 44, 4826 (1991).
[24] M. Araujo, S. Havlin, H. Larralde, and H.E. Stanley, Phys. Rev. Lett. 68, 1791 (1992).
[25] E. Ben-Naim and S. Redner, J. Phys. A 25, L575 (1992).
[26] H. Larralde, M. Araujo, S. Havlin and H. E. Stanley, Phys. Rev. A 46, 855 (15 July 1992).
[27] D. Avnir and M. Kagan, Nature, 307, 717 (1984)
[28] S. Havlin, M. Araujo, H. Larralde, H.E. Stanley and P. Trunfio, Physica A. 191, 143 (1992).
[29] M. Araujo, H. Larralde, S. Havlin and H.E. Stanley, (preprint).
[30] H. Larralde, M. Araujo, S. Havlin and H. E. Stanley, Phys. Rev. A. 46, R6121 (1992)

[31] H. Larralde, Y. Lereah, P. Trunfio, J. Dror, S. Havlin, R. Rosenbaum, and H. E. Stanley, Phys. Rev. Lett. 70, 1461 (1993)

[32] J. Crank, *Free and Moving Boundary Problems* (Clarendon Press, Oxford, 1984).

[33] H. Larralde, S. Havlin and H. E. Stanley, to be published.

[34] See, e.g., A. Bunde and S. Havlin (eds.), *Fractals and Disordered Systems* (Springer-Verlag, Berlin, 1991).

[35] S. Havlin, H. Larralde, P. Trunfio, J. E. Kiefer, H. E. Stanley and G. H. Weiss, Phys. Rev. A 46, R1717 (1992).

NUCLEAR SPIN RELAXATION IN AEROGELS

AND POROUS GLASSES

L. Malier, J.P. Boilot, F. Chaput, F. Devreux

Laboratoire de Physique de la Matière Condensée
URA CNRS n°1254
Ecole Polytechnique
91128 PALAISEAU Cedex
FRANCE

INTRODUCTION

It has been known for two decades that most of the physical properties are drastically different in amorphous materials as compared to their crystalline counterparts, specially when examined at low temperatures[1,2]. These specific behaviors have been accounted for by the phenomenological two-level-system (TLS) model [3,4] or its more recent generalization through soft modes[5]. One of the properties specific to the glassy state concerns the nuclear magnetic relaxation, which presents common features over numerous inorganic glasses[6,7,8]. Though different mechanisms have been debated, all of them require the TLS model to account for the weak temperature-dependance of the relaxation. The lack of structural model for these TLS prevents precise estimation of their coupling with the spins and then an absolute evaluation of the relaxation. In the case of electron spin resonance in bio-polymers, relaxation involving fractal-structure vibrations (fractons) have been considered[9]. Primarily introduced by S.Alexander and R. Orbach[10] on a theoretical basis, fractons have been experimentally tracked during these last years, with peculiar attention paid on aerogel dynamics[11-14]. Aerogels are materials whose fractal structure has been evidenced by small angle scattering[15,16] in the reciprocal space and by nuclear spin relaxation in the direct space[17]. Their density of state, deduced from inelastic light and neutron scattering, presents a weak energy dependance in part of the giga-Hertz range (typically 3 to 100 GHz). Nevertheless, this density cannot be described by a single spectral dimension as initially proposed, and the relevance of scalar or tensorial models is still a subject of debate. Moreover, the characteristic structural length determined from small angle scattering and the one deduced from the dynamics measurements present disrepancies which are not yet understood.

We measured the temperature-dependance of the spin-lattice relaxation time, for various alumino-silicate aerogels, corresponding porous glasses and crystalline counterparts. The purpose of these experiments is threefold : (i) to compare the relaxation response of these very porous amorphous materials to the general one of more 'classical' glasses, (ii) to see whether fractons, whose vibrationnal amplitudes are large, contribute to relaxation mechanisms, (iii) to follow - through variations of the density - the dependance of this dynamical property on the structural parameters, (iv) to test the theoretical predictions about relaxation in disordered systems proposed by R. Orbach and S. Alexander.

Soft Order in Physical Systems, Edited by Y. Rabin
and R. Bruinsma, Plenum Press, New York, 1994

EXPERIMENTAL

As previously stated, the adequate probe for these NMR experiments is a quadrupolar nucleus. Silicon being a spin 1/2 and oxygen presenting a low sensibility, we chose alumino-silicate aerogels rather than the much more studied silicate ones ; the nucleus of interest is then aluminum (spin 5/2). The synthesis and small angle scattering characterization of these materials has been detailed previously[16]. In the fractal frame, the structure of such aerogels is described by the two characteristic lengths bounding the fractal domain and the fractal dimension. The lower bound -a- corresponds to the size of dense particles, its typical value lies between 10 and 15 angströms.The upper one -ξ- is the length over which the material is homogeneous ; it also represents the mean size of the aggregates formed by the condensation of the particules. Most of the porosity lies in the fractal domain, which explains the scaling of ξ with the density. In the alumino-silicate aerogels, the fractal dimension values 2.1.

The aerogels and porous glass were obtained from the same organic precursor solution, submitted to different dilutions. Hypercritical drying for the aerogels and ambiant air sintering and drying for the glass led to densities of 0.095, 0.19 and 1.2. Heating at 500°C during several hours was achieved for all these samples, in order to complete condensation and to remove carbon groups. The crystalline sample was obtained by heating a porous glass for 3 hours at 1200°C. This heating procedure leads to a non-homogeneous material. It is structurally composed of crystallites of mullite embedded in a silica matrix. Prior to NMR experiments, pumping at 200°C during one day was performed, in order to remove water molecules adsorbed on the surface of the samples. It was checked that further pumping did not influence the relaxation data. We also verified by EPR that no paramagnetic impurity could provide an alternative relaxation path.

RESULTS

The spin-lattice time measurements were taken by the saturation-recovery method. Care was given to ensure that no heating occured through the radio-frequency pulses. As often encountered for quadrupolar nuclei, the relaxation profile is not exponential. An intrinsic reason arises from the fact that spins I are (2I+1) level systems[18]. Another source of non-exponentiality is that the disorder proper to the material might cause a distribution of relaxation times. A detailed analysis of the different mecanisms leading to departure from exponential recovery is out of scope of this paper but can be found elsewhere[6]. Nevertheless, the absorption line does not change in shape upon partial recovery. It might indicate that partial spin diffusion smears out the structural disorder effects and non-exponentiality would rather be due to magnetic effects. Two curve-fitting procedures were used to obtain a single relaxation time : a simple exponential on the first part of the recovery (from 1 to 1/e), and a stretched-exponential. The two times show relative difference close to 5%. This discrepancy does not affect either the temperature, or the density dependances.

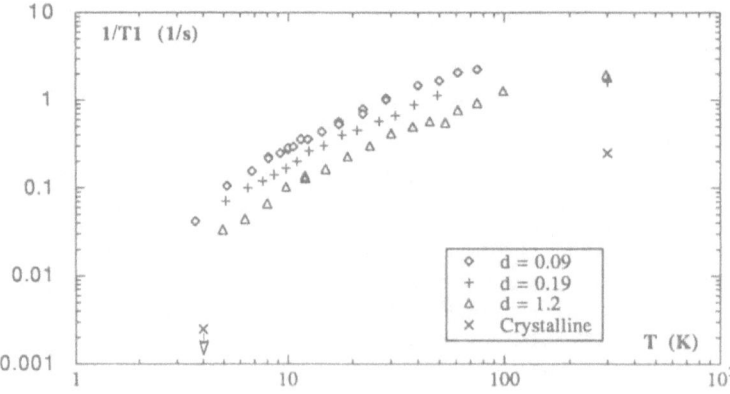

Figure 1. Relaxation rates as a function of the temperature. The value for the crystalline sample at 4K is an overestimation.

Figure 1 presents the different relaxation behaviors for both aerogels, the porous glass and the crystalline sample. For the first three samples, the relaxation rates presents stricking similarities : (i) a T-dependance very close to linear for temperature between 4K and 60 to 80K (a log-log plot leads to the following law : $1/T_1 \approx T^{1+\alpha}$, with $\alpha=0.2\pm.1$), (ii) a beginning of saturation or even decrease of the rates for T>100K, (iii) a much more efficient relaxation process for amorphous samples as compared to crystalline one.

The totally new experimental result lies in the strong dependance of the relaxation rates upon density. The effectiveness of the relaxation raises with porosity, more rapidly in the low density range than in the high one.

DISCUSSION

The temperature-dependance of the nuclear-spin relaxation in our very porous amorphous materials is qualitatively the same as in most of the explored classical glasses[6,7,8]. In the latters, the relaxation rates behave as $T^{(1+\alpha)}$, with α lying between 0.1 and 0.5. As showed by previous studies, the faster ionic conductor the glass is, the faster the relaxation. When compared to non ionic-conducting amorphous materials, our oxydes present particularly short relaxation times.

For quadrupolar nuclei, the most efficient relaxation mechanism is due to the coupling between the nuclear spin and the electrical field gradient which is modulated by lattice vibrations[18]. In crystals, the relaxation rate varies as T^7. Different authors[6,7] applied this formalism to TLS-type vibrations, and succeeded to account for $T^{1+\alpha}$ dependance, owing a slight dependance in energy of the density of modes of the TLS (proportionnal to $E^{\alpha/2}$) The temperature range described by these models corresponds to k_BT values lower than the maximum TLS energy. Reinecke et al. modified the classical two-phonon relaxation and considered the mixed and the two-TLS mechanisms. Relaxation then occurs through simultanous excitation and deexcitation of overlapping excitations (TLS or phonons) at the spin site. The energy difference of the two involved excitations has to equal the nuclear zeeman energy. Only the two-TLS mechanism accounts for the experimental results, and this is the most widely accepted model. A strong objection one can oppose to this mechanism is the expected low number of spin sites for which two TLS have the required energies and a non vanishing overlap. Partly answering that point, Ngai refined this model by taking the TLS correlation time (or energy-width) into account.

Szeftel and Alloul published a detailed analysis of the different mechanisms which led them to the conclusion that the interaction between TLS modulates the electric field gradient and provides the dominant relaxation.

Using this frame seems natural as it accounts for the temperature-dependance of our results ; the question is whether they can also account for the density-effect we observed. When lowering the density of the aerogels, the sole consequence is an increase of the length ξ[14]. There is an apparent difficulty to make such a local property as the interaction between TLS and nuclear spin depend on the variation of ξ which values several hundreds of angströms. The simplest application of the above models to the density-effect requires a dependance of the number of TLS upon density. Thermal and acoustic properties at low temperatures, which are peculiarly sensitive to TLS, have been performed on aerogels[11,12]. They indicate that the number of TLS does not increase with decreasing density. Szeftel model allows an other way to take density into account. The interaction between the TLS of energy E rises from indirect coupling through the phonons of the same energy E. The development of that coupling in case of usual phonons leads to a nuclear spin relaxation proportionnal to $d^2 v^4$ with d being the density, and v the sound velocity. According to the evolution of the sound velocity with the density found in aerogels, we can infer that such a mechanism leads to a decrease of the relaxation rate with decreasing density. Nevertheless, in the case of aerogels, we cannot just consider usual phonon bath. The physical reason for the interaction betwen TLS and lattice vibrations is the strain modifications in the TLS environment. In aerogels, the vibrations corresponding to lengths between a and ξ presents strickingly high amplitudes, dependant on the frequency. Neutron measurements of the mean-square displacement at 300K give for $\langle r^2 \rangle^{1/2}$ values of 0.04Å for the high-frequency limit, and 1Å for the low-frequency boundary[14]. As the most interacting TLS control the relaxation and the largest strain is due to lattice-vibrations in the ξ-frequency range, it is reasonnable to postulate that this mechanism applied to aerogels is sensitive to density variation, in the same qualitative way as we experimentally observed.

In a different theoretical approach, Orbach et al. analysed the spin relaxation on a fractal structure[19], taking the peculiar density of states and the localization of the vibrations into account. They predicted non-exponentiality of the recovery, and power-law dependance of the relaxation rates, with a lower boundary of 2 for the exponent, in the case we are concerned with. This obviously fails to account for our experimental results.

CONCLUSION

We report first measurements of nuclear spin-lattice relaxation in aerogels, and porous glass. Though the temperature dependance of this relaxation is very similar to the classical behavior encountered in 'classical' glasses, we observed a variation of the relaxation with the density. On the one hand, the usual models for relaxation in glasses fail to account for this density-effect, while on the other hand Orbach et al. model for relaxation on fractal network provides too steep temperature dependance. A mechanism of relaxation involving interacting-TLS seems promising provided that large amplitude vibrations found in aerogels are considered. Additionnal experiments on samples presenting other densities are under progress.

REFERENCES

1. R.C. Zeller and R.O. Pohl, 'Thermal conductivities and specific heat of non-crystalline solids', *Phys. Rev. B* 4 p2029 (1971).
2. R.B. Stephens, 'Low-temperature specific heat and thermal conductivity of non-crystalline dielectric solids', *Phys. Rev. B* 8 p2896 (1973).
3. P.W. Anderson, B.I. Halperin, C.M. Varna, 'anomalous low-temperature thermal properties of glasses and spin glasses', *Phil. Mag.* 25 p1 (1972)
4. W.A. Phillips, 'Two-level states in glasses', *Rep. Prog. Phys.* 50 p 1657 (1987).
5. Y.M. Galperin, V.G. Karpov, V.I. Kozub, 'Localized states in glasses', *Adv. in Physics* 38 p669 (1989).
6. J. Szeftel, H. Alloul, 'Nuclear spin lattice relaxation in the amorphous state : towards an understanding', *J. of Non-Cryst. Solids* 29 p 253 (1978), and references therein.
7. T.L. Reinecke, K.L. Ngai, 'Low-temperature nuclear spin-lattice relaxation in glasses', *Phys. Rev. B* 12 p3476 (1975), and references therein.
8. S. Estalji, O. Kanert, J. Steinert, H. Jain, K.L. Ngai, 'Uncommon nuclear-spin relaxation in fluorozirconate glasses at low temperatures', *Phys. Rev. B* 43 p7481 (1991).
9. J.T. Colvin, H.J. Stapleton, 'Fractal and spectral dimensions of biopolymer chains', *J. Chem. Phys.* 82 p4699 (1985).
10. S. Alexander, R. Orbach, 'Density of states on fractals : "fractons"', *J. Phys. (Paris)* 43 pL-625 (1982).
11. A.M. de Goer, R. Calemczuk, B. Salce, J. Bon, E. Bonjour, R. Maynard, 'Low-temperature energy excitations and thermal properties of silica aerogels', *Phys. Rev. B* 40 p8327 (1989).
12. A. Bernasconi, T.Sleator, D. Posselt, J.K. Kjems, H.R. Ott, 'Dynamic properties of silica aerogels as deduced from specific-heat and thermal-conductivity measurements',*Phys. Rev. B* 45 p10363(1992).
13. E. Courtens, J. Pelous, J. Phalippou, R. Vacher, T. Woignier, 'Brillouin scattering measurements of phonon-fracton crossover in silica aerogels', *Phys. Rev. Lett.* 58 p128 (1987).
14. R. Vacher, E.Courtens, G. Coddens, A. Heidemann, Y. Tsujimi, J. Pelous, M. Foret, 'Crossovers in the density of states of fractal silica aerogels', *Phys. Rev. Lett.* 65 p1008 (1990).
15. R. Vacher, T. Woignier, J. Pelous, E. Courtens, 'Structure and self-similarity of silica aerogels', *Phys. Rev. B* 37 p6500 (1988), and references therein.
16. F. Chaput, J.P. Boilot, A. Dauger, F. Devreux, A. De Geyer, 'Self similarity in alumino-silicate aerogels', *J. of Non-Cryst. Solids* 116 p133 (1990).
17. F. Devreux, J.P. Boilot, F. Chaput, B. Sapoval, 'NMR determination of the fractal dimension in silica aerogels', *Phys. Rev. Lett.* 65 p614 (1990)
18. E.R. Andrew, D.P. Tunstall, 'Spin-lattice relaxation in imperfect cubic crystals andin non-cubic crystals', *Proc. Phys. Soc.* 78 p1 (1961).
19. R. Orbach, S. Alexander, O. Entin-Wohlman, 'Relaxation and non-radiative decay in disordered systems : II. 2-fracton inelastic scattering', *Phys. Rev. B* 33 p3935 (1986).

FARADAY EFFECT IN THE MULTIPLE SCATTERING OF LIGHT: A MONTE CARLO SIMULATION

A.S. Martinez and R. Maynard
Laboratoire d'Expérimentation Numérique and
Centre de Recherche sur les Très Basses Températures
Maison des Magistères CNRS
BP 166
38042 Grenoble Cedex 9 France

ABSTRACT

A Monte Carlo simulation is used to obtain the intensity correlation function in the multiple scattering of an incident linearly polarized light in a magneto-optically active medium. The scatterers are finite spheres and each single scattering is calculated by the Mie theory. For the diffusion regime, the results predicted by a simple stochastic theory are verified. On the other hand, in the intermediate regime, the correlation function is described by the one-dimensional model, which explains the origin of the unexpected oscillations of the correlation function.

The observation of the weak localization in multiple scattering of waves in disordered media such as the enhanced back scattering of light[1] or the universal conductance fluctuation in mesoscopic conductors,[2] has stimulated research in this field.[3] However, our knowledge on the wave propagation in disordered media remains mainly restricted to scalar waves, point-like (isotropic) scatterings and independent Feyman paths.[3] The level of this last approximation to describe the multiple scattering of nth order is analogous to the problem of the conformation of n monomers in a polymer chain.[4] A sequence of n scatterings corresponds to a trajectory in real space with broken lines of mean length l: the photon mean free path. These lines are the bond lengths between two monomers in the polymer. The anisotropy of the scattering corresponds to a rigidity of the polymer described by the persistence length and the diffusion constant of the photons is proportional to the gyration radius. However, the fact that the multiple scattering sequences are only the partial scattered waves, which must be recomposed to build up the scattered wave, provides an important difference between this two fields.

Moreover, for the case of light, the vectorial nature of the field must be taken into account, as well as the size of the scatterers. The combination of the polarization of light and the finite size of the scatterers, which are supposed to be spherical, leads to the anisotropic Mie scattering.[5] This is a difficult problem for the following reason: during the successive multiple scatterings, the transverse electric field is changed by the scattering amplitudes and in each step it is given in a local basis. The difficulty occurs when the electric field is projected into the laboratory frame, in order to compare the different polarization states and to take into account the boundary conditions. The simplest experimental configuration for measuring the intensity correlation is a slab confining the scatterers where the incident and emergent wavevectors are perpendicular to its planes.

The weak localization of light is observed throughout the enhancement factor 2 in the reflection experiment. The presence of a magnetic field B in a magneto-optically active media produces a Faraday rotation of the polarization vectors between successive scatterings. This effect is proportional to $V\hat{k}.B$, where V is the Verdet constant and $\hat{k}$ the unitary wavevector. It breaks down the time reversal symmetry[6] and destroys the weak localization effect by decreasing the enhancement factor of the coherent back scattering cone.[7] This was recently observed experimentally as well as the intensity correlation under magnetic field.[8-10]

Several problems occur in the context of the multiple scattering of electromagnetic field and call for a deepening of the analysis. For a very large number of scatterings, corresponding to thick slabs, the general frame of the theory is the diffusion equation. This "pure" diffusion regime corresponds to complete depolarization of the light. But the approach to the diffusion regime, or still, the cross-over between the single scattering and the "pure" diffusion regime, that we call intermediate regime, raises important problems concerning the characteristics lengths of the system. It has be shown[11,12] that the polarization decays for the two basic incident polarized light - linear and circular - are exponential but with different characteristic lengths which depend on the particle size. The five characteristic lengths (slab thickness, wavelength, particle size, mean free path and transport mean free path) of the problem are not enough to describe the vectorial properties of the medium.

In this paper we will consider the approach towards a pure diffusive regime when incident linear polarized light is multiply scattered by a magneto-optically active random medium. Motivated by recent experiments[8-10] we are interested in the intensity along the x-axis: $i_x = |E_x|^2$, where E is the electric field as a function of the magnetic field B. The correlation function for the transmitted light through an infinite slab parallel to the x-y plane and of thickness L along the z-axis, in the approximation of independent paths, is given by:

$$G_{xx}^{(2)}(B) = \frac{<i_x(0)i_x(B)>}{<i_x(0)><i_x(B)>} - 1 = \frac{|<E_x(0)E_x^*(B)>|^2}{<i_x(0)><i_x(B)>} , \qquad (1)$$

where the symbol <...> stands for the ensemble average. Notice that $G_{xx}^{(2)}(B) = 1$ for a homogeneous medium.

Let us recall the simple stochastic[7] which can be considered as a first approach to the law of variation of $G_{xx}^{(2)}(B)$. Consider isotropic (Rayleigh) scattering regime: the wavevector $\hat{k}$ is randomized in the length scale of the mean free path l. It is assumed that in the same length scale, the polarization is randomized by a simple random process of helicity flip (of "Ising" type) in the circular wave representation. This is described by an external second random variable $\eta_\kappa = \pm 1$ with equal probability, with the index κ being the scattering label. The phase difference between right- and left-handed helicity $\delta\phi_\kappa = \eta_\kappa \alpha_\kappa$ with $\alpha_\kappa = VBl \cos\Omega_\kappa$, where Ω_κ, the angle between B and $\hat{k}_\kappa$, is uniformly distributed in the interval $[0,\pi]$. The phase difference is a random variable with zero mean value. For a given sequence, the total phase difference Φ between the circular states is $e^{j\Phi} = exp(j \sum_{\kappa=1}^{n} \delta\phi_\kappa)$ since $\delta\phi_\kappa$ are independent random variables of zero average. By virtue of the central limit theorem $<e^{j\Phi}> = exp(-n<\delta\phi^2>) = exp(-n<\alpha^2>)$. Confining the scatterers in a slab of thickness L, from the diffusion theory $(L/l >> 1)$, we use the path length (s) distribution function $P(s)$. The transmitted light is completely depolarized $<i_x(0)> = <i_x(B)> = \frac{1}{2}$ and from Eq. 1, one gets for $<\alpha^2> << 1$:

$$G_{xx}^{(2)}(\xi) = [\int_{L}^{\infty} ds\, P(s)\, exp(-\frac{s<\alpha^2>}{l}))\,]^2 \cong [\frac{\xi}{sh(\xi)}]^2 \approx 1 - \frac{\xi^2}{3}\,, \qquad (2)$$

where $\xi = L\sqrt{<\alpha^2>}\,/l = VBL$ The variable ξ couples exactly B and L through the factor BL. It is believed that this theory can be applied for anisotropic scatterings by changing the mean free path into the transport one l^*.[8,9]

By the rotation of the polarization states between the scatterings events, the Faraday effect is taken into account in the Monte Carlo simulation that we have described in reference 12. The effect of the magnetic field inside the scatterers is ignored. We have considered incident linear polarized light along the x axis with a wavelength in the vacuum $\lambda_{vac} = 0.4579\ \mu m$. The Verdet constant of the medium is $V = 157.1\ rad/(mT)$ and its refraction index $n_m = 1.69$. The spheres have a radius $a = 0.1\ \mu m$, refraction index $n_s = 1.45$ and are diluted, representing $\Phi = 1\%$ of the volume. The magnetic field is applied along the z axis and we have considered 10000 sequences for each slab thickness. From the Mie theory ($ka = 2\pi n_m a/\lambda_{vac} = 2.32$ and $n_s/n_m = 0.858$) we obtain a total scattering cross section $\sigma_t = 0.00441\ \mu m^2$, $l = 1/(\sigma_t\ \Phi) = 0.0951\ mm$, the mean value of the cosines of the scattering angle is $<cos\theta> = 0.669$ leading to a Boltzmann transport mean free path $l^*/l = 1/(1 - <cos\theta>) = 3.02$.

For transmission, the intensity correlation functions are shown in figure 1. In the diffusion limit (inset of figure 1) $L / l^* = 7,8,9,10$ and 11, VBL, is the good reduced variable, according to the stochastic theory for small values of VBL. For larger values of VBL, $G^{(2)}_{xx}(B)$ cannot be written only in terms of VBL. This is apparent from figure 1 where oscillations are observable above the noise. For the intermediate regime, we have plotted $G^{(2)}_{xx}$ as a function of VB for $L / l^* = 1,2,3,4,5$ and 10. It is striking that for small values of L/l^* the correlation function displays very net oscillations (Fig. 1) which do not exist in a homogeneous slab (no scatterings) where $G^{(2)}_{xx} = 1$. Still more striking are the amplitudes of these oscillations, which are no more damped for larger VB (see curves for $L / l^* = 3$ and 4).

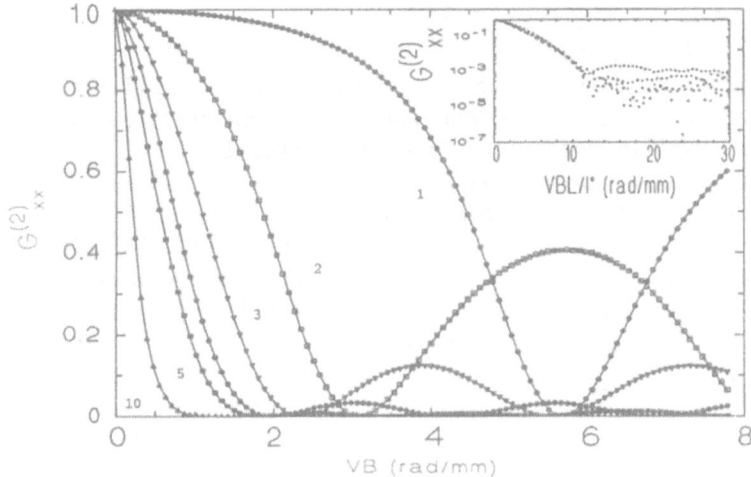

Figure 1. Intensity correlation (Eq. 1) as function of VB for a slab of thickness $L / l^* = 1,2,3,4,5$ and 10. Inset: Intensity correlation (in log scale), for $L / l^* = 7,8,9,10$ and 11, as a function of VBL/l^*, i.e. ξ since l^* is a constant. The lines are only a guide to the eyes.

For thinner slabs, by reason of anisotropy, the scattering paths are almost straight lines. This is relevant of the Faraday effect in a chain, where the photon execute a one dimensional random walk. In this model, the polarization states rotate in the same sense between the succesive scatterings, which can be either forwards or backwards, leading to an accumulation of the rotation angle proportional to the total length of the path. This simple idea gives for an incident linear polarization $G^{(2)}_{xx}(B) = <cos(VBs)>^2/<cos^2(VBs)>$ where the average is performed over the path length distribution $P(s)$. In transmission, $P(s)$ is characterized by a dispersion, which can be pictured as a window along the s axis. Hence, for narrow distributions corresponding to thin slabs, oscillations arise from the average values of $cos(VBs)$ and $cos^2(VBs)$ inside this window. The oscillations attenuation occurs for strong magnetic field and for thicker slabs. Actually the curves in figure 1 can be understood in a simplified theory of Rayleigh scattering in three dimensions.

In conclusion, we have developed a Monte Carlo simulation in order to obtain the intensity correlation function in the multiple scattering regime in a magneto-optically active medium. For the diffusion regime, the results predicted by the simple stochastic theory are qualitatively verified. In the intermediate regime, the correlation function can be described by the one-dimensional random walk model, which explains the origin of the unexpected oscillations of the correlation function.

We thank F. Erbacher, R. Lenke and G. Maret for very stimulating discussions. One of us (A.S.M.) also wish to thank CAPES for financial support.

REFERENCES

1. Y. Kuga and A. Ishimaru, J. Opt. Soc. Am. **A8**, 831 (1984).
   M.P. van Albada and A. Lagendijk, Phys. Rev. Lett. **55**, 2692 (1985).
   P.E. Wolf and G. Maret, Phys. Rev. Lett. **55**,2696 (1985).
   E. Akkermans, P.E. Wolf and R. Maynard, Lett. **56**, 1471 (1986).
2. *"Mesoscopic Phenomena in Solids,"* ed. B.L. Al'shuler, P.A. Lee and R.A. Webb, North Holland, Amsterdam, (1991).
3. *"Classical Wave Localization,"* ed. P. Sheng, World Scientific, Singapore, (1990).
4. P.J. Flory, *"Statistical Mechanics of Chain Molecules,"* John Wiley & Sons, New York, (1969).
5. H.C. van de Hulst, *"Light Scattering by Small Particles,"* Dover Publications, New York (1981).
6. A.A. Golubenstev, JETP **59**, 26 (1984).
7. F.C. MacKintosh and S. John, Phys. Rev. **B37**, 1884 (1988).
8. F. Erbacher, R. Lenke and G. Maret, Europhys. Lett. **21**, 551 (1993).
9. F. Erbacher, R. Lenke and G. Maret, in: *"Localization and Propagation of Waves in Random and Periodic Structures,"* ed. C.M. Soukoulis, Plenum Publishing Corporation (to appear).
10. R. Lenke and G. Maret, this issue.
11. F.C. MacKintosh, J.X. Zhu, D.J. Pine and D.A. Weitz, Phys. Rev. **B40**, 9342 (1989).
12. A.S. Martinez and R. Maynard, in: *"Localization and Propagation of Waves in Random and Periodic Structures,"* ed. C.M. Soukoulis, Plenum Publishing Corporation (to appear).

FRACTON DIMENSIONS FOR ELASTIC AND ANTIFERROMAGNETIC PERCOLATING NETWORKS

Tsuneyoshi Nakayama

Department of Applied Physics
Hokkaido University
Sapporo 060, Japan

INTRODUCTION

It has been ten years since a celebrated paper by Alexander and Orbach[1] introduced the concept of a "fracton" to represent quantized excitations of fractal networks. Since that time, a great deal of activity has been concentrated on understanding the nature of fractons, where the fracton (or spectral) dimension plays a key role.

This article presents the evaluation of the fracton dimensions via the calculations of the densities of states(DOS) for both percolating *elastic* and *antiferromagnetic* networks ($d = 2\text{-}4$). The latter belongs to a different universality class that for scalar elasticity. We claim the fracton dimension $\tilde{d}_{AF}$ for antiferromagnetic fractons to be very close to *unity* independent of the Euclidean dimension d.

THE FRACTON DIMENSION $\tilde{d}$ FOR PERCOLATING ELASTIC NETS

Let us consider bond-percolating(BP) networks consisting of N atoms with unit mass and linear springs connecting two atoms in the nearest neighbor. The calculated DOS and the integrated DOS are given in Figs. 1 and 2 for $d = 2$ percolating networks formed on 1100×1100 square lattice ($N = 657426$) with periodic boundary conditions(bottom). These data are taken from *one sample* at $p_c = 0.5$. The fracton dimension $\tilde{d}$ defined by $\mathcal{D}(\omega) \propto \omega^{\tilde{d}-1}$ is obtained as $\tilde{d} = 1.33 \pm 0.01$ from Fig. 1, while the data in Fig. 2 indicates the value of $\tilde{d} = 1.325 \pm 0.002$. The DOS and the integrated DOS for $d = 3$ bond-percolating networks are given in Figs. 1 and 2 by filled triangles (middle). These data show the averaged DOS and the integrated DOS over three samples at the percolation threshold $p_c(= 0.249)$. The networks, formed on $100 \times 100 \times 100$ cubic lattices, have 155385, 114303, and 143026 atoms. The fracton dimension $\tilde{d}$ is obtained as $\tilde{d} = 1.31 \pm 0.02$ from the least squares fitting by using the data of Fig. 1 obeying the power law, while $\tilde{d}$ takes the value of 1.317 ± 0.003, when using the data of Fig. 2.

Soft Order in Physical Systems, Edited by Y. Rabin
and R. Bruinsma, Plenum Press, New York, 1994

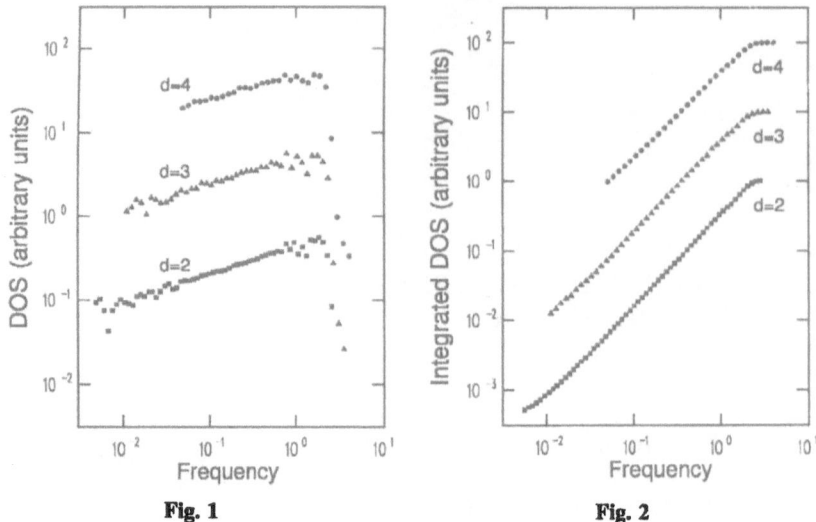

Fig. 1. The density of states (DOS) per atom for $d = 2$, $d = 3$, and $d = 4$ bond-percolating networks at $p = p_c$. The angular frequency ω is obtained in units of mass $m = 1$ and force constant $K = 1$. The networks are formed on $1100 \times 1100(d = 2)$, $100 \times 100 \times 100(d = 3)$, and $30 \times 30 \times 30 \times 30(d = 4)$ lattices with periodic boundary conditions, respectively.
Fig. 2. The integrated DOSs per atom at $p = p_c$ for $d = 2$, $d = 3$, and $d = 4$ bond-percolating networks. The symbols correspond to those in Fig. 1.

The DOS and the integrated DOS of bond-percolation networks at $p_c = 0.160$, formed on $30 \times 30 \times 30 \times 30$ hypercubic lattices, are shown in Figs. 1 and 2 by filled circles (top), respectively, which were obtained by averaging over 15 samples. The network sizes are $N = 8410 \sim 64648$. The DOS in the frequency region $0.12 < \omega < 0.9$ shows clearly the power law as well as in the $d = 2$ and 3 case. The fracton dimension $\tilde{d}$ is estimated as $\tilde{d} = 1.31 \pm 0.03$ from the least squares fitting using the data of Fig. 1. Simulated results suggest that the fracton dimension $\tilde{d}$ is very close to 4/3 independent of the Euclidean dimension d as conjectured by Alexander and Orbach,[1] but slightly smaller than 4/3.

THE FRACTON DIMENSION $\tilde{d}_{AF}$ FOR ANTIFERROMAGNETS

The linearized equations of motion for transverse spin deviations S_i^+ from antiferromagnetic Néel order are expressed in the unit of $2S/\hbar = 1$ by

$$i\frac{\partial S_i^+}{\partial t} = \sigma_i \sum_j J_{ij}(S_i^+ + S_j^+) \ , \tag{1}$$

where $S_i^+ \equiv S_i^x + iS_i^y$, and J_{ij} are chosen as $J_{ij} = 1$ if both sites i and j are occupied and $J_{ij} = 0$ otherwise. In eq. (1), σ_i is +1 for the site i belonging to the A-sublattice and -1 to the B-sublattice. The type of these equations are different from the equations of motion for ferromagnetic spin-waves or elastic vibrations with scalar displacements due to the factor σ_i in addition to the sign of the second term of the right hand side of eq. (1), indicating that the universality class of percolating antiferromagnets is different.

The calculated DOSs and the integrated DOSs for BP *antiferromagnets* at p_c are shown in Figs. 3 and 4, respectively. Filled squares in Figs. 3 and 4 represent the results for $d = 2$ percolating antiferromagnets, in which 11 BP networks have been

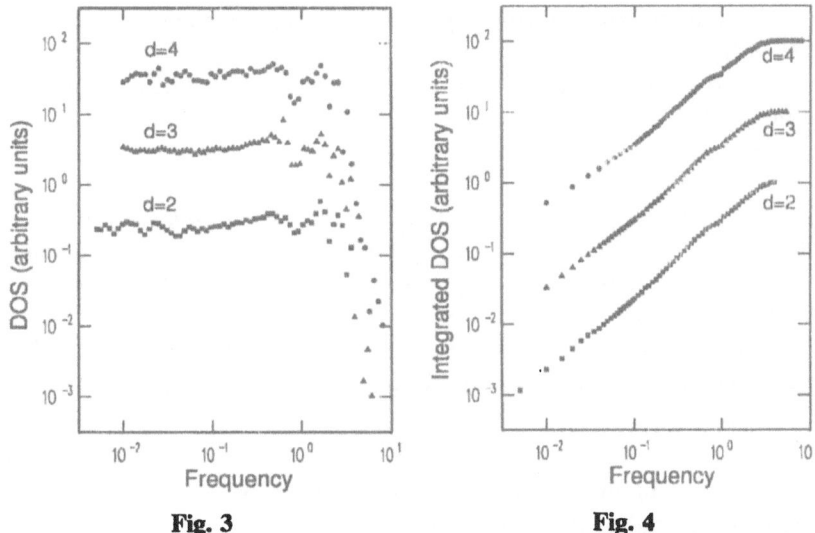

Fig. 3. The DOSs per one spin for $d = 2$ (squares), $d = 3$ (triangles), and $d = 4$ (circles) bond-percolating antiferromagnets at $p = p_c$. The results have been obtained by averaging over 11, 13, and 12 realizations of bond-percolating networks formed on $1,000 \times 1,000$, $96 \times 96 \times 96$, and $28 \times 28 \times 28 \times 28$ cubic lattices for $d = 2$, $d = 3$, and $d = 4$, respectively. For graphical reasons, the triangles ($d = 3$) and the circles ($d = 4$) are shifted upwards.
Fig. 4. The integrated DOSs per one spin for $d = 2$, $d = 3$, and $d = 4$ bond-percolating antiferromagnets at $p = p_c$. The symbols correspond to those in Fig. 3.

prepared on $1,000 \times 1,000$ square lattices at $p_c (= 0.5)$ with periodic boundary conditions. The largest network has 605,544 spins. The almost constant DOS for $\omega \ll 1$ is found in Fig. 3. The least squares fitting for filled squares in Fig. 3 leads to $\mathcal{D}(\omega) \propto \omega^{\tilde{d}_{AF}-1}$, with $\tilde{d}_{AF} = 0.97 \pm 0.06$.

The DOS and the integrated DOS for $d = 3$ percolating antiferromagnets at $p_c (= 0.2488)$ are shown in Figs. 3 and 4 by filled triangles, respectively. The BP networks of 13 realizations (the largest network has 177,886 spins) are formed on $96 \times 96 \times 96$ cubic lattices with periodic boundary conditions for all directions. The calculated DOS at low frequencies is constant as a function of frequency as well as the case of $d = 2$ percolating antiferromagnets. The value of $\tilde{d}_{AF} = 0.97 \pm 0.03$ is obtained by the least squares fitting from data in Fig. 3.

The DOS and the integrated DOS of spin-waves in $d = 4$ percolating networks are given by filled circles in Figs. 3 and 4, respectively. The BP networks of 12 realizations (the largest network has 26,060 spins) are formed on $28 \times 28 \times 28 \times 28$ hypercubic lattices at $p_c (= 0.160)$. The data of the DOS indicate $\tilde{d}_{AF} = 0.98 \pm 0.09$ for $d = 4$ antiferromagnetic fractons.

Our remarkable findings ($\tilde{d}_{AF} \approx 1$ independent of Euclidean dimension d) can be analyzed in terms of a dynamic scaling argument. One can express the scaling form of the dispersion relation of spin-waves as $\omega = k^{z_{AF}} f(k\xi)$, where z_{AF} is the dynamical exponent, and $f(x)$ is the scaling function.[2,3] For magnetic excitations in percolating antiferromagnets *above* p_c, there should be the characteristic frequency ω_c corresponding to the wavelength λ equal to the finite correlation length ξ. The *magnon* dispersion below ω_c behaves as normal, but with the p-dependent stiffness constant $C(p)$, i.e., $\omega = C(p)k$. The criticality of $C(p)$ in hydrodynamic limit is given[4] as $C(p) \propto (p - p_c)^{(\mu+\tau)/2}$, where

μ is the conductivity exponent and τ the exponent of the transverse susceptibility $\chi_\perp$ defined as $\chi_\perp(p) \propto (p - p_c)^{-\tau}$, respectively.

For $\lambda \gg \xi$, one has $f(x) \propto x^{1-z_{AF}}$ for $x \ll 1$, reflecting the linear dispersion relation. Combining these, the dynamic exponent becomes

$$z_{AF} = \frac{\mu + \tau}{2\nu} + 1 \ , \tag{2}$$

where ν is the exponent for the correlation length. The scaling function $f(x)$ for $x \gg 1$ should behave as $f(x) = const.$ because the dispersion relation does not depend on the correlation length ξ, namely, the dispersion relation for $x \gg 1$ becomes $\omega \propto k^{z_{AF}}$, This dispersion relation leads to the form of the DOS proportional to $\omega^{\tilde{d}_{AF}-1}$, where $\tilde{d}_{AF} = D_f/z_{AF}$ and D_f the fractal dimension of the percolating network, respectively. From eq. (2), the fracton dimension $\tilde{d}_{AF}$ of antiferromagnetic fractons is given by

$$\tilde{d}_{AF} = \frac{2\nu D_f}{\mu + \tau + 2\nu} \ . \tag{3}$$

Using the lower bound of the susceptibility exponent $\tau \geq \mu - \beta + (2 - d)\nu$,[4] eq. (3) yields the upper bound of the spectral dimension as

$$\tilde{d}_{AF} \leq \frac{2\nu D_f}{2\mu - \beta + (4 - d)\nu} \ . \tag{4}$$

By substituting the Alexander-Orbach conjecture $\tilde{d} = 4/3$ or equivalently $\mu = [(3d - 4)\nu - \beta]/2$ into eq. (4), we find that the upper bound of $\tilde{d}_{AF}$ of eq. (4) is *unity*, Our numerical results strongly support $\tilde{d}_{AF} \cong 1$. We claim the value of $\tilde{d}_{AF}$ to be very close to *unity* independent of the Euclidean dimension d.[5]

CONCLUSIONS

We have investigated the dynamic properties of $d = 2\text{-}4$ *elastic* and *antiferromagnetic* percolating networks. Antiferromagnetic fractons belong to the different universality class from that of vibrational or ferromagnetic fractons. The fracton dimension $\tilde{d}_{AF}$ of antiferromagnetic fractons are very close to unity ($\tilde{d}_{AF} \approx 1$) independent of the Euclidean dimension d.

I would like to thank S. Alexander and K. Yakubo for fruitful discussions. This work was supported in part by a Grant-in-Aid of Scientific Research of the Special Project, "Computational Physics as New Frontiers in Condensed Matter Research", of the Japan Ministry of Education, Science and Culture.

REFERENCES

1. S. Alexander and R. Orbach, J. Phys. (Paris) **43**: L625 (1982).

2. E. F. Shender, Zh. Eksp. Teeor. Fiz. **75**: 352 (1978), [Sov. Phys. JETP, **48**: 175 (1978)].

3. A. Christou and R. B. Stinchcombe, J. Phys. C **19**: 5895 (1986); *ibid*. **19**: 5917 (1986).

4. A. B. Harris and S. Kirkpatrick, Phys. Rev. B **16**: 542 (1977).

5. K. Yakubo, T. Terao, and T. Nakayama, J. Phys. Soc. Jpn. **62**: 2196 (1993).

FRACTONS IN COMPUTER AND LABORATORY EXPERIMENTS

René Vacher

Laboratoire de Sciences des Matériaux Vitreux
Université de Montpellier II
34095 Montpellier Cedex 5, France

Eric Courtens

IBM Research Division
Zurich Research Laboratory
8803 Rüschlikon, Switzerland

The current status of computer simulations and laboratory experiments on fractal aerogels is reviewed and compared to predictions based on dynamical scaling assumptions that have been made by Alexander and collaborators. The experimental results are so far in remarkable agreement with the single-length-scale postulate for fractons and with scaling hypotheses.

INTRODUCTION

The fractons considered here are the vibrational eigenmodes of strongly disordered fractals. Their properties have been analyzed theoretically on the basis of two central assumptions: a single-length-scale postulate (SLSP), and dynamical scaling hypotheses.[1-5] The postulate is that fractons of frequency ω are characterized by a single average dynamically relevant length scale, $\lambda(\omega)$. This length plays simultaneously the role of the wavelength, the scattering length, and the localization length, and no relevant length can be defined which does not scale like λ. The main dynamical scaling hypothesis is that the density of fracton states, $N(\omega)$, obeys a power law

$$N(\omega) \propto \omega^{\tilde{d}-1} , \qquad (1)$$

where $\tilde{d}$ is the "spectral" or "fracton" dimension.[1-3]
Coupled with the SLSP, it leads to[2,3]

$$\lambda(\omega) \propto \omega^{-\tilde{d}/D} , \qquad (2)$$

where D is the fractal dimension of the vibrating object.

A second scaling hypothesis concerns the modulation of the density in the embedding space by the vibration. This relates to the strain e, whose coherent part should only be sensitive to the average relation between the vibrational amplitude u_j, and the positions r_i within a given fracton. Averaged over many fractons of frequency ω, the strain is expected to scale as[4,5]

$$e(\omega) \propto u(\omega) / [\lambda(\omega)]^\sigma. \tag{3}$$

This defines a new scaling index σ.

The main purpose of this brief review is to examine the interpretation of experimental evidence in the light of the above assumptions. One finds that the consequences of these assumptions are so far in remarkable agreement with the results found both in computer and laboratory experiments.

EXPERIMENTAL SYSTEMS

To obtain valid experimental tests, it is necessary to study systems that are "good" disordered fractals, meaning that they should at least be characterized by a well-defined constant D over a significant range of lengths. Various methods can be applied to check for this, one of the simplest being to measure the static structure factor, $S(q)$. A necessary condition for a good fractal is that over a sufficient range of the wave vector q one obtains $S(q) \propto q^{-D}$.

Among the numerous computer models of fractals, the ones that have been most extensively studied are derived from percolation.[6] In particular, the "infinite" percolation cluster that remains after all finite clusters have been removed from a simulation is expected to be a mass fractal, the mass m within a sphere of radius r centered on an occupied site scaling on the average as $m \propto r^D$, at least for sufficiently large values of r. It is not the random filling of sites or bonds that generates this fractality, but it is the removal of the finite clusters that exist at all length scales. For that reason, bond-percolation clusters form good mass fractals already at small r, as there is no initial random filling of sites in the preparation of these objects. The latter characterizes site-percolation clusters and it produces a white-noise background in $S(q)$. This makes site-percolation clusters unsuited for a critical test of scaling laws up to very large cluster sizes.[7]

Several laboratory systems have been used to test dynamical properties of fractals, among which sintered powders,[8] fractal aggregates,[9] epoxy resins,[10] or diluted antiferromagnets.[11] It is however on silica aerogels that the most extensive measurements could be performed so far. These materials are demonstrated fractals,[12,13] they are suited for optical, x-ray and neutron-scattering studies, and they can be reproducibly prepared with a broad range of structural characteristics. In particular, with appropriate preparation, it has been possible to obtain series of materials of different densities which are not distinguishable from each other when observed at scales where they are fractal. The materials only differ by the extent of their fractal range, the so-called persistence length ξ.[13] Such series, which we named *mutually self-similar*, are particularly useful as they allow the investigation of scaling laws by just measuring macroscopic properties of samples as function of ξ.

The fractal dimension of percolation networks is $D = 91/48$ for $d = 2$ (d is the Euclidean dimension of the embedding space) and $D \approx 2.48$ for $d = 3$.[6] For silica aerogels, depending on preparation, D ranges from ~1.7 to ~2.5. In particular for the mutually self-similar series[13] one finds $D = 2.40 \pm 0.03$. Dynamically these materials are however very different from percolation networks. At percolation, the connecting path is the backbone, of fractal dimensions lower than D. This implies that the mass is entirely contained in dangling arms, which makes percolation network very floppy. Owing to their preparation, by reaction in a wet state, aerogels are not expected to have such a high fraction of floppy ends. Their higher connectivity leads to higher values of $\tilde{d}$ and lower values of σ as compared to percolation.

THE SINGLE-LENGTH-SCALE POSTULATE

The most direct test of various definitions of fracton lengths is achieved by simulations. As prerequisite one needs a good fractal support and a calculation of all fracton eigenmodes on the support. The latter can be done by diagonalization of the dynamical matrix which is the most direct procedure, but which is numerically expensive. An alternate approach is to simulate the dynamics by excitation of the network at various frequencies,[14] but this requires then considerable caution in separating nearby eigenmodes.[15]

Five different definitions of fracton size have been calculated on bond-percolation clusters using diagonalization of the dynamical matrix in the model of scalar elasticity.[7] These definitions are: 1) the square root of the second moment, $\Re \equiv [<|\mathbf{r} - \bar{\mathbf{r}}|^2>]^{1/2}$, where $\bar{\mathbf{r}}$ is the center of gravity of the fracton as defined in Ref. 16, and the angular brackets indicate averaging over the wave function; 2) the localization length, defined by the Thouless criterion;[5,17] 3) the root-mean-square spread of localization lengths; 4) the wavelength defined by q_{max}^{-1}, where $q_{max}(\omega)$ is the position of the maximum in the structure factor $S(q, \omega)$; 5) the scattering length given by the width of the structure factor.

All these definitions are found to be in agreement with (2), giving the values 0.70 and 0.53 for the exponent $\tilde{d}/D$, for $d = 2$ and 3, respectively.[7] This provides a very strong support for the single-length-scale postulate, at least for that particular model.

DYNAMICAL SCALING

The first point to check is the scaling of the density of states, Eq. (1). Early simulations[18] confirmed the conjecture[2] that $\tilde{d} \approx 4/3$ for percolation clusters in all dimensions and for scalar elastic models. The corresponding fracton dimension is now written $\tilde{d}_s$, with s for "stretch" or "scalar," as vibrations of tenuous objects dominated by *stretching* motion map to scalar elastic models. When *bending* dominates, the fracton dimension $\tilde{d}_b$ is smaller than $\tilde{d}_s$ for a given support. This was also recognized rather early.[18] A crossover from bending to stretching can occur as the observation frequency is increased,[19] and this was recently confirmed in simulations.[20] The value $\tilde{d}_b$ for percolation networks is ≈ 0.8.

In laboratory experiments, $N(\omega)$ is accessible rather directly by incoherent neutron-scattering measurements. On silica aerogels, and at low frequencies near the phonon-fracton crossover, one finds $\tilde{d}_b \approx 1.3...1.4$,[21] a value significantly larger than that of percolation networks. At higher frequencies, a crossover is observed near 10 GHz, beyond which $\tilde{d}_s \approx 2.2$,[22] much larger than 4/3. The scaling law (2), which results from the combination of the SLSP with (1) could be verified on a mutually self-similar series to which the above samples also belong. This was achieved by Brillouin scattering determination of the phonon-fracton crossover frequency and length.[23] From the scaling of frequency with length one obtains $\tilde{d}_b/D \approx 0.53$, and from the scaling of length with density, $D \approx 2.46$, from which $\tilde{d}_b \approx 1.3 \pm 0.1$, in agreement with the neutron determination.[21] In accordance with the SLSP, the crossover length determined in Brillouin scattering is approximately proportional to the fractal persistence length determined from the static structure factor,[13] and the proportionality constant relating these two quantities can be well explained.[24] The Brillouin spectra amount of course to a measurement of $S(q, \omega)$, at small values of q.[23] The fact that they could be very well fitted without taking into account the scaling relation (3) might already indicate that (3) is trivial in this particular case, meaning $\sigma = 1$.

A determination of σ has been performed by scaling $S(q, \omega)$ for bond-percolation simulations.[7] Excellent scaling is obtained with $\sigma = 1.05$ for $d = 2$, and $\sigma = 1.11$ for $d = 3$. Intuitively, one expects that for $d = 3$ the network is more floppy than for $d = 2$, and this corresponds to the higher value of σ. For aerogels, which are much more connected than percolation clusters as already shown by the much larger values of $\tilde{d}$, it is not surprising that one might find $\sigma = 1$ for all practical purposes. That value agrees with the scaling of the depolarized Raman scattering intensity[25,26] assuming it follows the predictions for the dipole-induced-dipole mechanism,[5]

$$I/n \propto \omega^{[(\tilde{d}/D)(2\sigma + 6) - 2\tilde{d} - 2]} \tag{4}$$

Here, I is the intensity and n is the Bose-Einstein factor. The values $\tilde{d}_b/D = 0.53$, $\tilde{d}_b = 1.3$, and $\sigma = 1$ lead to the observed exponent $- 0.36$.[25] That the bending fracton dimension is the proper one to use in (4) is consistent with the fact that it is the bending modes that depolarize the scattered light.

CONCLUSIONS

So far all observations are consistent with the fundamental assumptions that needed to be made to formulate a scaling theory of fractons. Experiments have revealed the dynamical exponent $\tilde{d}$ and σ and the remarkable scaling of $S(q, \omega)$. Clearly, the new exponents are not "universal," but provide additional information describing the dynamical properties of fractal objects.

ACKNOWLEDGMENTS

The authors have greatly benefitted from enlightening discussions with many colleagues, and in particular with Prof. Shlomo Alexander to whom this article is dedicated. Without his contributions to the field this subject would not have known the current development. The authors would also like to thank the numerous colleagues who participated in the measurements and simulations, in particular the co-authors whose names appear in Refs. 7, 13, 21, 22 and 24.

REFERENCES

1. S. Alexander, Percolation Structures and Processes, G. Deutscher, R. Zallen and J. Adler, eds., *Am. Isr. Phys. Soc.* 5:149 (1983).
2. S. Alexander and R. Orbach, *J. Phys. (Paris) Lett.* 43:L-625 (1982).
3. R. Rammal and G. Toulouse, *J. Phys. (Paris) Lett.* 44:L-13 (1983).
4. S. Alexander, *Phys. Rev. B* 40:7953 (1989).
5. S. Alexander, E. Courtens, and R. Vacher, *Physica A* 195:286 (1993).
6. D. Stauffer, "Introduction to Percolation Theory," Taylor and Francis, London and Philadelphia (1985).
7. E. Stoll, M. Kolb, and E. Courtens, *Phys. Rev. Lett.* 68:2472 (1992).
8. M.C. Malieppard, J.H. Page, J.P. Harrison, and R.J. Stubles, *Phys. Rev. B* 32:6261 (1985).
9. T. Freltoft, J.K. Kjems, and D. Richter, *Phys. Rev. Lett.* 59:1212 (1987).
10. A.J. Dianoux, J.H. Page, and H.M. Rosenberg, *Phys. Rev. Lett.* 58:886 (1987).
11. Y.J. Uemura and R.J. Birgeneau, *Phys. Rev. B* 36:7024 (1987).
12. D.W. Schaefer and K.D. Keefer, *Phys. Rev. Lett.* 56:2199 (1986).
13. R. Vacher, T. Woignier, J. Pelous, and E. Courtens, *Phys. Rev. B* 37:6500 (1988).
14. M.L. Williams and H.J. Maris, *Phys. Rev. B* 31:4508 (1985).
15. K. Yakubo, T. Nakayama, and H.J. Maris, *J. Phys. Soc. Jpn.* 60:3249 (1991).
16. E. Courtens, R. Vacher, and E. Stoll, *Physica D* 38:41 (1989).
17. D.J. Thouless, *Phys. Rep. C* 13:94 (1974).
18. G.S. Grest and I. Webman, *J. Phys. (Paris) Lett.* 45:L-1155 (1984); I. Webman and G.S. Grest, *Phys. Rev. B* 31:1689 (1985).
19. S. Feng, *Phys. Rev. B* 32:5793 (1985).
20. K. Yakubo, K. Takasugi, and T. Nakayama, *J. Phys. Soc. Jpn.* 59:1909 (1990).
21. E. Courtens, C. Lartigue, F. Mezei, R. Vacher, G. Coddens, M. Foret, J. Pelous and T. Woignier, *Z. Phys. B* 79:1 (1990).
22. R. Vacher, E. Courtens, G. Coddens, A. Heidemann, Y. Tsujimi, J. Pelous, and M. Foret, *Phys. Rev. Lett.* 65:1008 (1990).
23. E. Courtens, R. Vacher, J. Pelous, and T. Woignier, *Europhys. Lett.* 6:245 (1988).
24. R. Vacher, E. Courtens, E. Stoll, M. Böffgen, and H. Rothuizen, *J. Phys. Condens. Matter* 3:6531 (1991).
25. Y. Tsujimi, E. Courtens, J. Pelous, and R. Vacher, *Phys. Rev. Lett.* 60:2757 (1988).
26. E. Courtens and R. Vacher, *Phil. Mag. B* 65:347 (1992).

SOME REMARKS ON THE ELASTIC PROPERTIES OF THE NEMATIC AND SMECTIC A PHASES OF LIQUID CRYSTALS

L. Benguigui

Solid State Institute and Physics Department
Technion - Israel Institute of Technology
32000 Haifa, Israel

INTRODUCTION

The purpose of this paper is to emphasize some analogy between the mechanical response of the two most studied phases of liquid crystals: nematics and smectics A. In particular, we show that in nematics a new acoustical wave could be observed. This new wave has some formal analogy with the second sound in the smectics, although the physical origin is completely different.

ELASTIC PROPERTIES OF THE SMECTIC A PHASE

One can find a well detailed analysis of the elastic properties of the smectics in the review article of Miyano and Ketterson [1]. The essential point is that there are two elastic quantities needed to describe the state of the sample – the density and the strain perpendicular to the layers. We note ρ the relative change of the density from a reference state chosen to be the nematic-smectic transition point, and x_3 the second elastic variable. x_3 can be also seen as the relative variation of the layer thickness.

If one wants to use the conventional language of the elastic theory one considers as the two basic quantities x_3 defined as above and $(x_1 + x_2) = \rho - x_3$ where x_1 and x_2 are the deformations along two perpendicular axis parallel to the layer planes.

Using these two approaches, one can write the elastic energy as follows:

$$F = \frac{1}{2}A\rho^2 + \frac{1}{2}Bx_3^2 - Cx_3\rho \tag{1a}$$

or

$$F = \frac{1}{2}C_{11}(x_1 + x_2)^2 + \frac{1}{2}C_{33}x_3^2 + C_{13}(x_1 + x_2)x_3 \tag{1b}$$

Soft Order in Physical Systems, Edited by Y. Rabin
and R. Bruinsma, Plenum Press, New York, 1994

The transformation formulas from one set of elastic constant $(A, B$ and $C)$ to the other $(C_{11}, C_{13}$ and $C_{33})$ are easily deduced from (1a) and (1b):

$$A = C_{11}, \quad B = C_{11} + C_{13} - 2C_{13}, \quad C = C_{11} - C_{13} \tag{2}$$

From the free energy (1a) or (1b), one can calculate the particular elastic properties of the smectics A. We shall recall two effects: the sound propagation and the pressure to apply in the case of a smectic A sample confined in a box.

A. Sound propagation [1] [2]

The most striking property is the existence of two types of sound waves. The first is the regular longitudinal wave. The velocity depends on the angle θ, between the wavevector and the director and is given by the following expressions:

$$V_1^2 = \frac{1}{\rho_1}[C_{11} - 2(C_{11} - C_{13}) \cos^2 \theta + (C_{11} + C_{33} - 2C_{13}) \cos^4 \theta] \tag{3a}$$

$$V_1^2 = \frac{1}{\rho_1}[A - 2C \cos^2 \theta + B \cos^4 \theta] \tag{3b}$$

ρ_1 is the density of the material. The second kind of wave is a transverse wave with a velocity,

$$V_2^2 = \frac{1}{\rho_1} B \sin^2 \theta \cos^2 \theta \tag{4}$$

One notes that the "second sound", as this wave was called by de Gennes [2] by analogy with Helium, can propagate only if the wave vector is not parallel to the director, nor perpendicular to it. The only coefficient which appears in (4) is B, showing the close connection between the second sound and the layer structure of the smectic A, since B can be seen as the compressibility of the layers.

B. Smectic A confined in a box

We consider a smectic A sample confined in a cylindric box and we suppose that it is possible to vary the volume by a mobile piston. For simplicity, the plane of the piston is parallel to the layers. If we increase the pressure applied on the piston by $P_\parallel$, what is the increase $P_\perp$ in the pressure applied by the walls parallel to the director? Maybe it is more transparent to use eq. (1A) and one can write the relation between the stress $P_\parallel$ and $P_\perp$ and the strains x_1, x_2 and x_3:

$$P_\parallel = \frac{\partial F}{\partial x_3} = C_{13}(x_1 + x_2) + C_{33}x_3$$

$$P_\perp = \frac{\partial F}{\partial(x_1 + x_2)} = C_{11}(x_1 + x_2) + C_{13}x_3 \tag{5}$$

Using the relations between the C_{ij}'s and A, B, C, one gets

$$\begin{aligned} P_\perp &= A\rho - Cx_3 \\ P_\parallel - P_\perp &= -C\rho + Bx_3 \end{aligned} \tag{6}$$

Using first the usual elastic formalism and writing (6) from (5) permits us to give the physical meaning to the derivative $(\partial F/\partial x_3)_\rho = P_\parallel - P_\perp$. In the present situation

$x_1 + x_3 = 0$ (no deformation parallel to the layer) and $\rho = x_3$, one has $P_\perp = (A-C)x_3$ and $P_\parallel = P_\perp (A+B)/(A-B)$. There is a difference between the pressure $P_\parallel$ parallel to the director and the pressure $P_\perp$ perpendicular to the director.

ELASTIC PROPERTIES OF THE NEMATIC PHASE

Since a nematic is a liquid, the only thing one can do by changing the pressure is to modify the volume. One can define as an elastic constant the bulk compressibility. Therefore the only propagating sound wave is the longitudinal one. The velocity is independent of the direction, perpendicular or parallel to the director. Of course, this is correct only at very low frequency. In the acoustic range of the megahertz, the measurement of the sound wave gives an isotropy of V. However, the absorption is anisotropic. Because of the Kramer-Kronig relation, this means that the velocity is also anisotropic but this anisotropy is too small to be measurable.

A. Sound propagation at high frequency

Experimentally the attention has been attracted principally by the absorption [3] for which the anisotropy is easily observed. Only recently, detailed studies of the velocity anisotropy were published [4] in the case of a polymer liquid crystal. The advantage of a polymer is that the relaxation time is high such that at 1 Mhz the velocity anisotropy is almost 1%. The disadvantage is that this anisotropy is already almost independent of the frequency.

The interesting result is the variation of V with θ (defined as above as the angle between the wavevector and the director). It can be described by

$$V_N^2 = \frac{1}{\rho_1}[A_N - 2C_N \cos^2\theta + B_N \cos^4\theta] \tag{7}$$

which is formally identical with (3b). In fact, Jahnig [5] has already made a theoretical analysis of the elastic coefficients at high frequency. From his results, one can get easily (7) if one supposes that the increments of the compressibility due to the frequency are much smaller that the compressibility itself. The experimental results confirm this assumption and they are the first indication of the formal analogy between nematics and smectics. The essential difference between the two phases is that the coefficients B_N and C_N go to zero with the frequency. This is not the case for smectics A, for which the de Gennes constant B was measured at low frequency (~ 20kHz) [6][7].

An interesting consequence of the preceding results is that the elastic energy of a nematic at finite frequency has exactly the same form as that of a smectic A.

$$F_N = \frac{1}{2}A_N\rho^2 + \frac{1}{2}Bx_3^2 - C_N\rho x_3 \tag{8}$$

The deformation x_3 has nothing to do with a layer deformation; it is the instant deformation in direction of the director. Because of the uniaxial symmetry, we cannot have another elastic variable. (This point is easily seen when using the C_{ij} coefficients).

We can draw an immediate consequence of (8): we expect a propagative transverse wave with the velocity

$$V_{2N}^2 = \frac{1}{\rho_1}B_N \cos\theta \sin\theta \tag{9}$$

We cannot call this wave a "second sound" keeping this term for the case of smectics, because the physical mechanism permitting the acoustic anisotropy is completely different in nematics and smectics. In nematics it comes from relaxation effects (probably relaxation of the nematic order parameter), i.e. it is a dynamical effect.

B. Transient effects

As above, we consider a nematic sample confined in a cylindric box with a mobile piston. We neglect also, as in the case of the smectic sample, the influence of the walls.

First, the sample is prepared such that the director is perpendicular to the piston surface. At a given time, the pressure transmitted by the piston is changed suddenly by a small amount and one follows the change in the volume. (To avoid propagation phenomena, the sample is supposed to be small enough.) The volume will adapt immediately a value $V_\parallel$ and will relax toward a final value V_0, at long time scale.

If one repeats the experiment with the director parallel to the piston, the immediate value $V_\perp$ will be different to $V_\parallel$, but the final value will be V_0 as in the preceding case. This experiment can help to understand what is anisotropic behavior in nematics

As in the smectic sample confined in a box, this experiment is idealized. Its interest is to show that the phenomena will be different at short time scale but not at long time scale, when the only static coefficient is the compressibility. Transient phenomena are anisotropic in accordance with the anisotropy at high frequency.

Acknowledgements

The author thanks S. Alexander, E. Kats, V. Lebeder and P. Martinoty for very useful discussions. This research was supported by the Technion Fund for the Promotion of Research at the Technion.

References

1. K. Miyano and J.B. Ketterson "Sound Propagation in Liquid Crystals" in "Physical Acoustics", Vol. 14, W.P. Mason and R.N. Thurston Ed., Academic Press, New York (1979).

2. P.G. de Gennes, J. Phys. (Paris). Colloq. **30** C4-65 (1969).

3. C.A. Castro and C. Elbaum, Mol. Cryst. Liq. Cryst. **41**, 169 (1978).

4. a) L. Benguigui, P. Ron, F. Hardouin and M. Mauzac, J. Phys. (Paris) **50**, 529 (1989).
   b) L. Benguigui and P. Ron, Mol. Cryst. Liq. Crystal, **153**, 241 (1987).

5. F. Jahnig, Z. Phys. **258**, 199 (1973).

6. M.R. Fisch, L.B. Sorensen and P.S. Pershan, Phys. Rev. Lett. **47**, 43 (1981).

7. M. Benzekri, T. Claverie, J.P. Marcerou and J.C. Rouillon, Phys. Rev. Lett. **68**, 2480 (1992).

PERMEABILITY OF A SOAP FILM

Sylvie Cohen-Addad, David Quéré

Laboratoire de Physique de la Matière Condensée, URA 792 du CNRS,
Collège de France
11, place Marcelin-Berthelot, 75231 Paris Cedex 05, France

INTRODUCTION

We study the permeability of a soap bubble inflated with a gas nearly insoluble in water: even in such an unfavourable case, the bubble leaks. The role played by the surfactant in the transport of gas is investigated.

PERMEABILITY OF A SOAP BUBBLE

This first experiment consists in making a bubble with air saturated with heptane. The gaseous mixture is brought to the center of a surfactant droplet pinned between two horizontal glass plates: in this way, a "flat bubble" is inflated (radius of such a circular bubble: 4 cm, distance between the plates: 1.5 mm, and mean thickness of the film: 0.56 mm). Then the concentration of heptane inside the bubble is measured as a function of time: gas samples of 100 µl (i.e. about 1% of the initial volume of the bubble) are periodically taken with a syringe, and analysed by gas chromatography.

Results of such an experiment are shown in figure 1. The bubble is made from a solution of fluorinated surfactant [CF_3-$(CF_2)_5$-$(CH_2)_2$-SO_2-NH-$(CH_2)_3$-$(N^+CH_3CH_3)$-CH_2-COO^-] in water ($c_S = 3 \cdot 10^{-4}$ g/g = 8 cmc).

Soft Order in Physical Systems, Edited by Y. Rabin
and R. Bruinsma, Plenum Press, New York, 1994

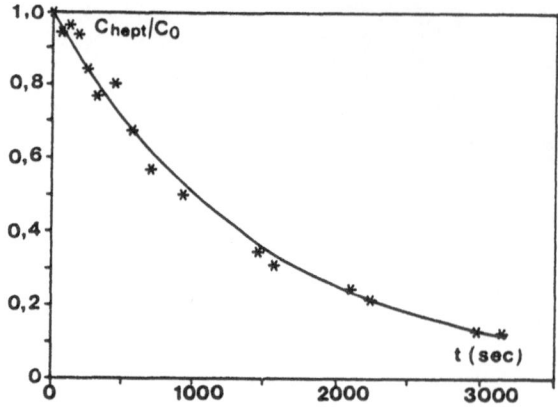

Figure 1. Concentration of heptane inside a bubble as a function of time.

The concentration can be fitted by a decreasing exponential (the line in the figure), from which a leak time τ can be deduced: we get $\tau = 1430 \pm 30$ s. If some hydrosoluble polymer (we used xanthan, of molecular weight $M = 2.5\ 10^6$ g) is added to the surfactant solution from which the bubble is inflated, similar curves are obtained: the bubble still leaks, but paradoxically faster than in the absence of xanthan. Besides, leak times decrease when the polymer concentration C_p increases (even though the solution becomes more viscous with increasing C_p): for $C_p = 10^{-3}$ g/g, we get $\tau = 830 \pm 50$ s and for $C_p = 2\ 10^{-3}$ g/g, $\tau = 630 \pm 50$ s. We do not understand this effect of the polymer.

A bubble made from a solution of Sodium Dodecyl Sulfate (SDS) is also permeable to heptane and the permeability is a function of the surfactant concentration. In order to analyze more precisely this process, and in particular the role played by the surfactant, we studied how heptane permeates soapy droaplets.

PERMEABILITY OF SOAPY DROPLETS

Because it is impossible to inflate long-lasting bubbles with a slightly soapy water $(0 < c_S < \text{cmc})$, we studied the permeability of thin droplets trapped inside small teflon tubes (diameter: 1.6 mm).

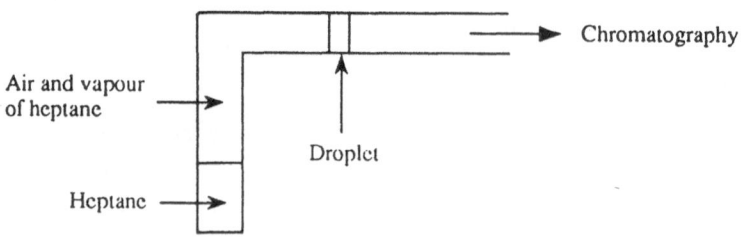

Figure 2. Experimental set-up to study the permeability of a soapy droplet.

The experimental set-up is drawn in figure 2: one side of the droplet is exposed to the vapour of heptane; air is periodically picked up on the other side, and analysed by gas chromatography. The volume of the gas sample (50 µl) exactly corresponds to the volume of the tube from the drop to its end. The droplet (2 mm long) is a solution of SDS. The results obtained for various SDS concentrations are gathered in figure 3.

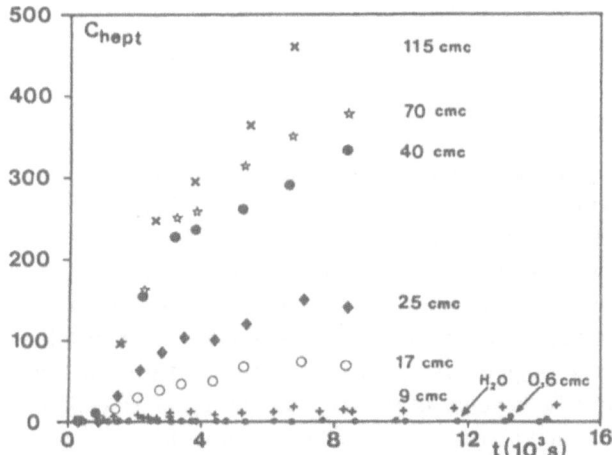

Figure 3. Quantity of heptane (in arbitrary unit) which permeates the droplet as a function of time, for various surfactant concentrations.

As shown in this figure, pure water and low concentrated (c_S = 0.6 cmc) droplets are heptaneproof: the concentration of heptane remains negligible. On the contrary, heptane permeates the droplet from the moment the SDS concentration exceeds the cmc: micelles are responsible for the permeation. Besides, a latent period τ_L is necessary for it to occur, τ_L being roughly independent of c_S (here we find that τ_L is about 15 mn). Then the heptane concentration increases with time, all the faster as c_S is large. Finally the curves seem to saturate: the value of the saturation also depends on c_S; it should indeed be proportional to the amount of heptane dissolved in solution. These remarks lead us to complementary experiments or comments.

Firstly we measured τ_L as a function of the length of the droplet L (at fixed concentration: c_S = 40 cmc). For 0 < L < 8 mm, we found a straight line passing through the origin. This fact probably proves that the micelles are carried inside the droplet by convection (convection which could be generated by small temperature differences on both sides of the droplet). To get rid of convection, we did the experiment with a droplet soaking an absorbent cotton: the time lag then tremendously increases (from one quarter of hour, it becomes of the order of one day!), revealing diffusion transport instead of convection.

Secondly it must be emphasized that at 20 °C the micellar phase of SDS solutions extends to 35 cmc. Beyond this concentration, other organized phases appear and they also should take part in the transport of heptane. Three curves in figure 3 belong to this high-concentration regime and possibly exhibit a particular behaviour (a kink at t ≈ 1 hour ?). These concentrated phases may imply other ways of carrying heptane - what we still have to understand.

Finally a similar study was performed with droplets of fluorinated surfactant. We found that contrary to the SDS case, their permeability is independent of the surfactant concentration and always remains equal to that of pure water. In this case, micelles cannot solubilize heptane (probably because heptane does not wet the fluorinated chains) and thus do not take part in the permeability process. Hence the leak of the fluorinated bubbles, described in the first paragraph, is simply understood in view of the (low) solubility of heptane in these solutions.

CONCLUSION

We have studied the permeability of soap films, starting from the observation that a bubble inflated with a mixture of air and heptane empties of the heptane molecules. The emptying process depends on the nature of the surfactant. In the SDS case, the micelles present in the solution are shown to be responsible for the transport of heptane through the film, what suggests as a possible scenario: heptane first wets the layer of surfactants exposed to the vapour; these surfactants exchange with naked ones coming from micelles; then heptane-loaded micelles are carried through the film, either by diffusion if it is thin, or by convection in the other case; and finally heptane is released on the other side of the film. The whole comprehension of such a scenario would allow to know how to formulate films of a given permeability. In the fluorinated case, the leak is much weaker, and simply explained by the solubility of heptane in water.

ACKNOWLEDGMENTS

We thank J.-M. di Meglio for fruitful help, D.H. Smith for an interesting suggestion and Elf-Atochem for financial support.

REFERENCES

See for instance: about the permeability of: (i) soap bubbles towards ether: C.V. Boys, *Soap Bubbles,* Dover, New-York, 1959; (ii) thin films towards small molecules: H.M. Princen, J.T. Overbeek and S.G. Mason, *J.C.I.S.,* 24, 125, 1967; about the solubilization of alcanes by surfactants: D.E. Clarke and D.G. Hall, *Colloid and Polymer Sci.,* 252, 153, 1974.

VESICLES OF HIGH TOPOLOGICAL GENUS

X. Michalet[1], D. Bensimon[1], B. Fourcade[2]

[1] Ecole Normale supérieure, Laboratoire de Physique Statistique
24 rue Lhomond , 75231 Paris Cedex 05, France
[2] Institut Laue Langevin, 156X Grenoble Cedex 38042, France

I. INTRODUCTION

We report the observation of phospholipid vesicles of high topological genus, exhibiting strong thermal fluctuations. By deeply affecting the global shape of the vesicle, they differ from the usual local thermal undulations of the membrane. They can be described as positional fluctuations of necks linking 2 nearby concentric membranes. We then present the results of a simple model for the shape of the necks based on an electrostatic analogy. This approach is corroborated by a numerical solution of the minimization problem [1].

II. EXPERIMENTAL OBSERVATIONS

Our vesicles are prepared using a phospholipid purchased from Avanti Polar Lipids, following a procedure already described in [2] and observed by phase contrast microscopy, which shows a cut of the membranes perpendicular to the focal plane.

Vesicles can be classified according to their volume (V) to surface (S) ratio, namely their dimensionless reduced volume: $v_{red} \equiv 6\sqrt{\pi}V/S^{3/2} \leq 1$.

We have already observed [2] that the shape of vesicles of high topological genus with $v_{red} \approx 1$ presents small thermal fluctuations at equilibrium, as may be expected from the simplest model for the curvature elastic energy (first introduced by Canham, Helfrich and Evans) [3]: $\mathcal{E} = \kappa/2 \iint H^2 dS$, where $\kappa/k_B T \approx 20$, H is the local mean curvature. The high value of $\kappa/k_B T$ indeed implies that the undulation modes of the membrane perturbe the overall shape by a few percents only. For small enough reduced volume v_{red}, however, strong fluctuations of the global shape are observed.

Fig.1 shows 3 snapshots of a vesicle taken at intervals of a few seconds. The vesicle can be viewed as two nearby and concentric spheres $\approx 4\mu m$ apart, with radii of $\approx$

$10\mu m$, connected by five necks with radii **a** of $\approx 1\mu m$. Their relative positions strongly fluctuate: the relative distance between the two necks, **L**, in the left-bottom corner of the pictures varies by a factor of two!

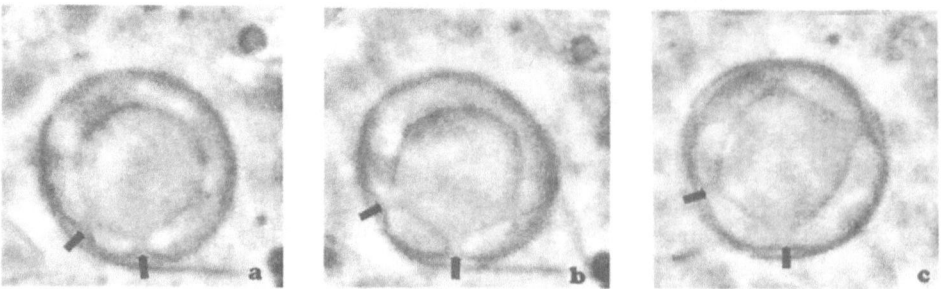

FIG. 1. fluctuating vesicle of genus 4

III. ANALYTICAL AND NUMERICAL RESULTS

To describe the experimental observations, we propose a simple model for the shape of the necks. For simplicity, we consider the shape of a neck linking two asymptotically flat parallel surfaces, and in order to have a finite volume,we take square parallel pieces of membrane of size $L \times L$ with periodic boundary conditions (PBC).

III.1. SHAPE

The solution minimizing the elastic energy with PBC on this cell is obtained using a boundary layer method. We shall use the Monge representation where the deviation of the membrane from the (x, y) plane is given by $\xi(x, y)$.

- Far from the neck ($a \ll r < L$), we expect a flat membrane, and may neglect the non-linear terms in H. Minimization of $\mathcal{E}$ then yields the usual bi-harmonic equation describing small fluctuations of a flat membrane, $\Delta H = \Delta^2 \xi = 0$.

  Its solution describing a neck and satisfying the PBC: $H(x \pm L, y \pm L) = H(x, y)$ can be obtained by solving the electrostatic (Poisson) problem obtained by integration of the last equation: $\Delta \xi = 4\pi\sigma = 2\pi a/L^2$, with a point charge $q = -a/2$ at the position of each neck, and PBC. Its solution is:

$$\xi_o(x, y) = Re\left\{a \ln\left(\frac{2L}{\pi a} \sin\left(\frac{\pi z}{L}\right) \prod_{m>0} [1 + \left(\frac{\sin \pi z}{\sinh \pi m}\right)^2])\right)\right\} - \frac{\pi a}{L^2} y^2.$$

- Near the neck ($a \leq r \ll L$), the non-linear terms in H cannot be neglected anymore. However the solution $H = 0$, which clearly minimizes $\mathcal{E}$ is the catenoid $\xi_i(r) = a \cosh^{-1}(r/a)$.

- These solutions matches logarithmically in the intermediate region and a uniform leading order approximation to the neck shape is: $\xi_{unif}(r) = \xi_o(r) + \xi_i(r) - a \ln(2r/a)$ in the vicinity of the $(0, 0)$ site.

200

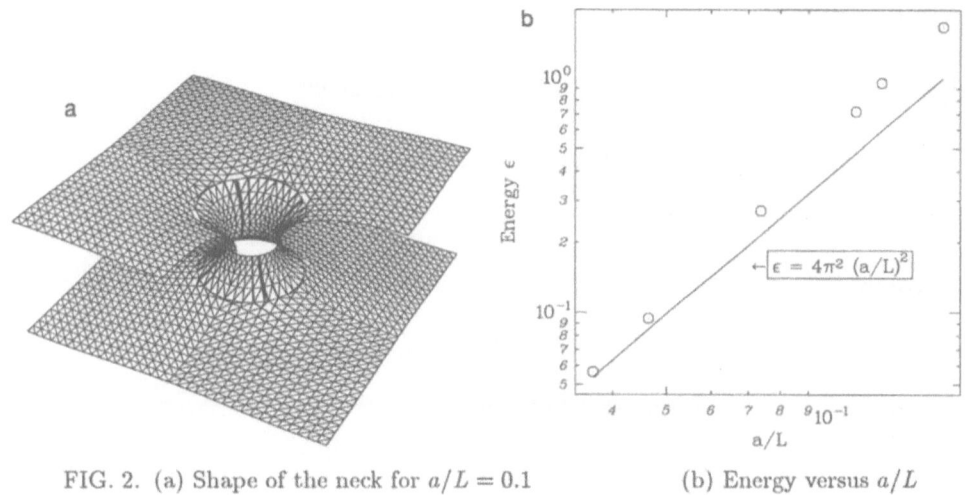

FIG. 2. (a) Shape of the neck for $a/L = 0.1$ (b) Energy versus a/L

In order to corroborate our previous discussion, we used a numerical minimization procedure for the preceding problem [1]. The shape of the neck is depicted in Fig.2(a) and is a surface whose curvature is roughly constant but varies very much in the near vinicity of the neck.

III.2. ENERGY

Finally, the energy of that neck is $\mathcal{E} = \kappa/2 \iint H^2 dS = 2 \times 2\pi^2 \kappa (a/L)^2$, which is minimal for $a = 0$. In fact, the stabilisation of a finite size neck ($a \neq 0$) is due to the (reduced) volume constraint.

Fig.2(b) shows a log-log plot of the energy per neck as a function of a/L, obtained numerically. The agreement with the preceding calculation (solid line) is very good as one approaches the domain of validity of the preceding method ($a/L \ll 1$).

III.3. INTERACTION

When two necks (or more) begin to overlap, the inner problem consists in determining a surface subtended by n disconnected contours ($n \geq 2$), with an asymptotic logarithmic behaviour matching the outer solution. A result due to R. Schoen [4] states that the only minimal surface of that kind is the catenoid: the inner solution is thus not a minimal surface anymore ($H \neq 0$), and the elastic energy of the system increases.

In conclusion, **to leading order**, the necks behave as **a gas of free particles** with **a hard core repulsion** whose range is $\approx \sqrt{aL}$, which is in qualitative agreement with our experimental observations.

REFERENCES

[1] Michalet X., Bensimon D., Fourcade B., *preprint, march 93*

[2] Fourcade B., Bensimon D., Mutz M., *Phys. Rev. Lett.* **68** (1992) 2551. Fourcade B., *J. de Phys. II (Fr.)* **2** (1992) 1705.

[3] Canham P.B., *J. Theor. Biol.* **26** (1970) 61. Helfrich W., *Z.Naturforschung* **28** c (1973) 693. Evans E.A., *Biophys. J.* **14** (1974) 923.

[4] Schoen R., *J. of Diff. Geom.* **18** (1983) 791.

AN INCOME-DISTRIBUTION DEMAND-SHIFT
MODEL OF INFLATION

Esther Alexander

The Ministry of Labor and Social Affairs
State of Israel
Kaplan St. 2, Jerusalem, Israel

"There are, of course, in principle, policies other than aggregate demand management to which we might turn and which are enticing in view of the unpleasant alternatives offered by demand management. The design of better alternatives is probably the greatest challenge presently confronting those interested in stabilization"

Modigliani (F.Modigliani, 1977)

I. INTRODUCTION

In the present paper we try to explore the dynamics which lie behind the coexistence and parallel development of inflation and unemployment. As a consequence of our explanation of that phenomenon we also try to answer the question of why conventional anti-inflationary measures of aggregate demand management are found in many cases to work in a direction opposite to that intended. We will use our findings to describe the principles of a policy, which fights inflation and unemployment by the same measures.

During the seventies and early eighties of this century, the world economy was plagued by the phenomenon called stagflation - the coexistence of inflation and unemployment. Throughout this period, inflation co-existed not only with unemployment, but also with harsh "inflation fighting" policy measures like high interest rates, severe budget cuts, low growth rates and shrinking real wages - all aiming to decrease aggregate demand. A priori there is no more reason to attribute the decrease of inflation at the end of this period to these anti-inflationary measures than to "credit" these measures with the high inflation and high unemployment with which they coexisted for the preceding eight years.

On the other hand, the decreasing rate of inflation - started in the U.S. in 1982 - is associated with expansionist policies of a higher budget deficit, with reduction in the interest rate, with increasing growth rate and with receding unemployment. These developments look like a deceleration course of stagflation rather than as a renewed trade off between inflation and unemployment.

As far as is known to us, no satisfactory answers exist to the questions raised by stagflation. Conventional theories run unto inconsistencies when they try to incorporate both inflation and unemployment happening at the same time and also when they are confronted with actual data. They run into difficulties in analyzing actual facts and in forecasting developments. They also have to assume a series of exogenous supply shocks in

Soft Order in Physical Systems, Edited by Y. Rabin
and R. Bruinsma, Plenum Press, New York, 1994

order to explain the continuity of the price rise - how a once-and-for-all price-rise is transformed into continuing inflation. In the eyes of these theories - except for the frequent increase of the money supply to accommodate price-jumps - recent inflations have no demand side at all.

We suggest that these theory failures happen because accepted theories - a) overlook a powerful economic factor which affects economic developments strongly, but does not affect economic theory, or affects it very little[1] and b) because conventional theories use the aggregates as their main tool of analysis. We suggest further that this overlooked factor is the income re-distribution process which occurs during inflation. We make income a function of the price level, besides the usual quantity demanded. We claim that this is a logical assumption, especially during inflation. Thus not only the shape of the demand curve is determined by the prices, according to the conventional analysis - but also its place in the coordinates P, Q which is usually attributed to exogenous factors. This extension enables us to disaggregate aggregate income, demand and consumption according to the income distribution and to analyze its effects on employment and on the price level with them.

Our model differs from the others in these aspects:

a) In the incorporation of the income redistribution, more precisely the process of income redistribution, in the theory and b) in the analysis of stagflation by disaggregated macro-economic variables and also c) in the coupling of them. The disaggregation of the aggregate variables is done in the paper according to the income re-distribution during inflation. Making the analysis by this kind of disaggregated variables results in a picture of the dynamics of inflation accompanied *necessarily* by recession and which also feeds itself intrinsically. We will also show that the inflation part of stagflation does have an excess demand side which is also endogenous to it, and which sustains and accelerates it intrinsically. We will show that it is not the *aggregate demand*.

Current theories consider the aggregates - and thus the aggregate income - as the decisive factors of the economy, and for this reason income redistribution is not incorporated into them as a major economic variable - even when they do acknowledge its existence. It is considered a moral issue, subject to value judgment, rather than as an economically meaningful force, and it hardly affects pure economic thought. It is often shown as a possible outcome of the interaction between different economic forces, but not as an active force in its own right. In those cases when economists do deal with the income distribution as an economic factor, it is considered only in a static way and in connection to its effect on the aggregate consumption function. Namely, they try to determine whose MPC is larger, that of the poor or that of the rich and accordingly, how the aggregate consumption is affected by the state of the income distribution in the economy. In either direction only slight effects were found (*H.Lubell*, 1947; *P.Davidson and E.Smolensky*, 1964; *A.S.Blinder*, 1975; *T.M.Stoker*, 1986). Instead, we deal with income distribution in its dynamics, with the income re-distribution process and its effect on differentiated consumption. Doing this we find that income redistribution plays a major role in the determination of employment and of the price-level.

---

[1]The feeling that some factor has been overlooked, has existed for some time in economics. The outcome is the recent emphasis placed on "Rational Expectations" as a major decisive economic factor. "Rational expectations" in their different forms are employed to solve the inconsistencies between the different theories and the actual facts. The "Rational Expectations Theory" regards inflation as a given process and examines its dynamics after it has already started, without considering its initial causes. In this respect, besides the attempt to solve inconsistencies, our "Income Redistribution Theory of Inflation" is similar to the expectations theories. Moreover, the inflationary anticipation is regard in the expectations theories as a phenomenon born in inflation, and which, at the same time , serves as the main cause for its acceleration. The income redistribution process in our theory plays a similar role.

Contrary to the expectation theories we do believe that expected inflation does redistribute incomes. We believe that not all the parts of the public, not all the owners of the different production factors, are able to act so as to cancel inflation's adverse effects on their income even when they expect it correctly.

In the work of S.Weintraub (1958, 1980)and of P.Davidson and E.Smolensky (1964) we also find disaggregated demand (consumption) functions. They also use the income effect of the price-rise in constructing an upward sloping demand curve - both are features of our model. Weintraub, Davidson and Smolensky constructed an aggregate demand (consumption) function which is the sum of separate demand functions of different social groups. This aggregate demand function depends on the level of employment in the economy and it is an upward sloping function of this variable - as usually perceived. Since they assume that rising prices are always a necessary and direct outcome of rising employment they also find that the aggregate demand in their model is an upward sloping function of prices. This function is not derived without using the income effect - through the rising employment - of the price rise (*Weintraub* 1958). In our model we do acknowledge the income effect - through income distribution - of the price rise on demand but we deny the notion that prices and employment necessarily move together and in the same direction. They maintain the view that rising prices are equivalent to rising employment even though they acknowledge that the price rise effects the incomes of the different groups, which compose their aggregate demand, differently.

According to their model the price-rise affects the incomes of rentiers and wage and salary earners adversely while the income of the profit receivers rises - a notion with which we agree. The separate demand of the different income groups comes into their analysis only through its effect on aggregate demand which is the variable in their analysis. In our model we use the separate (partial) demands of the different groups as the variables in the macroeconomic analysis.

The so called "*Historical Demand Curve*" is also an upward sloping aggregate demand function of the price level because of the growth of income over the years.

Income re-distribution is assumed to be a very possible consequence of inflation. We also assume that the gains of the gainers in inflation flow to the final consumption market. We deal only with the final product consumption-market[2]. In our Income Redistribution (IRE) Model of Inflation we assume that there are two groups, the group of the gainers and the group of the losers in inflation, which divide the national income between them, and each of which controls a substantial part of the national income. This means, that we assume that a much larger category of the public than the fixed-pension earners and the rentiers - according to the traditional perception - lose in inflation. It includes wage and salary earners who were not able to incorporate a full cost-of living compensation into their contracts and also farmers who receive the same price for their products while their prices rise sharply in the supermarkets.

II. THE MODEL

The model presented here is a model of demand-shift inflation where the shift in the existing demand structure results from the income distribution caused by inflation itself. Inflation is defined as a continuous rise in the price level. We concentrate on the analysis of an ongoing inflation and not on its initial cause. The first jump in the price level might occur because of an exogenous shock to the system, a shock to which inflation is often related, but according to the model the observed continuing inflationary process is endogenous to

---

[2]We consider inflation, according to Keynes, as a phenomenon of the final market. It is logical to assume that if the gains of the gainers - or most of them - are invested back into the production process, stagflation might turn into growth with slow inflation. If they flow out of the country or are accumulated in the financial markets, causing a huge liquidity of the system which is used for driving up stock and real-estate prices, stagflation may turn into a non-inflationary recession which we witness these days. If they are used for LBO practices, the non-inflationary recession may be accompanied by high interest rates - because of the huge demand for these liquid funds - which is a not less contradictory phenomenon than the existence of inflation in times of unemployment. The model here does not deal with these cases and assumes that the gains are spent in the consumption market.

inflation, it is due to the dynamics of income redistribution during inflation. We shall be able to show how stagflation comes into existence and to demonstrate a stagflationary spiral which drives itself.

1. Assumptions
1.I. Assumptions Concerning Income Distribution
a. Inflation redistributes income between the different production factors. This means that inflation changes the *real* income, because the changes in the money income of the different factors relate differently to the rate of inflation. There is no law, economic, social or other, according to which the rate of inflation would be incorporated equally into every factor's income[3].

b. Income is a function of the price level. This follows from assumption (a).

c. Production factors are divided into two groups: 1. The gainers in inflation (**g**) whose money income increases more rapidly during inflation than the rate of inflation; 2. The losers in inflation (**w**) whose money income lags behind the rate of inflation.

d. The two groups, **g** and **w**, constitute the economy. Each group controls a substantial part of the National Income.

e. Inflation redistributes income in a systematic way and not randomly. This means that gains and losses do not cancel each other in a certain factor's aggregate income. The same group always gains and the other group always loses during the process of the price rise.

1.II Assumptions Concerning the Composition of Demand
f. The income-gains of the gainers in inflation (**g**) flow to the final consumption market and are spent there (see comment 2 above).

g. The composition of demand is a function of the level of real income, even at constant relative prices.

In accordance with these assumptions the inflationary process is described as shown in *figure 1*.

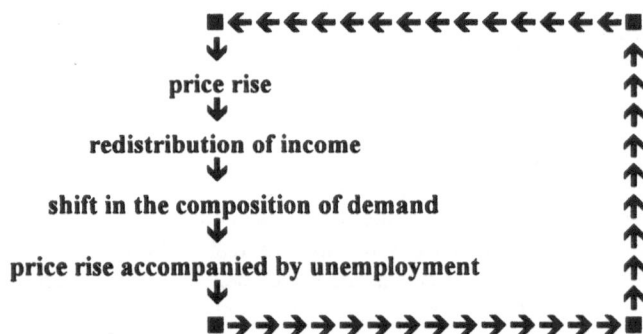

Figure 1. Schematic description of the loop of the dynamic process.

---

[3] One of the clearest recognitions of the redistribution of income by inflation can be found in Keynes'' article: "How to Pay for the War" (Keynes, 1930) where he discusses how a profit inflation financed World-War I. Another hint is the popular notion that inflation is a "tax" which recognizes income redistribution between the public and the government during inflation, but still assumes a uniformly reduced income for all members of that public, corresponding to the government's gain.

In our model one part of the public pays the tax of inflation to the other part of the public (according to a comment by J. Hirschleifer).

Other proofs of income redistribution during inflation we leave to statistics.

The outcome is a second round of price rises which starts the whole process again.

1.III. Additional Assumptions. The following assumptions are made for convenience. They are not necessary for the working of the model.

    h. The **GNP** is constant. This means that one group's gain is the other group's loss. We emphasize that this assumption is not essential, one could also have a declining or an increasing **GNP** during inflation.

    i. The Money Supply is rising with inflation. The total volume of transactions remains constant while money income grows and prices are rising so the growth of the money supply is implied by h. (There is no specific reference in the model to the changes in the money supply).

    j. The Marginal Propensity to Consume (**MPC**) is independent of income (*Blinder*, 1975; *Stoker* 1986).

    k. All goods are normal goods in respect to income and prices.

The world of the model is a closed economy where the production of final goods is considered as the aim of the activities of all production factors (in accordance with the Keynesian concept that inflation is a consumption-market phenomenon).

2. Introducing The Income-Effect Of A Price-Rise On The Demand Function - Besides The Price-Effect

In this section we shall analyze the overall effect of changes in the price level on demand. We introduce the **income-effect** of price changes on demand besides the usually considered **price-effect**. This will allow us to construct a new type of demand function representing the overall effect of prices on demand. These functions are different from the usual demand functions which only consider the direct effect of prices at constant nominal income. Because of the general slope of the usual demand function as a function of prices, the price effect - the move along the demand curve - effects all demand in a uniform downward way in case of a price rise, (All commodities are assumed normal - see assumption k.). In contrast to that, the direction of the income effect is not determined. It can affect demand upward or downward in case of a price change in either direction.

We assume that the change in the price level causes a change in the money income and take this into account in analyzing the overall response of demand. We derive two qualitatively different types of demand curves, one for the net gainers, g, whose real income increases with inflation (their money income rises faster than the rate of inflation, see assumption a), and one for the net losers, w. These demand curves respond differently to price rises.

While changes in the money income and changes in the price-level are often linked together the **income effect** on demand is still neglected in the analysis of inflation. It is frequently assumed that the *redistribution of income*, even if it occurs, cancels out in the aggregate and therefore has no relevance to macroeconomic analysis. The purpose of our analysis in this chapter is to show that a systematic - as distinct from a random - income redistribution does lead to a systematic shift in demand patterns. Demand for a specific commodity (or group of commodities) can either rise or fall with the general price level.

Figure 2 presents a family of parallel demand curves, which differ from each other only in the money income they assume. The demand curves to the right assume a higher money income than the demand curves to the left. In time period "*0*" the positions of the individuals **g** and **w** are equal. Both have the same demand curve $d_0^{w,g}$. Both are in equilibrium at the price level p_0. and demand a quantity $q_0^{w,g}$ at this price. On that picture we project an inflation. Both prices and money income are rising. The money income of **g** rises faster with the price level than that of **w**. At time "*1*", at the price level p_1 **w** has moved to the demand curve d_1^w and his consumption of q is now q_1^w, while **g** moved to the demand curve d_1^g and his consumption of q is now q_1^g.

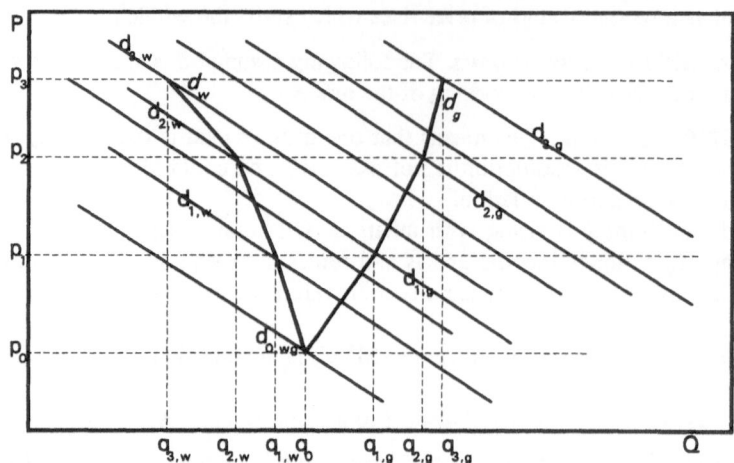

Figure 2. Developing the two Demand Functions. One for the Gainers, d_g, and one for the Losers in Inflation, d_w. In Figure 2 there is: money income y, two individuals: w the loser; g the gainer, demand is d. There are one price, p, and one commodity, q.

With the move of p_0 to p_1:

$$
\begin{aligned}
&\text{price effect for } \mathbf{w}: && q_0^{w,g} - q_1^{w,g} \\
&\text{price effect for } \mathbf{g} && \mathbf{q}_0^{w,g} - q_1^{w,g} \\
&\text{income effect for } \mathbf{w}: && \mathbf{q}_1^{w} - q_1^{w,g} \\
&\text{income effect for } \mathbf{g}: && \mathbf{q}_1^{g} - q_1^{w,g}
\end{aligned}
$$

Real consumption change for **w**:
income effect - price effect

$$(\mathbf{q}_1^w - q_1^{w,g}) - (\mathbf{q}_0^{w,g} - q_1^{w,g}) = \mathbf{q}_1^w - \mathbf{q}_0^{w,g};$$
$$\mathbf{q}_1^w < \mathbf{q}_0^{w,g}$$

Because the income effect is smaller than the price effect the change in real consumption is negative. Real loss of **w**:

$$- (\mathbf{q}_1^w - \mathbf{q}_0^{w,g}) = (\mathbf{q}_0^{w,g} - \mathbf{q}_1^w)$$

Real consumption change for **g**:
income effect - price effect

$$(\mathbf{q}_1^g - q_1^{w,g}) - (\mathbf{q}_0^{w,g} - q_1^{w,g}) = \mathbf{q}_1^g - \mathbf{q}_0^{w,g};$$
$$\mathbf{q}_1^g > \mathbf{q}_0^{w,g}$$

Because the income effect is smaller than the price effect the change in real consumption is positive. Real gain of **g**:

$$\mathbf{q}_1^g - \mathbf{q}_0^{w,g}$$

The move of the price p_1 to the level p_2 and the move of **w** to the demand curve $\mathbf{d}_2^w$ and of **g** to the demand curve $\mathbf{d}_2^g$ result in:

price effect for **w**:	$\mathbf{q}_1{}^w - q_2{}^w$	price effect for g	$\mathbf{q}_1{}^g - q_2{}^g$
income effect for **w**:	$\mathbf{q}_2{}^w - q_2{}^w$	income effect for g:	$\mathbf{q}_2{}^g - q_2{}^g$

Real change for **w**:

income effect - price effect $\qquad (\mathbf{q}_2{}^w - q_2{}^w) - (\mathbf{q}_1{}^w - q_2{}^w) = \mathbf{q}_2{}^w - \mathbf{q}_1{}^w;$

$$\mathbf{q}_2{}^w < \mathbf{q}_1{}^w$$

Thus the real loss of **w** is: $\qquad \mathbf{q}_1{}^w - \mathbf{q}_2{}^w$

The real change for **g** is: $\qquad (\mathbf{q}_2{}^g - q_2{}^g) - (\mathbf{q}_1{}^g - q_2{}^g) = \mathbf{q}_2{}^g - \mathbf{q}_1{}^g;$

$$\mathbf{q}_2{}^g > \mathbf{q}_1{}^g$$

The real gain of **g** is $\qquad \mathbf{q}_2{}^g - \mathbf{q}_1{}^g$

Figure 2 demonstrates that any time the income effect is larger than the price effect real gain occurs. If we connect the points corresponding to that definition we get the demand function for the gainers in inflation (d^g). The characteristic of this demand function is that it **rises** with the price level.

Similarly, whenever the income effect is smaller than the price effect, real loss occurs. If we connect the points corresponding to that definition we get the demand function of the losers in inflation (d^w). The characteristic of this demand function is that it declines with the rise of the price level.

Figure 2 assumes growth in the money supply with the rise in the price level but this is not a necessary assumption to get these results. The initial demand function d_0 is not necessarily the last one to the left. It could be one in the middle. Demand curves to the left of d_0 would assume decreasing money income. Then the re-distribution occurs between the group with decreasing money income and the group with a sufficiently increasing money income.

It is useful to express the same results in mathematical terms. We assume that the demand (**D**) is a function of the money income (y) and of the price (p), while the money income itself is a function of the price level:

$$\mathbf{D} = D(\mathrm{y,p}) \quad ; \quad \mathrm{y} = \mathrm{y(p)} \qquad (1)$$

We can therefore write

$$\mathbf{D} = D(\mathrm{y(p),p}) \qquad (2)$$

We differentiate with respect to p:

$$(\mathrm{d}D/\mathrm{d}p) = [(\mathrm{d}D/\mathrm{d}y)_{\mathrm{p=const}} \cdot (\mathrm{d}y/\mathrm{d}p) + [(\mathrm{d}D/\mathrm{d}p)]_{\mathrm{y=const}} \qquad (3)$$

The first term on the right hand side of equation 3 is the income effect, **IE** (of a change in price), and the second one represents the price effect, **PE**.

Quite generally we can assume that the price effect on demand:

$$\mathbf{PE} = [(\mathrm{d}D/\mathrm{d}p)]_{\mathrm{y=const}} < 0 \qquad (4)$$

is always negative. The income effect:

$$\mathbf{IE} = [(\mathrm{d}D/\mathrm{d}y)_{\mathrm{p=const}} \cdot (\mathrm{d}y/\mathrm{d}p) \qquad (5)$$

can have either sign. One can assume quite generally that demand rises with income (at constant price), so that:

$$(\mathrm{d}D/\mathrm{d}y)_{\mathrm{p=const}} > 0 \qquad (6)$$

Thus the sign of the income effect depends on the sign of dy/dp. If this is positive the overall effect of a price rise on demand depends on the relative magnitude of the two terms on the right hand side of equation 3.

If the income effect is larger than the price effect then dD/dp is positive. That means that the demand is an upward sloping function of the price level. If the price effect is larger than the income effect, then dD/dp is negative and the demand is a decreasing function of prices. Real increases and decreases in consumption (as a function of the price change) are represented by the difference between the income and the price effects.

The simplest assumption[+] is that only real income affects demand:

$$(dD/dp)_{y_{re}=const} = 0 \quad ; \quad y_{re} = y/p \tag{7}$$

or equivalently from equation 3:

$$p.(\partial D/\partial p)_{y=const} + (\partial D/\partial y)_{p=const} = 0 \tag{8}$$

With this assumption the overall dependence of demand on price becomes:

$$dD/dp = [1 - (p/y).(dy/dp)].(\partial D/\partial p)_{y=const} \tag{9}$$

where $.(\partial D/\partial p)_{y=const}$ is the slope of the usual demand function. Equation 9 shows clearly that the overall dependence of demand can have either sign.

We can now divide the consumers in the economy into two groups. The gainers in inflation (**g**) increase their real income, y^g_{re}, as a result of the rise in the price level.

$$dy^g_{re}/dp > y^g/p > 0 \tag{10}$$

and therefore their demand increases in real terms:

$$dD^g_{re}/dp > 0 \tag{11}$$

The second group are the losers (**w**):

$$dy^w_{re}/dp < 0 \tag{12}$$

$$dD^w/dp < (D^w/p)_{y_w=const} < 0 \tag{13}$$

By introducing the income effect we have demonstrated in figure 2 and in equations 1 - 13 that some individuals' demand rises with the prices and as a result of the price rise while that of others declines. We have to point out that figure 2 and related equations do not say anything about what happens to the total aggregate demand. The aggregate may increase, decrease or stay constant. The main result, the establishment of two differently sloping demand curves would not be invalidated by the changes in their sum once income redistribution has been assumed and the income effect is introduced.

We conclude: a) A rising price level does not decrease everybody's demand; b) Even when total demand is decreasing there can be a substantial increasing demand in the economy.

3. The Partial Demand Functions

In this section we move from money income to real income and from the micro to the macro level of the economy. The analysis will be done in terms of disaggregated macroeconomic variables instead of the conventional aggregate ones.

3.I Definitions: If the growth of different money incomes relates differently to the rate of inflation, an assumption we used in constructing figure 2, it follows that inflation redistributes real income. In that respect we divide aggregate real income (Y) into two parts: Y_g is the sum of the incomes of all the gainers in inflation, y_g, whose income increases faster than inflation.

$$Y_g = \Sigma\, y_g \tag{14}$$

and Y_w, which is the sum of the incomes of all the losers in inflation, y_w, whose income increases slower than inflation.

$$Y_w = \Sigma\, y_w \tag{15}$$

where Y is the national income.

$$Y = Y_g + Y_w \tag{16}$$

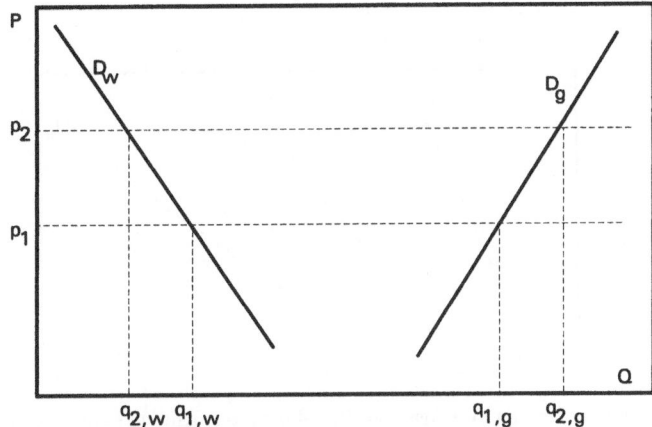

Figure 3. The figure shows the **Partial Demand Functions** D_g and D_w in the particular case where total demand D is constant. At every price p

$$D_w + D_g = D = \text{const.}$$
$$Q_{1,w} - Q_{2,w} = Q_{2,g} - Q_{1,g}$$

The loss of **w** equals the gain of **g**. Both increase with the increase of p referred to p_1

$$D_{w-} = D_w(Y_w(p),\, p)$$
$$D_g = D_g(Y_g(p),\, p)$$

We now introduce the **Partial Aggregate Demand Functions**. We divide the aggregate demand **(D)** into two parts: The partial aggregate demand of the gainers, D_g, is the sum of the demands of all the gainers in inflation, d_g, for all the goods in the market, **Q**.

$$D_g = \Sigma\, d_g \tag{17}$$

and the partial aggregate demand of the losers, D_w, is the sum of the demand of all the losers in inflation, d_w, for all goods in the market, **Q**.

$$\mathbf{D_w} = \Sigma \, \mathbf{d_w} \tag{18}$$

D is the aggregate demand

$$\mathbf{D} = \mathbf{D_g} + \mathbf{D_w} \tag{19}$$

As we have shown $\mathbf{D_g}$ rises with the price level

$$d\mathbf{D_g}/dp = \Sigma(dd_g/dp) > 0 \tag{20}$$

and $\mathbf{D_w}$ decreases

$$d\mathbf{D_w}/dp < 0 \tag{21}$$

In itself this does not tell us anything about the behavior of the total aggregate demand, it can either rise or fall with the prices.

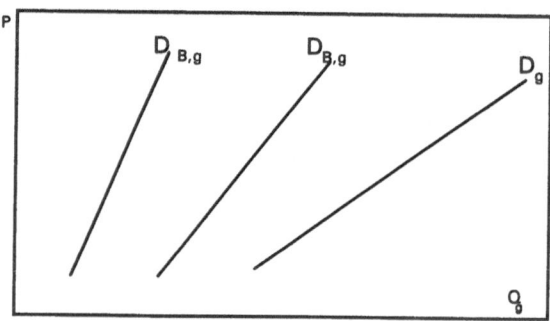

Figure 4. Decomposition of the **Partial Demand, $\mathbf{D_g}$**, of the group gaining in inflation into its components - the **Differentiated Partial Demands** in the luxury market, $\mathbf{D_{L,g}}$, and in the basic goods market, $\mathbf{D_{B,g}}$. $\mathbf{Q_g}$ represents the total goods and services demanded by g. It is composed of a larger part from the luxury market, $\mathbf{Q_{L,g}}$, and of a smaller part from the basic goods market, $\mathbf{Q_{B,g}}$.

4. The Differentiated Demand Functions

As we already noted we assume that inflation does not re-distribute incomes in a random way. There is an economic mechanism, inherent to the way the economy works, which systematically redistributes incomes between different social groups when prices are raised. We assume that $\mathbf{Y_g}$ and $\mathbf{Y_w}$ are the incomes of different social groups with different average incomes (y_g, y_w). Thus if one would differentiate between the demands of the two groups one would find that the demand of the gainers (group **g**) has a different structure than that of the losers. To describe this we divide the commodities (final products) market into two. The luxury market, **L**, is defined as the market to which demand is transferred when real income increases. We can assume that the majority of the buyers in the **L** market are the gainers in inflation, **g**. The second market is the market for basic goods, **B**, where the majority of the consumers are the losers in inflation, **w**. Together they constitute the whole market:

$$\mathbf{D} = \mathbf{D}_g + \mathbf{D}_w \qquad (22)$$

where D is the aggregate demand and $\mathbf{D}_L$ and $\mathbf{D}_B$ are the **Differentiated Demands** in the respective markets. Since we have two groups of consumers we also have:

$$\mathbf{D}_L = \mathbf{D}_{L,g} + \mathbf{D}_{L,w} \qquad (23)$$

and

$$\mathbf{D}_B = \mathbf{D}_{B,g} + \mathbf{D}_{B,w} \qquad (24)$$

Consumers whose real income increases will tend to devote a larger fraction of their income to luxury goods and therefore:

$$d\mathbf{D}_{L,g}/dp > d\mathbf{D}_{B,g}/dp \qquad (25)$$

The reverse will be true for the losers:

$$d\mathbf{D}_{L,w}/dp < d\mathbf{D}_{B,w}/dp \qquad (26)$$

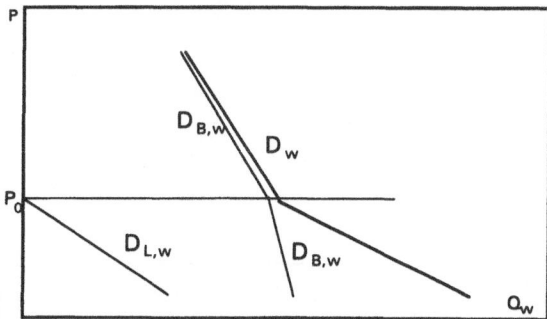

Figure 5. Decomposition of the **Partial Demand, $\mathbf{D}_w$**, of the group losing in inflation into its components - the **Differentiated Partial Demands** in the luxury market, $\mathbf{D}_{L,w}$, and in the basic goods market, $\mathbf{D}_{B,w}$. Q_w represents the total goods and services demanded by w. It is composed of a larger part of Q_B and of a smaller part of Q_L.

We emphasize the asymmetry between the two groups. Since Y_g is rising with prices both $\mathbf{D}_{L,g}$ and $\mathbf{D}_{B,g}$ should rise. Since the demand of this group for basic goods, $\mathbf{D}_{B,g}$, is essentially satisfied one expects $\mathbf{D}_{B,g}$ to rise only slowly. Most of the increase in Y_g is therefore reflected in the increase of the demand for luxury goods, $\mathbf{D}_{L,g}$. This is shown in *figure 4*. The situation for the group losing in inflation is quite different. Real income, $Y_{w,re}$, is declining. Thus both $\mathbf{D}_{L,w}$ and $\mathbf{D}_{B,w}$ decrease as prices rise. Since the consumption of basic goods is more important there is, however, a transfer effect in the demand - transferring demand from L to B. This will partially compensate for the effect of the decline in real income on $\mathbf{D}_{B,w}$. This compensation is possible only until, at some price level p_0, $\mathbf{D}_{L,w}$ vanishes. At this price the group w is eliminated from the market for luxury goods. The result is a kink in the demand for basic goods, $\mathbf{D}_{B,w}$, at the price level p_0. This is shown in figure 5.

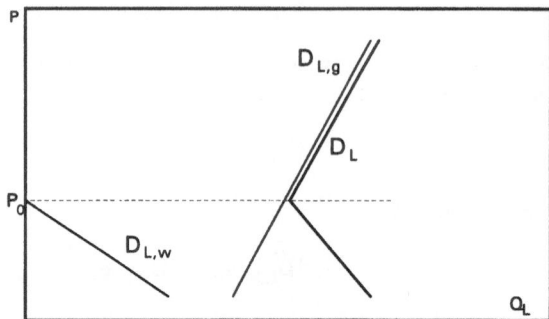

Figure 6. The **differentiated demand** for luxury goods, D_L resulting from the combination of the **differentiated partial demands** of the two income groups, $D_{L,w}$ and $D_{L,g}$, in this market (equation 23). One notes the kink in D_L at the price level p_0 where $D_{L,w}$ vanishes.

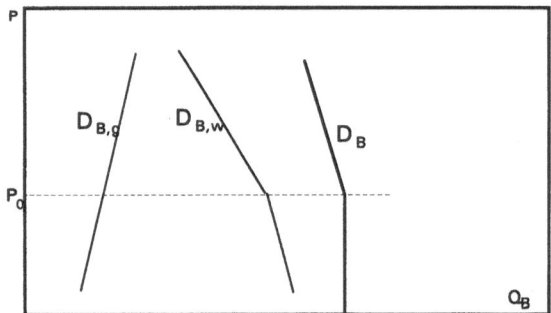

Figure 7. The **differentiated demand** for basic goods, D_B, resulting from the combination of the **differentiated partial demands** of the two income groups, $D_{B,w}$, and $D_{B,g}$, in this market (equation 24). Again there is a kink in D_B reflecting the kink in $D_{B,w}$ at the price level p_0. Above this price level the real demand for basic goods falls rapidly as prices rise.

Finally we combine these results to obtain the **Differentiated (aggregate) Demand** in the two markets. For the luxury market one has:

$$D_L = D_{L,g} + D_{L,w} \tag{27}$$

The result, obtained by combining the relevant curves in *figures 3* and *4* is shown in *figure 6*.

Similarly we have for the **Differentiated Demand** for basic goods, D_B.

$$D_B = D_{B,w} + D_{B,g} \tag{28}$$

Again, combining the curves in *figures 3* and *4* we obtain the result illustrated in *figure 7*.

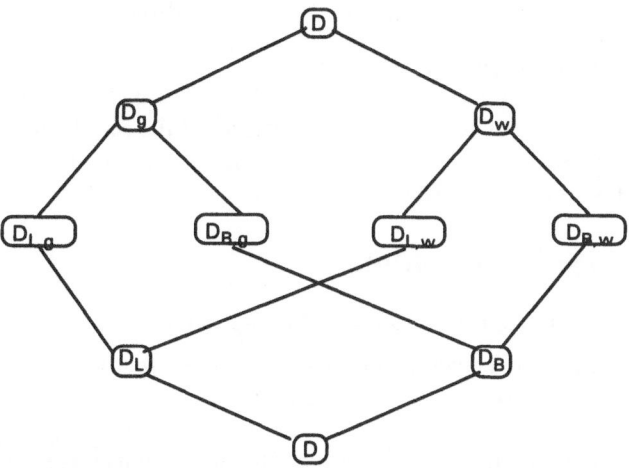

Figure 8. The disaggregation of the **aggregate demand** into two **Partial Demands** (D_g, D_w) and into two **Differentiated Demands** (D_L, D_B).

<table>
<tr><td align="center">**Partial Demands**</td><td align="center">**Differentiated Demands**</td></tr>
<tr><td align="center">$D = D_g + D_w$</td><td align="center">$D = D_L + D_B$</td></tr>
<tr><td align="center">$D_g = D_{L,g} + D_{B,g}$</td><td align="center">$D_L = D_{L,g} + D_{L,w}$</td></tr>
<tr><td align="center">$D_w = D_{L,w} + D_{B,w}$</td><td align="center">$D_L = D_{B,g} + D_{B,w}$</td></tr>
</table>

Figure 8 describes the ways aggregate demand is disaggregated. Starting with the upper **D**, aggregate demand is divided into two **Partial Demands**, one of the gainers, D_g, and one of the losers in inflation, D_w. D_g is divided further into the gainers' **Differentiated Partial Demand** in **L** - which is $D_{L,g}$ - and in **B** - which is $D_{B,g}$. The losers' **Partial Demand**, D_w, is divided in the same way into $D_{L,w}$ and $D_{B,w}$. If we combine $D_{L,g}$ and $D_{L,w}$ the result is the , D_L, which is the luxury market's aggregate demand. Similarly the **Differentiated Demand** D_B of the basic goods market is the sum of $D_{B,g}$ and $D_{B,w}$. Finally D_L and D_B are the components of the aggregate demand, **D**, into which we divide it starting from the lower **D** in *figure 8*.

We thus find that income redistribution will lead to a shift in demand at least above the threshold price p_0. Initially, at low prices, there are compensation mechanisms. The demand for luxury goods is transferred from the losers to the gainers. This is the dominant effect. The overall effect on the demand for these goods is not obvious.

$$dD_L/dp = dD_{L,g}/dp + dD_{L,w}/dp \qquad (29)$$

could have either sign, depending on the actual values of $dD_{L,g}/dp$ and $dD_{L,w}/dp$. The same holds for

$$dD_B/dp = dD_{B,g}/dp + dD_{B,w}/dp \qquad (30)$$

Thus the result depends on the specific details. The situation is quite different above the threshold price p_0. The losers have dropped out of the luxury market. They now spend all their income on basic goods. Thus when income is transferred from group w to group e one is essentially transferring demand from the basic goods market to the luxury market and:

$$dD_L/dp > 0 \quad ; \qquad dD_B/dp < 0 \qquad (31)$$

This is seen clearly in *figures 5* and *6*.

One could of course replace the sharp threshold assumed in figure 4 by a smoother curve but this would not change the qualitative features of the result. On a more formal level we note that we are assuming that the relative demand for the products, Q_L and Q_B depends explicitly on the level of income. This seems realistic.

III. IMPLICATIONS OF THE MODEL

1. The Joint Appearance Of Inflation And Unemployment

We now assume initial equilibrium in the two markets followed by an initial rise in the price level. According to our model a rising price level creates a **Differentiated Excess Demand (DED)** in the luxury market L, because D_L is an increasing function of the price level. This results in rising prices in that market. The same original rise in the price level also created a demand deficiency in the basic goods market **B** because D_B is a decreasing function of the price level.

If the price rise in **L** is transferred to **B** causing there stable, non-falling or actually rising (downward rigid) prices despite the decreasing demand, then we can conclude that the **Differentiated Excess Demand (DED)** which exists in one part of the market, as a result of the income redistribution during inflation, will bring about a general rise in the price level.

That new rise in the price level starts the whole process of income redistribution once again with all its consequences and thus leads to a new additional rise in prices. We thus have a dynamic process of continuing income redistribution, demand shift and inflation.

We have assumed that the price rise due to the excess demand in **L** is transferred to **B** in spite of the declining demand in this market. This means that **B** reacts to the **Declining Differentiated Demand (D_B)** by quantity adjustment rather than by price adjustment. Excess capacity and unemployment will thus be generated in **B**. Thus as a result of the Income Redistribution Effect (**IRE**) of inflation we get an inflation which not only naturally coexists with recession but necessarily generates and sustains itself as well as recession. It follows from the same mechanism that the dominant effect of the demand shift is a change in the general price level and not in the relative prices in the two markets. We therefore assume only one general price level in the economy (p) which measures the overall inflation.

We have seen that the effect of income redistribution in creating a demand shift is not quite clear before prices reach the price level p_0. Up to that price level the losers move out

216

of $\mathbf{L}$, thus effecting a decrease in $\mathbf{D_L}$ and, in a way, offsetting the upward effect of $\mathbf{D_{L,g}}$ on $\mathbf{D_L}$. They move over to the basic goods market offsetting the downward pressure of the decrease in $\mathbf{D_w}$ and of the price level on $\mathbf{D_B}$ - up to the price level p_0. There is a kink in the $\mathbf{D_L}$ and $\mathbf{D_B}$ curves at p_0 (see *figures 5* and *6*) where the effect of $\mathbf{D_{L,w}}$ on $\mathbf{D_L}$ and $\mathbf{D_B}$ stops. Beyond the point p_0, **IRE** exercises its full upward pressure on prices in $\mathbf{L}$ and downward effects on employment in $\mathbf{B}$ and then the stagflationary spiral gains its full momentum.

One would also expect a "kink" in the reaction of $\mathbf{B}$ to the demand shift. For small shifts, or sufficiently slow ones, the market can react, at least partially, by a price adjustment. This is not feasible in the short run when the demand shift is large. For a rapid dynamical demand shift process the market $\mathbf{B}$ cannot reach long term equilibrium. One therefore expects it to react to such shifts by quantity adjustments.

It is therefore intrinsic to our model that it can show two different modes of behavior depending on the size of the initial shock. If the initial shock was small and resulted in a price level ($p_{initial} < p_0$), the market forces are still effective. Thus the economy can return to equilibrium following the shock. For a large shock ($p_{initial} > p_0$) the initial demand shift is large and the $\mathbf{B}$ market reacts by quantity adjustment. This leads to a stagflationary instability of the economy. Leijonhufvud (*A.Leijonhufvud*, 1981) has described this type of situation as a corridor. Inside the corridor the classical rules of economics work. Outside one has different rules. "*The system is likely to behave differently for large than for moderate displacements from the "full coordination" time path. Within some range from the path (referred to as "corridor" for brevity), the system's homeostatic mechanisms work well, and deviation-counteracting tendencies increase in strength. Outside that range these tendencies become weaker as the system becomes increasingly subject to effective demand failures....*"(*A.Leijonhufvud*, 1981, pp.109). The specific model we have developed is of course different from the one discussed by Leijonhufvud.

There is also another factor which prevents the $\mathbf{B}$ market from returning to equilibrium. Because of the high share of quasi-rent in the price-structure of luxury goods, the $\mathbf{L}$ market absorbs much less real production factors than the $\mathbf{B}$ market in order to produce the same addition to the GNP. Thus a substantial part of them will remain unemployed when demand is transferred from $\mathbf{B}$ to $\mathbf{L}$.

2. Price Rise Transfers From L To B The Double Effect Of The DED

In our model, the double effect of the **Differentiated Excess Demand (DED)**: 1) pulling up prices in $\mathbf{L}$, while 2) at least preventing them from falling in $\mathbf{B}$, is the main factor in activating the stagflationary spiral.

The appearance of the **DED** means that the system is displaced sufficiently "far out" in the Leijonhufvud sense (*Leijonhufvud* , 1981) and it is subjected to an effect - the double effect of **DED** - which prevents its return to equilibrium. **DED** means, that a substantial excess demand market exists in the economy in parallel with a declining demand market. In this excess demand market, as pointed out above, the prices are rising.

It is sufficient to assume the existence of the well known and well documented phenomenon of the downward rigidity of prices in the deficient demand market, $\mathbf{B}$, in order to get a rise in the general price level of the economy, p - it is a natural consequence of the price rise in $\mathbf{L}$ and the price rigidity downward in $\mathbf{B}$ - and also a recession in $\mathbf{B}$, where demand decreases at the prevailing prices. In the framework of our model we do not need more in order to get a stagflationary spiral.

Nevertheless we think that the existence of a substantial excess demand market beside a declining demand market, the **DED**, is responsible for the frequently observed phenomenon of the "downward rigidity of prices", which in its turn is responsible for the quantity adjustment of the low-demand markets. (Historically it is demonstrated that when

all incomes lose, and decrease in a real sense, then prices also fall as in the Great Depression. (*J.F.Walker and H.G.Vatter*, 1986)). It is logical to assume that the existence of an excess-demand-market in the same economy keeps costs, the prices in the factor markets and also the expected rate-of-return in the deficient-demand-markets higher than would be appropriate to these markets on their own, and thus prevents prices from adjusting. The effect of **DED** might well explain *Nordhouse*'s finding (*Nordhouse*, 1976) of widespread cost-plus pricing practices which are not responsive to demand.

The failure of prices to adjust in the declining-demand market also affects the economy in two ways: **1)** It causes quantity adjustment in this market, which means unemployment and decreased production; **2)** because of it the development of prices in **B** does not moderate the increase in the price level caused by rising prices in

The final result of the **Double-Effect of the Differentiated-Excess-Demand (DED)** is the appearance of inflation accompanied by recession. The feedback of the price-rise on income re-distribution (**IRE**) and the link of the latter to demand shift, turns that phenomenon into a stagflationary spiral.

3. Linking Systematic Income Redistribution To Shift In Demand

The spiraling process of stagflation, in our model, is brought about by a demand-shift inflation between two well defined markets (**L, B**) linked to a systematic income redistribution by inflation between two well defined income groups (**g, w**). The demand whose structure changes, the income which is redistributed and the markets which are effected are macroeconomic magnitudes. It is a model of demand shift inflation sustained through the feedback effect of the income redistribution on the structure of demand. This kind of inflation not only can, but necessarily does coexist with unemployment.

Demand shift inflation is known in the literature (*L.C.Schultze*, 1959; *J.H.G.Olivera*, 1964; *E.A.Cardoso*, 1981). While this concept acknowledges inflation caused by a partial demand (and not by the aggregate demand) the demand involved is limited, its shift is random and not connected to income re-distribution. Even so it brings about inflation.

The income redistribution process during inflation is also acknowledged and widely discussed in the literature (*I.Fisher*, 1913; *J.M.Keynes*, 1930, 1935, chp.I; *S.Weintraub*, 1958, 1978; *P.Davidson and E.Smolensky*, 1964; *A.Babeau, A.Masson and D.Strauss-Kahn*, 1975, *L.C.Thurow*, 1975; *E.Alexander*, 1975, 1978, 1990; *M.Allais and J.Allais*, 1976; *J.G.Williamson*, 1977; *R.G.Ramos*, 1980 and many others).

Our results were obtained by linking the income redistribution by inflation with demand-shift inflation in a systematic way in a macroeconomic model through the disaggregation of aggregate demand. This way the, so far overlooked effects of the income redistribution by inflation appear as decisive economic factors in the determination of the price level and of the level of employment.

The increase in the money supply by itself does not play a central role in our model of inflation. In this aspect it is considered as a means to redistribute income (see figure 2). Even if excessive and frequent increases in the money supply do cause inflation, it is often thought in the literature to be meaningless if it does not redistribute income. It is thought that the only important real effect of inflation is the redistribution of income, the **IRE** in the language of our model. Without this inflation is nothing more than rescaling of prices or renaming of the money. It is neutral to the economy and might cancel itself or could easily be stopped (*Milton Friedman*, 1974). We only add to this that our model shows that **IRE** of inflation not only gives "meaning" to it and makes it difficult to stop but it also accelerates inflation and brings recession along.

The initial income redistribution may occur for many other reasons. A regressive budgetary policy is one possibility; another would be the sudden rise of the oil prices (*R.G.Gordon*, 1977), and will then be followed by a demand shift with its subsequent inflationary and recessionary consequences. We predict stagflationary effects with or without accommodating such shocks by an increase in the money supply. One might think, nevertheless, that the recessionary effect will be more severe when there is no monetary accommodation. But the effect of easing the recession is canceled if the increasing money supply deepens the income redistribution - as our model shows and as often happens.

IV. FORECASTING AND POLICY CONSEQUENCES

According to our income redistribution demand shift (**IDDSMI**) model the information one needs in order to forecast economic developments success- fully is information on the magnitude and trend of the ongoing income redistribution. Today, information of this type which can be extracted from published statistical data is very limited and published with considerable delay (see *A.Babeau*, 1975 and *A.Babeau, A.Masson and D.Strausskahn*, 1975). The data should be provided more extensively and frequently similar to data on changes in the money supply in the 80's which were used to estimate the magnitude and trend of inflationary expectations - given the assumption that an increase in the money supply means more inflation - and thus to forecast coming economic developments. The data on income redistribution should refer to the changes in the income of the different factors due to inflation and/or to the government's discretional policy. We have to watch the movement of the **Differentiated Excess Demand**, given the assumption that it is a function of real income. Using the model developed here we can predict, that its increase means more inflation and more recession and its decrease means moderation in both developments.

The aim of an effective stabilization policy should be to offset the effects of the **DED** in the economy. Curbing the increase of the **DED** means fighting inflation and recession at the same time and by the same means.

The management of **DED** differs from the well known aggregate demand management in that it specifies the social groups, or production factors, which should carry the burden of the adjustment of the economy, once the gainers in inflation are identified. As far as aggregate demand management is concerned, the question where put the burden in order to fight inflation, is a political one without economic meaning. After all the decrease in the aggregate demand, the aim of aggregate demand management during inflation, will be achieved no matter whose demand will decrease, that of the gainer or that of the looser. In **DED** management the question where to put the burden is crucial and strictly economic. According to it the tax of the readjustment should be paid exclusively by the gainers; otherwise the policy will not be effective. According to our income redistribution demand shift (**IDDSMI**) model, on which the **DED** management is based, if the burden is put on the groups which loose anyway in inflation, inflation will not affected even in case of a decline in the aggregate demand.

More than that, in case the government causes even worse loss to the losers in inflation, it increases at the same time and by the same means also the gains of the gainers because one factor's loss is, almost inevitably, another factor's gain. Thus such a policy increases **DED** and at the same time, and because of this, it also increases, instead of decreasing, inflation and recession. A policy which puts the burden of aggregate demand management on the losers, strengthens - instead of weakening - the income redistribution effect (**IRE**) of inflation which is the very heart of the stagflationary spiral.

The repeated failure of aggregate demand management in affecting current inflations and recessions is easy to understand within our model when one investigates the effect of the specific policies employed within its framework on the **DED**. Using this type of analysis the author was the only one to predict the consequences of the *"new economic policy"* (*"the right economics"*) of Finance Minister Aridor in Israel in the spring of 1981 (*E.Alexander*, 1981b). The actual developments contradicted all predictions based on aggregate demand management or monetarist considerations.

In a stagflationary situation an economic policy which is based on **DED** management will always try to counteract the **IRE** of inflation, which means fighting inflation and recession at the same time and by the same means, with a more equitable income policy. On the other hand aggregate demand management, because it disregards the **IRE** and because its means mostly are inequitable, is only able to design policies which try to fight inflation or recession one at a time, and one at the expense of the other, even when they co-exist. Thus its failure to achieve stabilization and full employment is intrinsic to that method.

REFERENCES

Alexander, E., 1975, **How Inflation Accelerates Itself and the Case of Stagflation**, (Hebrew) Jerusalem *Emda*, November

Alexander, E., 1978, **"The Inflationary Effect of the Income Redistribution Process during Inflation, - the Case of Israel and France"**, Los Angeles, U.C.L.A. Working paper, presented at the conference of the Western Economic Association, Honolulu, June.

Alexander, E., 1981a, **"National Income, Private Consumption, Disposable Wages and Inflation in the Israeli Economy 1968-1974"** Jerusalem. The Ministry of Energy and Infrastructure. Working paper presented at the Conference of the Israeli Economic Association, January.

Alexander, E., 1981b, **Why inflation will come down**, (Hebrew) Tel-Aviv, *Haolam Hazeh*, March.

Alexander, E., 1990, **"The Power of Equality in the Economy - The Israeli Economy in the 80's, the True Picture"**, (Hebrew) 336 pages, Hakibbutz Hameuhad Publishers, Tel Aviv, September 1990.

Allais, M. and Allais, J., 1976, **"Inflation, income distribution and indexation with reference for the French economy, 1947-1975"**, Paris, Centre d'Analyse Economique C.N.R.S., March.

Babeau, A., Masson, A. and Strauss-Kahn, D., 1975, **"Inflation et Partage des Surplus: le Case des Menages"**, Paris, Editeur Cupas.

Blinder, A.S., 1975, **Distribution effects and the aggregate consumption function**, *Journal of Political Economy*, June.

Bruno, M. and Sachs, J., 1985. **"Economics of Worldwide Stagflation"**. Harvard University Press, Cambridge Mass.

Cairncross, A., 1985, **Economics in theory and practice**, Richard T. Ely Lecture, *A.E.R.*, May.

Cambridge Economic, Policy Group 1977, **General review and policy analysis**, *Economic Policy Review*, March.

Cardoso, E.A., 1981. **Food supply and inflation**, *Journal of Development Economics*. vol.8 p.269.

Davidson, P. and Smolensky, E., 1964. **"Aggregate Supply and Demand Analysis"**, Harper and Row, Publishers, New York, Evanston and London.

Fields, G.S., 1977. **Who benefits from economic development?. a reexamination of Brazilian growth in the 1960's**, *A.E.R.* February.

Fisher, I., 1913, **"The Purchasing Power of Money"**, Augustus M. Kelley, Publishers, Fairfield, N.J.

Friedman, M., 1974, **L'indexation un moyen de degonfler les governmentals**, *Fortune* and *Le Figaro*, October.

Gordon, R.J., 1977, **The theory of domestic inflation**, *A.E.R.* February.

Hansen, B., 1951, **"A Study in the Theory of Inflation"**, Allen and Anwin, Chapter 7.

Hedlund, J.D., 1983, **Distribution theory revisited: an empirical examination of the Weintraub synthesis**, *Journal of Post Keynesian Economics*, Fall.

Kessel, R.A. and Alchian, A.A., 1960. **Meaning and validity of the inflation - induced lag of wages behind prices**, *A.E.R.* March.

Keynes, J.M., 1930. **"A Treatise on Money"** vol.II, London, MacMillan, pp. 181-198.

Keynes, J.M., 1935. **"The General Theory of Employment, Interest and Money"**.

Kaldor, N., 1976, **Inflation and recession in the world economy**, *The Economic Journal*, December.

Leijonhufvud, A., 1975. **Cost and consequences of inflation**, U.C.L.A. Discussion paper No. 58. Written for the International Economic Association Conference on the Microfoundation of Macroeconomics. S'Agro, Spain, April.

Leijonhufvud, A., 1981. **"Information and Coordination"**, Oxford University Press.

Lubell, H. 1947, **"Effects of Redistribution of Income on Consumers' Expenditures"**, A.E.R. 37, March.

Modigliani, F., 1977. **The monetarist controversy or should we forsake stabilization policies"**, *A.E.R.* March.

Nordhouse, W.D., 1973. **The effect of inflation on the distribution of economic welfare**, *Journal of Money, Credit and Banking*, February.

Nordhouse, W.D., 1976. **Inflation, theory and policy**, Cowles Foundation Paper No 436.

Olivera, J.H.G., 1964. **On structural inflation and Latin-American Structuralism**, *Oxford Economic Papers* p.321.

Ramos, R.J., 1980. **The economics of hyper stagflation**, *Journal of Economic Development*, vol. 7 p.467.

Robinson, J., 1971. **The second crisis of economic theory**, American Economic Association Meeting, December. General Learning Press, Morristown, N.J.

Schultze, L.C., 1959. **"Recent Inflation in the United States"**, Joint Economic Committee, Congress of the United States, Washington, D.C., September.

Stoker, T.M., 1986. **Single tests of distributional effects on macroeconomic equations**, *Journal of Political Economy*, August.

Taylor, J.B., 1984, **Recent changes in macro-policy and its effects: some time-series evidence**, *A.E.R.* May.

Thurow, L.C., 1975, **"Generating Inequality: Mechanisms of Distribution in the U.S. Economy"**, Basic Books Inc., New-York.

Tremblau, R., 1975, **La transflation ou l'inflation par le transferts de pouvoir d'achat**, *L'Actualite Economique*, Avril-Juin.

Walker, J.F. and Vatter, H.G., 1986. **Stagnation-performance and policy: a comparison of the depression decade with 1973-1984**, *Journal of Post Keynesian Economics*, Summer.

Weintraub, S. 1958, **"An Approach to the Theory of Income Distribution"**, Chilton Co. Publishers, Philadelphia.

Weintraub, S. 1978, **"Capitalism's Inflation and Unemployment Crisis: Beyond Monetarism and Keynesianism"**, Addison Wesley Publishing Co. Reading Mass.

Weintraub, S. 1980. **Comment on aggregate demand and price level diagrammatics**, *Journal of Post Keynesian Economics*, Fall.

Williamson, J.G., 1977. **Strategic wage goods, prices and inequality"**, *A.E.R.* March.

NEO-DARWINIAN PROCESSES IN THE EVOLUTION OF SCIENCE AND OF HUMAN SOCIETIES

Yuval Ne'eman

Wolfson Chair Extraordinary
Raymond and Beverley Sackler Faculty of Exact Sciences
Tel-Aviv University
Tel-Aviv, Israel 69978
and
Center for Particle Physics*
Physics Department
University of Texas
Austin, Texas 78712

THE AGE OF RANDOMNESS

Determinism dominated the early Nineteenth Century as a central paradigm.Laplace assumed that given the positions and momenta at a given instant for all the particles in the universe, plus extensive computing capabilities (nowadays we would say "given a Cray supercomputer" ..) one could reconstruct the entire past and predict the future of the physical world. A contemporary interest in precisely such a problem is provided by meteorology: Given all data about temperature, pressure, wind velocity, cloud coverage (including the heights, i.e. a 4-dimensional problem), earth topographies etc. all over the world, predict tomorrow's weather (or next month's). However, atmospheric scientists who work on this problem at the end of the XXth Century know that some questions cannot be answered, due to so-called "chaotic" behavior. As a matter of fact, "chaos" was first identified by E.N. Lorenz in this area.

There have been four inroads into determinism, leaving very little behind: (1) Boltzmann and Gibbs, explaining thermodynamics through statistical mechanics - presented first as an information deficiency (the difficulty in knowing coordinates and momenta for $> 10^{24}$ particles) but now redefined by Kolomogorov and Chaitin; (2) Quantum Indeterminacy - fundamental, as displayed in the experimental checks of Bell's inequalities; (3) Chaos, i.e. instability in the dependence on initial conditions; (4) Neo-Darwinian evolutionary processes. Although it was first noted in the biological kingdom, we now know that evolution is at work in a much broader domain, all areas of which similarly require production of randomized mutations, plus a selective mechanism following certain criteria of stability.

Soft Order in Physical Systems, Edited by Y. Rabin
and R. Bruinsma, Plenum Press, New York, 1994

With the discovery of the genetic code, we now know that DNA simply undergoes "copyists' error's" in its reproduction and that this is the mechanism producing mutated organisms. The mutated DNA molecules themselves are first exposed to the direct test of their own physical stability. Once this is passed, the Darwinian selective process at the level of the organism gets its chance to test for "fitness", i.e. surviving in the environment in which chance and the fitness of the parent species have landed it. The mutational machinery is random - within limitations due to physical initial conditions. At each step, one does not expect "big jumps." The old claim that there hasn't been enough time for evolution to make man was based on the assumption that 10^{27} molecules suddenly had to arrange themselves into a human being. This has been likened by Dawkins to wondering how the Mississippi manages, in its meanderings, to pass so precisely underneath all its bridges.

SERENDIPITY-DRIVEN EVOLUTIONARY EPISTEMOLOGY

Karl Popper and Donald Campbell [1] have initiated the formulation of a Neo-Darwinian scheme for the description of how science evolves. Popper's famous contribution to the philosophy of science had consisted in the emphasis on **"falsification"**, i.e. that no theory is worth the title unless it can be tested experimentally and failed on some of its predictions. A theory which fails a test is replaced by a better one, or one whose domain of validity is more extensive. Example: The replacement of Newtonian gravitation by Einstein's, which also explains the precession of the perihelion of Mercury, which could not be explained by Newtonian mechanics. The development of the system of science is clearly an evolutionary process and falsification already provides us with the necessary selective machinery. What we still need is the (random) mutational mechanism. Popper and Campbell therefore added a hypothesis of "blind variation". In a recent publication [2], Kantorovich and I have offered a more structured identification for the mutation-generating mechanism. I was drawn into this question following a previous interest in the role of scientific research in the evolution of humanity [3,4] and in the interaction between evolution and moral choices [5]. The main issue we had to address is an apparent paradox: If science evolves randomly, how come we treat research as an orderly methodical activity? Our epistemological evolutionary thesis identifies **serendipity** as the driving mutational mechanism. Another important point to bear in mind is that it is in so-called **revolutionary science** [6] that the key mutations occur. In **normal** science, the scientist solves problems by intentional action, almost as in an engineering problem. Considering the limitations on the size of manuscripts for these Proceedings, I refer the interested reader to a tractate in which I have recently put together many examples proving our thesis [7].

The more revolutionary the discovery, the more unexpected the results: They could not - by their very unexpectedness, have been aimed at by the researcher. Example: The discovery of nuclear fission, in which Fermi, bombarding Uranium nuclei with neutrons, thought he was producing elements 93 and 94. Instead, as realized by Lise Meitner and Otto Friesch, he had split the Uranium nucleus in two and was observing known elements from the middle of the periodic table. In this example, chance intervened "surreptitiously". In other cases, the scientist **knowingly** plays the odds. Example: In the discovery of High T_c Superconductivity, Bednorz and Mueller originally had been testing such oxide metallic ceramics as could be "improved" (raising their critical temperatures) within the context of the existing paradigm - BCS theory - by increasing the density of states in the Fermi electron levels, e.g. changing the Pb:Bi ratio in $BaPb_{1-x}Bi_xO_3$, etc. After three years of this kind of **normal** research, they played the dice, perhaps as the result of a hunch; they tested similar ceramic materials which the Caen chemists had investigated (BaLaCuO combinations), even though these did not fit the ideas based on BCS. They hit the jackpot - and there is as yet no new paradigm replacing or modifying BCS to explain their results..

It is very important to understand that what is mutating is not the static status of science, but rather the **dynamical** research mechanism. There has to be first an ongoing research mechanism, a search for "A". This search must be on, for anything to occur. Without the scientist being already searching, there will be no mutation, no serendipitous discovery, no accidental finding of "B". Remember - genetic mutations do not occur in static DNA; they require a reproduction process, in which the error will slip in. "Normal" science - the normal willed and programmed research process, whose aim is generally directed at a further elucidation of previous results - this is where the mutational "error" will occur. Under such conditions, we give evolution its chance, i.e. to find B even though we were searching for A.

The two great scientific revolutions at the beginning of this Century were ushered in by serendipitous advances. For Relativity, it was Michelson's lifetime interest in the velocity of light, that made him apply his skills to measure the velocity of the earth in the ether. Instead, he found a zero answer, unexplainable within classical physics, which was finally resolved by Einstein. The geometrical interpretation, however, was only supplied in 1908, by Einstein's former teacher, Hermann Minkowski (see pp. 80-85 of [8]), hence our use of the Minkowski metric. This was essential in the next step, the lesser known story of the discovery of the General Theory of Relativity. Alfred Kastler provided the missing link in a contribution at the Einstein Centennial Conference in Jerusalem [9]. In "The Quarterly Journal of Legal Medicine and Public Health" (in German) (in the context of a Festschrift such as this one..) Einstein tested the behavior of light in a Newtonian gravitational field, using $E = Mc^2$ "backwards". There is an acceleration, a change in his sacrosanct constant "c". However, having just read Minkowski (and dismissed the geometrical interpretation at first as unimportant). Einstein could now inquire with his former classmate Marcel Grossmann, now an expert in Differential Geometry, whether there exist geometries with varying metrics. The mutation here was provided by Minkowski's remark (read at the Cologne meeting of the German Association of scientists and physicians..).

For Quantum Mechanics, Planck's interest in the Second Law of Thermodynamics made him attempt to fit a formula to the spectrum of black- body radiation. The only formula he could find was one that would have resulted from an assumption that radiation is emitted in quanta of action. He treated his formula as an ad hoc temporary measure. Later on Bohr introduced yet another ad hoc temporary treatment in his atomic theory. Einstein's treatment of a particle-like photon in 1905 was another such hunch, sticking his neck.. It took till 1925 to find a new paradigm. The mutations here were wild guesses, completely unjustified by the existing theories.

An example from a relatively young field in medicine, chemopsychotherapy. The fist serious tranquilizers, the benzodiazepines, (valium, librium, etc.) were found in 1963, discovered serendipitously. In 1977, two research teams identified specific receptors for these drugs in the brain. The implication, of course, is that nature had developed them eons ago on its own..[10]. The first effective drugs against depression were discovered in the mid-fifties. Monoamine oxidase inhibitors (MAOI) typify searching for A and finding both A and B: iproniazid, a drug developed against tuberculosis, was observed to have euphoric effects. After a while, this brought about the idea of using the drug to relieve depressions. An impediment as a drug's side-effect was turned into a new medical tool. This reminds us of Fleming's discovery of penicillin as a result of his observation of the death of the bacteria in the culture he had prepared, when a mold had contaminated it. The preventive action of lithium salts, when used to fight manic-depression, was discovered in 1949 by F.J. Cade, an Australian psychiatrist who was investigating the effects of uric acid, said to cause excited behavior. Injecting lithium urate to wild guinea pigs, he observed them getting very tame rather than excited. Cade deduced that lithium was the calming agent. He then administered it to ten hospital patients suffering from manic depression, six from schizophrenia and three from depression. The manics were greatly helped - and there was no effect on the others..

THE UNIVERSALITY OF EVOLUTION

A Century and a half after Darwin, evolution has become a universal "doctrine". In almost every area of natural phenomena, there is a role for evolution. The evolution of non-living matter is tightly bound to that of the stars and is known as nucleosynthesis. Progress in the ever more complex structure of atomic nuclei - and as a result, of the chemical elements, the building up of Mendeleyev's periodic table from $Z=1$ to $Z=92$ - can be compared to the development of the DNA molecule's structure from that of a virus to man's. At the same time, the formation, life-cycle and collapse of stars (into either white dwarfs, neutron stars or black holes) parallel that of the living organisms and undergoes evolutionary processes regulated by the stage reached in the nuclear reactions inside it: They represent the "blueprint" or "program" for the star, a control mechanism, in the same sense that DNA controls biological evolution.

Evolutionary processes are characterized by a random mechanism generating stochastic mutations, plus a selective machinery which decides which mutations are "stable", introducing a new species. The random mutation mechanism in nucleosynthesis within the stars consists in very-high-energy collisions between atomic nuclei, i.e. temperatures of hundreds millions degrees, resulting from the gravitational energy due to the star's huge mass. This pulls it inwards and is counterbalanced by pressure generated in the nuclear reactions thereby ignited. Stable levels are the various nucleides created in these collisions. Other features are the dissipative nature of the system, i.e. a metabolism in a generalized sense: Stars absorb low-Z atomic "dust" and make higher atomic elements from it. Apparent teleonomy, both "strategic" and "tactical" are also characteristic of evolutionary systems. Tactically, the organism appears to serve a purpose, making helium from hydrogen, for instance. Strategically, it is as if the entire evolution is meant to lead to a "final" target ("from microbes to man"). Entropy-wise, evolutionary processes generate order and decrease entropy by pushing it into their environment, whose entropy thereby increases faster than otherwise.

THE ROLE OF SCIENTIFIC RESEARCH IN THE EVOLUTION OF SOCIETY

In the last million years, except for the transition between Neanderthal and Sapiens, the scene has been dominated by the evolution of human societies. There have been no major genetic changes in homo sapiens, and yet he has evolved tremendously, through his grouping in societies.

In the evolution of societies, the stable evolutionary levels are determined by technologies. Paleo-archeology has given us a description in which we go from the Paleolithic to the Mesolithic and Neolithic, following the evolution of the stone-made tools. Next we go from copper tools (the Chalcolithic) to the Ages of Bronze and of Iron. We entered the Industrial Age two hundred years ago and we are now in the Age of Information technology, according to many analysts. The latter divisions are not clear cut, because science has also entered the equation. The technological changes of Prehistory did not follow a scientific advance, whereas the new technologies since the XVIIIth Century do represent applications of advances in science: Watt and Stephenson follow Boyle and Mariotte, etc.

Analyzing the social history of mankind, we realize that technology, and more recently science, represent the control mechanism, the code or blueprint for the social organism's evolutionary stage. It is the DNA of the social species, just as the nuclear composition determines the star's evolutionary fate. With the arrival of science and its rapid progress, society is evolving much more rapidly.

In any evolutionary process, there has to be first and foremost a random mutational mechanism. What is it here? In 1977, I suggested [3] that since the XVIIIth Century, it is scientific research that fulfills the role of the randomized mutational machinery for social evolution. Any important advance implies a surprise discovery, otherwise it would not be considered important, it would not represent a new departure.

This is my thesis, thereby throwing new light on the role of serendipity: It drives the revolutionary mutations in science, like the errors in DNA reproduction - and it thereby also triggers the mutational advances in the evolution of human society.

One lesson is that basic research should not be "directed". You can only "direct" towards targets you know, but real progress can only come from the ones you cannot know of. Concentrating on "relevant" basic research implies stopping progress, staying at the same evolutionary level for ever. It then becomes like a Peter's Principle result. There were such attempts to direct science into "useful" channels in the USSR in the Fifties, in China during the so-called "Cultural Revolution" and in the USA around 1965-70, though in a milder fashion, due to the democratic structure. I have the impression that there is a renewed such threat in the present mood of the democratic structures in the USA

There is also a lesson concerning the methodology of grants for scientific research. Generally, it requires the submission of a proposal, which is supposed to include a research plan and targets the researcher is aiming at. Obviously, any serendipitous result lurking in the shadows cannot be anticipated. Thus, the most important possible results will never appear in the proposal. Therefore, the grantor should not take proposals too seriously. It is important, because if you will not set out properly and methodically to find A, you will also not discover B serendipitously. However, the actual program in itself is less important. What matters mostly is whether this researcher will recognize B when he encounters it. Past performance might sometime be a guide for this selection. For a new researcher, there is still no way of judging except by the way he has planned his project; again, what is important is less the aims stated than the method presented.

Using the criterion of "relevance" can be a tragic joke. Perhaps the most ironic of all such developments is to be found in the evolution of mathematical logic and of the theory of sets. These used to be regarded as the most abstruse and esoteric branches of mathematics, rather more of philosophy than of mathematics. Then came the Russell-Whitehead paradox, Goedel's proofs of the completeness of the Predicate Calculus (1930) and of the incompleteness of arithmetic (1931). He used computable functions - which made some mathematicians wonder if the functions he used included all "computable" functions. This required a redefinition of "computable". Alan Turing invented a "gedanken" machine, to define "computation". Then came the Second World War, Turing was put to work in cryptography. There was a need for extensive computations and Turing developed his machine - leading directly to the invention of the modern computer, developed independently by J. von Neumann in the USA, to help in counter-battery artillery fire. Imagine having to justify one hundred years of set theory by promising that some day it would usher in the most useful machine ever? Would any funding agency set on "relevance" have bought the argument?

REFERENCES

1. D.T. Campbell, Evolutionary epistemology, in "The Philosophy of Karl Popper", Vol. I, P.A. Schilpp, ed., La Salle (Open Court), pp. 413-463.

2. A. Kantorovich and Y. Ne'eman, Serendipity as a Source of Evolutionary Progress in Science, "Studies in the History and Philosophy of Science", Vol. 20, pp. 505-529 (1989).

3. Y. Ne'eman, Science as evolution and transcendance, Caltech Fairchild Colloquium (1977), unpub; reproduced in "Acta Scient. Venezuelana", 31 (1980) 1.

4. Y. Ne'eman, in "Metabolic, Pediatric and Systemic Ophthalmology", 11 (1988) 12.

5. Y. Ne'eman, Darwin, Nietzsche and the Judeo-Christian Ethic, "Jour. of Social and Evolutionary Systems" (1993).

6. T.S. Kuhn, "The Structure of Scientific Revolutions", Un. of Chicago Press, Chicago (1962).

7. Y. Ne'eman, "Serendipity, Science and Society - An Evolutionary View", Emil Starkenstein Foundation, Rotterdam (1993).

8. E. Segre, "From X-Rays to Quarks", W.H. Freeman and Co., New York (1980).

9. A. Kastler, in "To Fulfill A Vision", Y. Ne'eman, ed., Addison- Wesley Pub., Reading, Mass. (1981), pp. 77-78.

10. M.E. Lickey and B. Gordon, "Drugs for Mental Illness", W.H. Freeman and Co., New York (1983).

INDEX